Quaternionic and Clifford Calculus for Physicists and Engineers

Quaternionic and Clifford Calculus for Physicists and Engineers

Klaus Gürlebeck
Bauhaus-University Weimar, Germany

Wolfgang Sprössig
Frieberg University of Mining and Technology, Germany

JOHN WILEY & SONS
Chichester • New York • Weinheim • Brisbane • Singapore • Toronto

National 01243 779777
International (+44) 1243 779777
e-mail (for orders and customer service enquiries): cs-books@wiley.co.uk
Visit our Home Page on http://www.wiley.co.uk
or http://www.wiley.com

Other Wiley Editorial Offices

John Wiley & Sons, Inc., 605 Third Avenue,
New York, NY 10158-0012, USA

Wiley-VCH Verlag GmbH, Pappelallee 3,
D-69469 Weinheim, Germany

Jacaranda Wiley Ltd, 33 Park Road, Milton,
Queensland 4064, Australia

John Wiley & Sons (Asia) Pte Ltd, 2 Clementi Loop #02-01,
Jin Xing Distripark, Singapore 129809

John Wiley & Sons (Canada) Ltd, 22 Worcester Road,
Rexdale, Ontario M9W 1L1, Canada

British Library Cataloguing in Publication Data

A catalogue record for this book is available from the British Library

ISBN 0 471 96200 7

Produced from camera-ready copy supplied by the authors
Printed and bound in Great Britain byBiddles Ltd, Guildford and King's Lynn
This book is printed on acid-free paper responsibly manufactured from sustainable forestation, for which at least two trees are planted for each one used

Contents

Preface

A monograph, particularly a textbook, should have messages. Our book is intended to give three messages:

1. One variable complex analysis has a pendant in higher dimensional Euclidean space the so-called CLIFFORD analysis.

2. Clifford and quaternionic analysis are natural approaches for the treatment of elliptic boundary value problems in higher dimensions.

3. There exists a discrete CLIFFORD calculus, which supports a corresponding numerical analysis.

Because CLIFFORD analysis covers a great part of the quaternionic analysis we will prefer the notation CLIFFORD analysis. CLIFFORD analysis counted from the very beginning (elementary results for quaternionic functions) is about 100 years old. Its history started with the paper "On the Newtonian potential" by A.C. DIXON from the University of Belfast . In the thirties of our century papers by G.N. MOISIL/N. TEODORESCU from Cluj and fundamental papers by R. FUETER and his followers from Zürich were published. There were physicists such as D. HESTENES [?], [?] and physically oriented mathematicians, as M. RIESZ [?], who wrote initially books, however, from a more algebraical point of view. A first closed representation on CLIFFORD analysis appeared in 1982 by the Belgian group F. BRACKX, R. DELANGHE and F. SOMMEN entitled "*Clifford analysis*" [?]. It followed monographs by R.P. GILBERT and J. BUCHANAN [?] in 1983, by the authors [?] in 1990, by B. GILBERT and A. MURRAY [?] in 1992, by R. DELANGHE, F. SOMMEN and V. SOUCEK [?] in 1993, by M. MITREA [?] in 1994, by J. RYAN (ed.) [?] in 1995, and by M.V. SHAPIRO and V.V. KRAVSHENKO [?] in 1996.

Beginning in 1985 International Conferences with the constant title *Clifford Algebras and Their Applications in Mathematical Physics* has been periodically organized (Canterbury (1985), Montpellier (1989), Deinze (1992), Aachen (1996)). Participants were mathematicians, physicists, and engineers. Moreover, a number of further conferences, e.g., Fayetteville (1993), Mexico-City (1994), Triest (1994) and Seiffen (1996) attracted many prominent people.

With a growing number of publications, intensified conference activities on this subject and the rapid increase of applications it becomes more and more important to have a book which reflects the main ideas in this field with large number of different applications on a level which makes quaternionic and CLIFFORD analysis accessible and applicable for a relatively broad readership. This book will also help to develop a corresponding curriculum such that CLIFFORD analysis can be better integrated into university calendars.

The book is addressed to mathematicians, physicists, and engineers who are interested in the analysis of partial differential equations together with their applications in mathematical physics. Our intention is to equip the reader (user) with some elementary tools for the analytical and numerical treatment of boundary value problems. Of course, we have to pay a certain price in realizing this intention. The price is a lack of generality, the restriction to special (but not too small) classes of equations and problems. Because we also explain the history of the used methods we hope to enable the reader to adapt the methods to his own problems. A corresponding part on numerical analysis is included. We have to consider that any multidimensional theory is not easy in general. We believe that CLIFFORD analysis enables us to enter into higher dimensions in a natural way, assuming that the reader is familiar with the first steps in one variable complex function theory, basic knowledge in functional analysis, and algebra. Parts of the book (e.g. Sections 1.1–1.3, 2.1–2.13) are also suitable for undergraduate students, too. We try to show the historical continuity of the subject. Many mathematicians of different fields, physicists and engineers were involved in the development, which we considered in this book. Meanwhile, there exists a maze of papers, names, directions, and results at very different levels. That is why we try to give the reader who is not involved in disciplinary research at the top level a guide through this maze.

There are two types of sections. On the one hand we have sections where our representation is relatively complete with many proofs (e.g. Sections 1.2, 2.3, 2.5, 3.4, 3.6, 4.2, 4.3, 4.4, 4.5, 4.12, 5.3) and on the other hand there are sections where we only give a rough outline of the material (definitions, notations, important statements, e.g. Sections 1.4–2.2, 3.7, 3.8, 4.8, 5.2, 5.8, 6.3). Altogether, we do not lay claim to completeness. Often we can only explain the main ideas by examples. Most of the recently interesting fields are mentioned. However, sometimes we can only refer to the literature. If it is possible then we stress the close relations among such basic mathematical disciplines as algebra, geometry and analysis including corresponding discrete mathematical methods and functional analysis in the frame of CLIFFORD structures.

In Sections 1.1–1.7 we present the necessary algebraical and geometrical background. Sections 1.4–2.2 are devoted to CLIFFORD regular functions, which replace the "holomorphic" functions in the one-dimensional complex function theory. We give only a very roughly survey on the hypercomplex FOURIER transform as well as the exterior calculus (Sections 2.1, 3.8). In detail we explain the generalization of the classic complex function theory to higher dimensions (Sections 2.1–2.6, 3.1–3–7). Introducing the operators of the TEODORESCU transform (Section 3.1), CAUCHY-type operators, corresponding singular integral operators (Sections 3.3, 3.4) as well as HILBERT space orthoprojections (Section 3.6) we formulate a calculus which will be applied in Sections 4.1–4.15. In particular, nonlinear boundary value problems form the contents of the Sections 4.8–4.10. Very different applications follow (Sections 4.13–4.16). A complete presentation of a discrete CLIFFORD calculus is the aim of Sections 5.1–5.4. In Section 6.1 we will show only that our methods also work in the case of parabolic equations. Section 6.2 is more a historical review of the application of CLIFFORD algebras and CLIFFORD analysis to field theories. It follows some remarks and hints about the state of the art in this field of research with comments on further investigations. In Appendix A we have listed an integral table for the TEODORESCU operator and in Appendix B one can find a table of discrete fundamental solutions. Together with the detailed description of the discrete Cauchy 'integral' operator in Appendix C we try to prepare some tools for practical applications of the CLIFFORD analysis up to numerical methods. Summarizing all the progress of recent years it seems to be the case that there are more and more requests from outside the small circle of specialists for information and work with methods of applied CLIFFORD analysis. This process is supported by data from the Appendix.

It is a pleasent duty for us to thank all our friends and colleagues who have helped us in writing this book. We are deeply indepted to S. BERNSTEIN and U. KÄHLER who read the whole manuscript in detail and suggested a number of corrections and improvements. Without their comments the book could not be as it is. In addition we thank J. RYAN (University of Arkansas) for stimulating discussions and valuable comments on parts of the material.

It is also a pleasure to thank Mrs. K. UHLEMANN in Freiberg and Mrs. G. SCHMIDT in Weimar for typing parts of the manuscript. Finally, many thanks go to Wiley & Sons Publishers especially to D. IRELAND and I. STONEHAM for the realization of this book.

Series Preface

Mathematics is a fundamental and enabling discipline in most of the applied sciences. More than ever, both students and research workers in many fields are finding that in order to meet the modern day demands of their discipline they need a sound and practical appreciation of the various mathematical methods which are available for their use.
By tradition, many topics in mathematics are first taught in pure mathematics courses, where they are developed theoretically, and then again in applied mathematics courses where practical applications are the main concern. This can often cause confusion for the applied scientist. The present series is a response to the demand to reconcile these two approaches and, in addition, is intended to meet the perceived and increasing demand for mathematics as applied to specific disciplines and industries.
The books in this series will each present a well motivated and persuasive account, involving many practical examples, of the mathematical techniques required within each title's special topic area. They aim to provide mathematical support for workers in a wide range of research environments.

Bruno Brosowski
Universität Frankfurt/Main, Germany

Gary Roach
University of Strathclyde, UK
1997

Chapter 1

Quaternions and multivectors

1.1 Real and complex quaternions

Quaternions were invented in 1843 by the Irish mathematician WILLIAM ROWAN HAMILTON. He had worked since 1830 in the field of complex numbers and in 1833 obtained the result that complex numbers form an algebra of pairs of real numbers. W.R. HAMILTON tried for more than ten years to extend this concept to triples of real numbers with one real and two imaginary units. He could himself well imagine suitable operations of addition and multiplication of triples (which he later defined as 'vectors'), but he was unable to find a method for the division of such vectors. W.R. HAMILTON was often asked by his elder son during breakfast: "Father, have you learned how to divide vectors?" After the rejection of the commutativity law of multiplication he could answer on October 16, 1843 : "Yes, I have!" This introduction of a noncommutative structure into this algebra led to the dispute over more then twenty years with other famous mathematicians of his time.

1.1.1 Elementary properties of real quaternions

In this section we assemble a number of well known properties of real quaternions, which we found in different classic books in this field.

Let $e_0 = 1, e_1, e_2, e_3$ the basic elements in $\mathbb{R}^4$ where the vector e_k is to identify with this 4-tupel which has at the $k+1$-th component the number one and is zero otherwise. An arbitrary element $x \in \mathbb{R}^4$ now has the representation $x = x_0e_0 + x_1e_1 + x_2e_2 + x_3e_3$. We will call the part $x_0e_0 =: \mathrm{Sc}\, x$ the *scalar part* of x and $\underline{x} = x - x_0e_0$ written by $\mathrm{Vec}\, x = \underline{x}$ is the *vector part* of x . Let $x, y \in \mathbb{R}^4$. Now we define a product in $\mathbb{R}^4$ which fulfils the following conditions:

(i) $e_1^2 = e_2^2 = e_3^2 = -1$,

(ii) $e_1e_2 = e_3, e_2e_3 = e_1, e_3e_1 = e_2$,

(iii) $e_ie_j + e_je_i = 0 \quad (i, j = 1, 2, 3\,;\, i \neq j)$.

If we write for short $(\underline{x}, \underline{y}) := x_1y_1 + x_2y_2 + x_3y_3$ and $\underline{x} \times \underline{y} := (x_2y_3 - x_3y_2)e_1 + (x_3y_1 - x_1y_3)e_2 + (x_1y_2 - x_2y_1)e_3$, then by using of the algebraic rules (i),(ii),(iii) the *quaternionic product* xy is given by

$$xy = x_0y_0 - (\underline{x}, \underline{y}) + x_0\underline{y} + y_0\underline{x} + \underline{x} \times \underline{y}.$$

After introduction of this product the tuple $(\mathbb{R}^4, \cdot)$ is called *algebra of real quaternions*. In honour of W.R. HAMILTON we signify $(\mathbb{R}^4, \cdot)$ by $\mathbb{H}$. If $x = \underline{x}$ then x is called *pure quaternion*. The subset of all pure quaternions is denoted by $\mathrm{Vec}\,\mathbb{H}$, while the subset of all scalars will be signified by $\mathrm{Sc}\,\mathbb{H}$. For $x \in$

Vec $\mathbb{H}$ we identify $x = \underline{x}$. The quaternion $\bar{x} = x_0 - \underline{x}$ is called the *conjugate* to x. The mapping $x \longmapsto \bar{x}$ is called *conjugation*. The number $|x|$ defined by $x\bar{x} = |x|^2$ is named *norm* or *absolute value* of x . If $|x| = 1$ then a quaternion is said to be a *unit quaternion*.

Example 1.1. *Let* $x = 2 + e_1 + 3e_2 + e_3$ *then* $\bar{x} = 2 - e_1 - 3e_2 - e_3$, $Sc\, x = 2$ $|x|^2 = 4 + 1 + 9 + 1 = 15$. *Let* $y = e_1 - 2e_3$ *be another quaternion then*

$$xy = 2 - (-1) + 2e_1 - 2e_3 + (-6)e_1 + 4e_2 - 3e_3 = 3 - 4e_1 + 4e_2 - 5e_3.$$

Lemma 1.2. *xy is an $\mathbb{R}$-bilinear and associative product, but it is not commutative.*

Remark 1.3. For $x, y \in \mathbb{H}$ and $\mu \in \mathbb{R}$ one has the relations $\overline{x + y} = \overline{x} + \overline{y}$, $\overline{\mu x} = \mu\overline{x}$, $\overline{\overline{x}} = x$ and $\overline{xy} = \overline{y}\,\overline{x}$.

Corollary 1.4. *Let* $x, y \in$ Vec $\mathbb{H}$

(i) $x \times y = \mathrm{Vec}(xy) = \frac{1}{2}(xy - yx)$

(ii) $(x, y) = \mathrm{Sc}\,(\overline{x}y) = \frac{1}{2}(\overline{x}y + y\overline{x})$.

Corollary 1.5. *From* $x^2 = y^2$ *it does not follow that* $x = \pm\, y$.

Proof. Take for instance e_1 and e_2. #

Lemma 1.6. *A quaternion x is a vector if and only if $x^2 < 0$ or $x = 0$.*

Proof. Let $x = \underline{x}$, then $x^2 = -x\overline{x} = -(x, x) = -(x_1^2 + x_2^2 + x_3^2)$. In the opposite direction we have $x^2 = x_0^2 - |\underline{x}|^2 + 2x_0\underline{x}$. Now only two cases are possible: (i) $x_0 = 0$ and then we are ready, (ii) $\underline{x} = 0$ and $x_0 \neq 0$ then x^2 would be positive contrary to the assumption. #

Lemma 1.7. *A quaternion x is real if and only if for each other quaternion y holds $yx = xy$.*

Proof. It is necessary to show only that from the commutativity relation with an arbitrary quaternion y it follows that x has to be a real number. Taking for $y = e_1$ we get

$$xe_1 = -x_1 + x_0e_1 - x_2e_3 + x_3e_2 = -x_1 + x_0e_1 + x_2e_3 - x_3e_2 = e_1x.$$

Hence $0 = 2(x_2e_3 - x_3e_2)$. It immediately follows that $x_2 = x_3 = 0$. After repeating this procedure with the quaternion $y = e_2$ we obtain that x_1 also has to be zero. #

Proposition 1.8. *For all $x, y \in \mathbb{H}$ the identity $|xy| = |x|\,|y|$ holds. If $x \neq 0$ then x^{-1} exists and we have $x^{-1} = \overline{x}|x|^{-2}$. Furthermore, we get $|x|^{-1} = |x^{-1}|$.*

Lemma 1.9. *Let $\underline{y}$ be an invertible vector. Then for any vector $\underline{x}$ the expression $\underline{y}\underline{x}\underline{y}^{-1}$ is also a vector.*

Proof. As $\underline{x}$ is a vector, $\underline{x}^2$ is real and non-positive. Using Lemma 1.6. we have $(\underline{y}\underline{x}\underline{y}^{-1})^2 = \underline{x}^2$ and so $\underline{y}\underline{x}\underline{y}^{-1}$ has to be a vector. #

Corollary 1.10. *The map $\rho_y : \mathbb{R}^3 \ni \underline{x} \to -\underline{y}\underline{x}\underline{y}^{-1} \in \mathbb{R}^3$ is a reflection in the plane which lies orthogonal to the vector $\underline{y}$.*

Proof. The map ρ_y is linear and $\rho_y(\underline{y}) = -\underline{y}$, while for any vector $r \perp \underline{y}$ it follows that $\rho_y(r) = -\underline{y}r\underline{y}^{-1} = r\underline{y}\underline{y}^{-1} = r$. #

Corollary 1.11. *Each rotation in $\mathbb{R}^3$ has for some non-zero $\underline{y}$ the form $-\rho_y$. Conversely, any such map can be seen as rotation in $\mathbb{R}^3$.*

Proof. We already know that the product of two plane reflections is just a rotation and vice versa. #

Lemma 1.12. *Let y be a quaternion. Then there exists a vector $\underline{a} \neq 0$, such that $y\underline{a}$ is also a vector.*

Proof. Let $\underline{a}$ be a vector in $\mathbb{R}^3$ orthogonal to the vector part of y. Then

$y\underline{a} = y_0\underline{a} + \underline{y} \times \underline{a} \in \mathbb{R}^3.$ #

Corollary 1.13. *Each quaternion can be described as a product of two vectors.*

Lemma 1.14. *An arbitrary unit quaternion can be represented as the product $yxy^{-1}x^{-1}$, where $x = \underline{x}$ and $y = \underline{y}$ are non-zero vectors.*

Proof. We know from Lemma 1.12. that for any unit quaternion e a non-zero vector $\underline{x}$ exists such that $e\underline{x}$ is a vector. Because of $|e| = 1$ we have $|e\underline{x}| = |\underline{x}|$. In this way $e\underline{x}$ has to be a rotation. Then there exists a vector $\underline{y} \neq 0$ with $e\underline{x} = \underline{y}\underline{x}\underline{y}^{-1}$ and so the statement follows. #.

1.1.2 Representation of real quaternions

We will group here some of the most important properties of the representation of real quaternions.

Theorem 1.15. *An arbitrary quaternion $x \in \mathbb{H}$, $\underline{x} \neq 0$ permits the representation*

$x = |x|(\cos\phi + \omega(x)\sin\phi),$

where $\phi = \operatorname{arc\,cot}(x_0/|\underline{x}|)$ and $\omega(x) = \underline{x}/|\underline{x}| \in S^3$.

Proof. It is well known that

$$\sin\phi = \frac{1}{\sqrt{(1+\cot^2\phi)}} \quad \text{and} \quad \cos\phi = \frac{\cot\phi}{\sqrt{(1+\cot^2\phi)}}.$$

We obtain under our assumption

$$\sin \operatorname{arccot}\frac{x_0}{|\underline{x}|} = \frac{1}{\sqrt{1+(\frac{x_0}{|\underline{x}|})^2}} \quad \text{and} \quad \cos \operatorname{arc\,cot}\frac{x_0}{|\underline{x}|} = \frac{\frac{x_0}{|\underline{x}|}}{\sqrt{1+(\frac{x_0}{|\underline{x}|})^2}}.$$

Then we get by a straightforward calculation $\cos\phi + (\underline{x}/|\underline{x}|)\sin\phi = x/|x|$ which verifies our theorem. #

Remark 1.16. For $x_0 = 0$ we get the decomposition mentioned in Proposition 1.8.

Example 1.17. Let $x = 3 + 2e_1 + 2e_2 + e_3$ then $|x| = 3\sqrt{2}$, $|\underline{x}| = 3, \phi = 45°$. Thus we obtain the representation

$$x = 3\sqrt{2}\left[\cos 45° + \frac{(2e_1 + 2e_2 + e_3)}{3}\sin 45°\right].$$

Corollary 1.18. (MOIVRE's formula.) *Let $x \in \mathbb{H}, \underline{x} \neq 0, n \in \mathbb{N}$, then the following formula is valid:*

$$(\cos\phi + \omega(x)\sin\phi)^n = \cos\, n\phi + \omega(x)\sin\, n\phi.$$

Lemma 1.19. *The $\mathbb{R}$-linear hull of the set $\{1,\ x\}$, where x is not real, forms a subalgebra which is isomorphic to $\mathbb{C}$.*

Theorem 1.20. *The algebra of real quaternions $\mathbb{H}$ can be represented by the matrices*

$$\begin{pmatrix} 1 & 0 \\ 0 & 1 \end{pmatrix}, \quad \begin{pmatrix} -i & 0 \\ 0 & i \end{pmatrix}, \quad \begin{pmatrix} 0 & -1 \\ 1 & 0 \end{pmatrix}, \quad \begin{pmatrix} 0 & -i \\ -i & 0 \end{pmatrix}.$$

Therefore each quaternion x permits the representation

$$x = \begin{pmatrix} \overline{z}_1 & -\overline{z}_2 \\ z_2 & z_1 \end{pmatrix}$$

with $z_1 := x_0 + ix_1$ and $z_2 := x_2 + ix_3$.

Corollary 1.21. *The algebra of real quaternions $\mathbb{H}$ can be generated by the real matrices*

$$\begin{pmatrix} 1 & 0 & 0 & 0 \\ 0 & 1 & 0 & 0 \\ 0 & 0 & 1 & 0 \\ 0 & 0 & 0 & 1 \end{pmatrix}, \qquad \begin{pmatrix} 0 & -1 & 0 & 0 \\ 1 & 0 & 0 & 0 \\ 0 & 0 & 0 & 1 \\ 0 & 0 & -1 & 0 \end{pmatrix},$$

$$\begin{pmatrix} 0 & 0 & -1 & 0 \\ 0 & 0 & 0 & -1 \\ 1 & 0 & 0 & 0 \\ 0 & 1 & 0 & 0 \end{pmatrix}, \quad \begin{pmatrix} 0 & 0 & 0 & 1 \\ 0 & 0 & -1 & 0 \\ 0 & 1 & 0 & 0 \\ -1 & 0 & 0 & 0 \end{pmatrix}.$$

Remark 1.22. In [9] the following good relations between quaternions and matrices are given. For this reason we introduce a vector $u = (u_0, u_1, u_2, u_3)^T \in \mathbb{R}^4$ and a corresponding matrix

$$T^{\pm}(u) = \begin{pmatrix} u_0 & -\underline{u}^T \\ \underline{u} & u_0 E_3 \pm K(\underline{u}) \end{pmatrix}$$

with

$$K(\underline{u}) = \begin{pmatrix} 0 & -u_3 & u_2 \\ u_3 & 0 & -u_1 \\ -u_2 & u_1 & 0 \end{pmatrix} \quad \text{and} \quad E_3 = \begin{pmatrix} 1 & 0 & 0 \\ 0 & 1 & 0 \\ 0 & 0 & 1 \end{pmatrix}.$$

It is clear that $K(\underline{v})\underline{w} = \underline{v} \times \underline{w}$, where $\underline{v}, \underline{w} \in \mathbb{R}^3$.

Lemma 1.23. *Let $u, v, w \in \mathbb{R}^4$. Then we have*

(i) $vw = T^+(v)w = T^-(w)v$,

(ii) $uvw = T^+(u)T^+(v)w = T^+(u)T^-(w)v = T^-(w)T^-(v)u.$

1.1.3 Complex quaternions

More than real quaternions, complex quaternions play an important role in theoretical physics. Let us now discuss the fundamental properties of quaternions with complex-valued coefficients. We will use the so-called PAULI matrices:

$$\sigma_0 = \begin{pmatrix} 1 & 0 \\ 0 & 1 \end{pmatrix}, \quad \sigma_1 = \begin{pmatrix} 1 & 0 \\ 0 & -1 \end{pmatrix}, \quad \sigma_2 = \begin{pmatrix} 0 & -i \\ i & 0 \end{pmatrix}, \quad \sigma_3 = \begin{pmatrix} 0 & 1 \\ 1 & 0 \end{pmatrix}.$$

PAULI matrices form an algebra P. Then we get the representation of an arbitrary element $x \in P$ in the form

$$x = x_0\sigma_0 + x_1\sigma_1 + x_2\sigma_2 + x_3\sigma_3 + x_4\sigma_2\sigma_3 + x_5\sigma_3\sigma_1 + x_6\sigma_1\sigma_2 + x_7\sigma_1\sigma_2\sigma_3.$$

We have to distinguish four classes of elements, namely complex linear combinations of scalars σ_0, vectors $\sigma_1, \sigma_2, \sigma_3$, bivectors $\sigma_2\sigma_3, \sigma_3\sigma_1, \sigma_1\sigma_2$ and pseudoscalars $\sigma_1\sigma_2\sigma_3$. PAULI matrices satisfy the conditions $\sigma_i\sigma_j + \sigma_j\sigma_i = 2\delta_{ij}\sigma_0$ $(i, j = 1, 2, 3)$, where δ_{ij} denotes KRONECKER's symbol. On account of $(\sigma_1\sigma_2\sigma_3)^2 = -\sigma_0$ we obtain that the linear space generated by $\{\sigma_0, \sigma_1\sigma_2\sigma_3\}$ is isomorphic to the field of complex numbers $\mathbb{C}$.

Remark 1.24. The centre of P is $\mathbb{C}$.

Lemma 1.25. *With $\varepsilon := i\sigma_0$ we get*

$$-\varepsilon\sigma_0 = \sigma_1\sigma_2\sigma_3, -\varepsilon\sigma_1 = \sigma_2\sigma_3, -\varepsilon\sigma_2 = \sigma_3\sigma_1, -\varepsilon\sigma_3 = \sigma_1\sigma_2, \varepsilon\sigma_1\sigma_2 = +\sigma_3,$$
$$\varepsilon\sigma_2\sigma_3 = +\sigma_1, \varepsilon\sigma_3\sigma_1 = +\sigma_2, \varepsilon\sigma_1\sigma_2\sigma_3 = +\sigma_0,$$

i.e. the multiplication by ε transforms scalars into pseudoscalars, vectors into bivectors, bivectors into vectors, and pseudoscalars into scalars.

Remark 1.26. The operator of multiplication by ε is often called HODGE *map* and marked by $*$.

Definition 1.27. *Let $x = x_1\sigma_1 + x_2\sigma_2 + x_3\sigma_3$ and $y = y_1\sigma_1 + y_2\sigma_2 + y_3\sigma_3$ be vectors in P. The formal determinant*

$$\det \begin{bmatrix} \sigma_1 & \sigma_2 & \sigma_3 \\ x_1 & x_2 & x_3 \\ y_1 & y_2 & y_3 \end{bmatrix}$$

is called cross product and denoted by $x \times y$.

Lemma 1.28. *The algebra P contains zero divisors.*

Proof. One has to take only the complex quaternions $(1 - \sigma_1)$ and $(1 + \sigma_1)$. The product of both is zero. #

We find for each element $x = p + q\varepsilon$ of the PAULI algebra with $p, q \in \mathbb{H}$ the conjugation (do not confuse with the notion of the complex-conjugation) acts as follows : $\bar{x} = \bar{p} + \bar{q}\varepsilon$.

Definition 1.29. *Let $x = p + q\varepsilon, \quad p = p_0 + \underline{p}, \quad q = q_0 + \underline{q}$. Then we call the complex magnitude $||x||_{\mathbb{C}}$ of the element x defined by*

$$||x||_{\mathbb{C}}^2 = x\bar{x} = |p|^2 - |q|^2 + 2\varepsilon[(p_0q_0 + (\underline{p}, \underline{q})]$$

complex-valued norm in P where $|p|^2 = p\bar{p}$ and $|q|^2 = q\bar{q}$.

Lemma 1.30. *There exists a unique real-valued norm $||x||$ of $x \in P$ which satisfies the conditions*

(i) $||xy|| = ||x||\,||y||$,

(ii) $||\mu\sigma_0|| = |\mu|\ (\mu \in \mathbb{R})$.

This norm is given by

$$||x||^4 = |\,||x||_{\mathbb{C}}^2|^2.$$

1.2 Quaternions and vector algebra relations

In most of the lectures on vector algebra and analytical geometry the unit vectors were denoted by i, j, k. This goes back to W.R. HAMILTON who called the pure quaternion $xi + yj + zk$, "vector", (in a paper in 1846). Up to this time three-component objects were already studied in a number of physical papers. It was for almost a hundred years not completely clear that HAMILTON's vectors are indeed nothing other than axial vectors (cf. [242], 1958) and can be finally identified with binary rotations (cf. [7]). Note here that vectors can be classified in polar and axial vectors, which are also called vectors and pseudovectors, respectively. Vectors are triples of scalars and pseudovectors are triples of pseudoscalars. We further note that a binary form of a quaternion x is written as

$$x = x_0 + \omega|\underline{x}| \qquad (\omega \in S^3).$$

This led in the past to confusion in the use of vectors. More on the eventful history of vectors, rotations, and quaternions and also a good overview on corresponding references can be found in ALTMANN's book [7].

1.2.1 Isomorphic embeddings in $\mathbb{H}$

Now let us consider relations between the quaternionic structure in $\mathbb{R}^4$ and the classical vector algebra. An arbitrary element $x = (x_0, x_1, x_2, x_3) \in \mathbb{R}^4$ corresponds with the quaternion: $x = x_0e_0 + x_1e_1 + x_2e_2 + x_3e_3$, where $e_0^2 = 1$ and $e_ie_j + e_je_i = -2\delta_{ij}$ $(i, j = 1, 2, 3)$. In this way $\mathbb{R}$ and $\mathbb{R}^3$ are embedded in $\mathbb{H}$. The corresponding embedding mappings are defined by

$$x_0 \longrightarrow x_0e_0, \quad (x_1, x_2, x_3) \longrightarrow x_1e_1 + x_2e_2 + x_3e_3, \tag{1.1}$$

respectively. The most natural realization of the elements e_0, e_1, e_2, e_3 is to identify them (as W.R. HAMILTON has also done) with the basis $e_0 = (1, 0, 0, 0)$, $e_1 = (0, 1, 0, 0)$, $e_2 = (0, 0, 1, 0)$, $e_3 = (0, 0, 0, 1)$ in $\mathbb{R}^4$ and define the multiplication between these basic elements by the CAYLEY table:

	e_0	e_1	e_2	e_3
e_0	e_0	e_1	e_2	e_3
e_1	e_1	$-e_0$	e_3	$-e_2$
e_2	e_2	$-e_3$	$-e_0$	e_1
e_3	e_3	e_2	$-e_1$	$-e_0$

Another realization is given by elements which are based on the PAULI matrices given in Section 1.1. We used there the basis $e_0 := \sigma_0$, $e_1 := -i\sigma_2$, $e_2 := -i\sigma_1$, $e_3 := \sigma_1\sigma_2$. Because of $e_1^2 = -1$ the elements $z_1 = x_0e_0 + x_1e_1$ and $z_2 =$

$x_2e_0 + x_3e_1$ can be considered as complex numbers. One obtains the following isomorphism between $\mathbb{C}^2$ and $\mathbb{H}$:

$(z_1, z_2) \longleftrightarrow z_1 + z_2e_2$

$\mathbb{H}$ has to be taken as a left vector space on $\mathbb{C}$.

1.2.2 Hamilton's vector algebra

In this section let $x, y, z \in \operatorname{Vec} \mathbb{H}$, i.e. $x = \underline{x} = x_1e_1 + x_2e_2 + x_3e_3$, $y = \underline{y} = y_1e_1 + y_2e_2 + y_3e_3$, and $z = \underline{z} = z_1e_1 + z_2e_2 + z_3e_3$.

Lemma 1.31. *We have the following relations:*

(i) $x^2 = -|x|^2e_0$,

(ii) $xy + yx = |x|^2 + |y|^2 - |x + y|^2 \in \operatorname{Sc} \mathbb{H}$.

Proof. Property (i) immediately follows from the conditions $e_ie_j + e_je_i = -2\delta_{ij}(i, j - 1, 2, 3)$. Indeed, $x^2 = (x_1e_1 + x_2e_2 + x_3e_3)^2 = -(x_1^2 + x_2^2 + x_3^2)e_0 = -|x|^2e_0$. Property (ii) results by $-|x + y|^2 = (x + y)^2 = x^2 + xy + yx + y^2 = -|x|^2 - |y|^2 + xy + yx$. #

Definition 1.32. *In accordance with the definition of the quaternionic multiplication in Section 1.1 we set*

$(x, y) := \frac{1}{2}(\overline{x}y + y\overline{x}) \in \operatorname{Sc} \mathbb{H}$

and call (x, yt) scalar product of x and y in Vec $\mathbb{H}$.

Corollary 1.33. *The scalar product is commutative, i.e., $(x, y) = (y, x)$.*

Corollary 1.34. (PYTHAGORAS' theorem) $(x, y) = 0$ *if and only if* $|x+y|^2 = |x|^2 + |y|^2$.

Proof. Clearly, that $(x, y) = 0$ is equivalent to $(\overline{x}, y) = 0$. Using Lemma 1.31. we prove the corollary. #

Corollary 1.35. *The scalar product permits the representation*

$(x, y) = |x||y| \cos(\overline{x}, y)$.

Proof. This follows from the definition and by the aid of the cosine theorem. #

Lemma 1.36. *This scalar product fulfils the relations:*

(i) $(\lambda x, \mu y) = \lambda\mu(x, y) \quad (\lambda, \mu \in \mathbb{R})$

(ii) $(x, y + z) = (x, y) + (x, z)$ *and* $(x + y, z) = (x, z) + (y, z)$.

Proof. The proof immediately follows from the definition. #

Example 1.37. Let $x = \cos\alpha e_1 + \sin\alpha e_2$, $\overline{y} = \cos\beta\overline{e_1} + \sin\beta\overline{e_2}$ then $|x| = |y| = 1$ and $(x, y) = \cos\alpha\,\cos\beta + \sin\alpha\,\sin\beta = \cos(x, \overline{y}) = \cos(\alpha - \beta)$.

Remark 1.38. Clearly, that in a fixed basis $\{e_1, e_2, e_3\}$ in Vec $\mathbb{H}$

$(x, y) = \sum_{i=1}^{3} x_i y_i$.

Lemma 1.39. *Let $a \in \text{Vec}\,\mathbb{H}, d \in \mathbb{R}$. Then any solution of the equation*

$(a, x) = d, \quad x \in \text{Vec}\,\mathbb{H},$

can be written in the form

$$x = \frac{\overline{a}}{|a|^2} d + u$$

with $u \in \text{Vec}\,\mathbb{H}$ and $(a, u) = 0$.

Note that J. Gibbs introduced in 1902 the cross-product, which we defined in Subsection 1.1.1.

Lemma 1.40. *We have $x \times y = \frac{1}{2}(xy - yx)$.*

Proof. The proof is a consequence of the decomposition

$xy = \frac{1}{2}(xy + yx) + \frac{1}{2}(xy - yx)$. #

Corollary 1.41. *The cross-product has the properties*

(i) $x \times y = -(y \times x)$, *hence* $x \times x = 0$,

(ii) $x \times (y + z) = x \times y + x \times z$.

Proof. It is sufficient to show (ii). We find $x \times (y + z) = \frac{1}{2}[x(y+z) - (y+z)x]$ $= \frac{1}{2}[xy + xz - yx - zx] = \frac{1}{2}[xy - yx] + \frac{1}{2}[xz - zx] = x \times y + x \times z$. #

Corollary 1.42. *The vector $x \times y$ is orthogonal to x and y, i.e.,*

$(x \times y, x) = (x \times y, y) = 0.$

Proof. A straightforward calculation leads to

$$\begin{aligned}(x \times y, x) &= -\tfrac{1}{4}\left([xy - yx]x + x[xy - yx]\right) \\ &= -\tfrac{1}{4}[(xy)x - yx^2 + x^2y - xyx] = 0.\end{aligned}$$

Similar considerations prove $(x \times y, y) = 0$. #

Proposition 1.43. *We have the following relations:*

(i) $|x|^2|y|^2 = |(x, y)|^2 + |x \times y|^2$ (Lagrange's identity),

(ii) $|(x,y)| \leq |x||y|$ (inequality of SCHWARZ).

Proof. The quaternonic multiplication rule provides that for any quaternion $x = x_0 + \underline{x}$ with $\underline{x} = x_1e_1 + x_2e_2 + x_3e_3$ holds $|x|^2 = x_0^2 + |\underline{x}|^2$. Therefore, we have only to square the identity $xy = -(x,y) + x \times y$. #

Corollary 1.44. *The norm of the cross-product permits the representation*

$$|x \times y| = |x||y| \sin(x,y).$$

Proof. This follows from Proposition 1.43. and Corollary 1.35. #

Corollary 1.45. *The cross-product is homogeneous according to both vectors, i.e.,*

$$(\lambda x) \times (\mu y) = \lambda\mu(x \times y)$$

for $\lambda, \mu \in \mathbb{R}$.

Proof It follows from the definition. #

There are cases where the mutually geometric position of two vectors can be expressed by algebraical means.

Proposition 1.46. *We get*

(i) *x, y are parallel (i.e. $x \times y = 0$!) if and only if $xy = yx$,*

(ii) *x is orthogonal to y if and only if $xy = -yx$.*

Proof. The identity $x \times y = \frac{1}{2}(xy - yx)$ shows (i). The relation $(x,y) = -\frac{1}{2}(xy + yx)$ verifies (ii). #

Proposition 1.47. *The following identities are true:*

(i) $(x, y \times z) = (x \times y, z)$,

(ii) $(x \times (y \times z)) = (x,z)y - z(x,y)$.

Proof. As $(xy)z = x(yz)$ we obtain $x \times (y \times z) - x(y,z) - (x, y \times z) = (x \times y) \times z - (x,y)z - (x \times y, z)$. This leads directly to (i). Setting now $h := x \times (y \times z)$ then h is orthogonal to x and $y \times z$. Therefore h has to be a linear combination of y and z. Similarly, $(x \times y) \times z$ is a linear combination of x and y. This leads now to

$$x \times (y \times z) + z(y,x) = (x \times y) \times z + (z,y)x = \lambda y.$$

Multiplying by x from the left we get

$$(x, x \times (y \times z)) = (x \times x, y \times z) = 0$$

and, therefore,

$$(x, z)(y, x) = \lambda(y, x).$$

In the case of $(y, x) \neq 0$ we get $\lambda = (x, z)$ and such our proposition. If $(y, x) = 0$, then one has to multiply by z from the left and in the case of $(y, z) \neq 0$ it follows also that $\lambda = (x, z)$. It only remains to consider the case where y is orthogonal to both x and z. Then we multiply the equation

$$x \times (y \times z) = \lambda y$$

by y and get by using (i) $(y \times x, y \times z) = \lambda |y|^2$. This leads to

$$|y|^2 |x||z| \cos(y \times x, y \times z) = \lambda |y|^2$$

and, finally, to

$$\lambda = |x||z| \cos(x, z) = (x, z). \qquad \#$$

Definition 1.48. *The real number*

$$(x, y \times z) = (y, x \times z) = (x \times y, z)$$

is called spatial product of the vectors x, y and z. Its denotation is (x, y, z).

Corollary 1.49. *Let $x, y, z, w \in \mathrm{Vec}\,\mathbb{H}$. Then we have*

$$(x \times y) \times (z \times w) = y(x, z, w) - x(y, z, w).$$

Special case:

$$(x \times y) \times (y \times w) = y(x, y, w)$$

(double vector rule).

Proposition 1.50. *Let $x, u \in \mathrm{Vec}\,\mathbb{H}$. Then x permits the decomposition*

$$x = (x, u)\frac{u}{|u|^2} + (u \times x) \times \frac{u}{|u|^2},$$

where $(u \times x) \times \frac{u}{|u|^2}$ is orthogonal to u.

Proposition 1.51. *Let $a, b \in \mathrm{Vec}\,\mathbb{H}$. Then under the necessary condition $(a, b) = 0$ the equation*

$$a \times x = b$$

has the general solution

$$x = \frac{b \times a}{|a|^2} + \lambda a \quad (\lambda \in \mathbb{R}).$$

Proof. It is an immediate consequence of Proposition 1.50. #

1.3 Rotations

As is well known the matrices

$$R(\phi) = \begin{pmatrix} \cos\phi & -\sin\phi \\ \sin\phi & \cos\phi \end{pmatrix}$$

with $\phi \in \mathbb{R}$ form a group. It is easy to verify that for $\phi,\ \psi \in \mathbb{R}$ the relation

$$R(\phi + \psi) = R(\phi)R(\psi)$$

holds. Any rotation in the plane can be characterized by such a matrix. The corresponding group of all rotations in the plane is called *special orthogonal group* $SO(2)$. Another realization of this group is the unit circle S^1 in the complex plane with the complex multiplication as the group operation. We can identify the matrix $R(\phi)$ with the complex number $e^{i\phi}$.

1.3.1 Rotations in $\mathbb{R}^3$

Every rotation $R(\phi)$ in $\mathbb{R}^3$ is given by an axis a and an angle ϕ of rotation around this axis. The group of all rotations is denoted by $SO(3)$. This group may be realized by linear transformations with corresponding matrices $R(\phi) = R$ which preserve the scalar product

$$(x, y) = (Rx, Ry)$$

and $\det R = 1$. Rotations around the coordinate axis can be expressed by the matrices

$$R_1(\alpha) = \begin{pmatrix} 1 & 0 & 0 \\ 0 & \cos\alpha & -\sin\alpha \\ 0 & \sin\alpha & \cos\alpha \end{pmatrix},$$

$$R_2(\beta) = \begin{pmatrix} \cos\beta & 0 & \sin\beta \\ 0 & 1 & 0 \\ -\sin\beta & 0 & \cos\beta \end{pmatrix},$$

$$R_3(\gamma) = \begin{pmatrix} \cos\gamma & -\sin\gamma & 0 \\ \sin\gamma & \cos\gamma & 0 \\ 0 & 0 & 1 \end{pmatrix}.$$

The angles α, β, γ are called EULER *angles*. Now, let us explain the role of quaternions for the representation of rotations in the space. We already know from Section 1.1.2 that a real quaternion x permits the binary representation

$$x = |x|\left(\frac{x_0}{|x|} + \omega\,\frac{|\underline{x}|}{|x|}\right)$$

with

$$\omega^2 = \left(\frac{\underline{x}}{|\underline{x}|}\right)^2 = -1.$$

Definition 1.52. *Let $\omega \in \mathrm{Vec}\,\mathbb{H}$, $\omega^2 = -1$. The exponential function $e^{\omega\theta}$ is defined by the formal expansion*

$$e^{\omega\theta} := 1 + \omega\theta - \frac{1}{2!}\theta^2 - \frac{1}{3!}\omega\theta^3 + \cdots,$$

where we have uniform convergence of the expansion

$$e^{\theta} = 1 + \theta + \frac{1}{2!}\theta^2 + \cdots, \theta \in \mathbb{R}$$

and the estimation

$$|e^{\omega\theta}| \leq 1 + |\theta| + \frac{1}{2!}|\theta|^2 + \cdots = e^{|\theta|}.$$

Proposition 1.53. *Let $\omega \in \mathrm{Vec}\ \mathbb{H}$ with $\omega^2 = -1$. Then*

(i) $e^{\omega\theta} = \cos\theta + \omega\sin\theta$

(ii) $|e^{\omega\theta}| = 1$

(iii) In the binary representation we have

$$\cos\theta = \frac{x_0}{|x|}, \quad \sin\theta = \frac{|\underline{x}|}{|x|}.$$

Proof. It is sufficient to compare separately the vector parts and scalar parts in the definition of $e^{\omega\theta}$. #

Corollary 1.54. *Let $x, y \in Vec\,\mathbb{H}$. Then*

(i) $xy = yx \longrightarrow e^x e^y = e^{x+y}$.

(ii) *From $e^x e^y = e^{x+y}$ it does not follow that $xy = yx$* .

Proof. We have only to verify (ii). For this reason we take $x := 3\pi e_1$ and $y := 4\pi e_2$. Then we obtain $e^{3\pi e_1} = -1, e^{4\pi e_2} = 1$ and $e^{3\pi e_1 + 4\pi e_2} = e^{((3/5)e_1 + (4/5)e_2)5\pi}$. The latter identity is minus one, because of $((3/5)e_1 + (4/5)e_2)^2 = -1$. On the other hand, $4\pi e_2 \cdot 3\pi e_1 = -3\pi e_1 \cdot 4\pi e_2 \neq 3\pi \cdot e_1 4\pi e_2$.
#

A similar definition can also be given in the algebra P.

Definition 1.55. *Let $\omega \in \mathrm{Vec\,P}$, $\theta \in \mathrm{R}$, $\omega^2 = 1$ then the exponential function can be defined by the series*

$$e^{\theta\omega} := 1 + \omega\theta + \frac{1}{2!}\theta^2 + \omega\frac{1}{3!}\theta^3 + \cdots .$$

The convergence of this series is ensured (cf. Definition 1.52.).

Proposition 1.56. *Let $\omega \in \mathrm{Vec\,P}$, $\omega^2 = 1$ then we have*

(i) $e^{\theta\omega} = \cosh\theta + \omega\sinh\theta$

(ii) $e^{\theta(j\omega)} = \cos\theta + (j\omega)\sin\theta,\ j^2 = -1$,

where $j\omega$ is a unit bivector in P.

Proof. It follows from the complex analysis in the plane. We have only to consider the relations $\cosh\theta = \cos j\theta$ and $\sin j\theta = -j\sinh\theta$. #

Remark 1.57. In physics the mapping $x \longmapsto c^{\theta e_1e_2e_3}x$ $(\theta \in \mathbb{R})$ with $(e_1e_2e_3)^2 = -1$, where $\{e_1, e_2, e_3\}$ is a basis in $\mathbb{R}^3$, is called the LARMOR–REINICH *transformation.*

Corollary 1.58. (EULER's formula.) *For the rotation of a vector $x \in \mathrm{Vec}\,\mathbb{H}$ we have the formula*

$$x' = x\cos\theta + (\cos\theta - 1)(x, \omega) + (\omega \times x)\sin\theta$$

with $\omega^2 = -1$ and ω stands for the rotation axis.

Proof. We start with the decomposition $x = u + v$, with $u \times \omega = 0$ and $(v, \omega) = 0$. We do the same for $x' = u' + v'$ with $u' \times \omega = 0$ and $(v', \omega) = 0$. Since ω commutates with u, we get

$$u' = u = -(\omega, x)\omega.$$

Furthermore,

$$v' = e^{\frac{1}{2}\theta\omega}ve^{-\frac{1}{2}\theta\omega} = ve^{-\theta\omega}.$$

Hence,

$$v' = v(\cos\theta - \omega\sin\theta) = v\cos\theta - (v \times \omega)\sin\theta$$

and

$$\begin{aligned} x' &= u' + v' = -(\omega, x)\omega + v\cos\theta - (v \times \omega)\sin\theta \\ &= -(\omega, x)\omega + x\cos\theta - u\cos\theta + (\omega \times v)\sin\theta \\ &= -(\omega, x)\omega - u\cos\theta + x\cos\theta + (\omega \times x)\sin\theta \\ &= (\cos\theta - 1)(\omega, x)\omega + x\cos\theta + (\omega \times x)\sin\theta. \end{aligned}$$ #

Corollary 1.59. *The coordinates of* $x \in \text{Vec}\,\mathbb{H}$ *transform as follows:*

$$x_1' = x_1 \cos\theta + \omega_1(\cos\theta - 1)[-(x_1\omega_1 + x_2\omega_2 + x_3\omega_3)] + (\omega_2 x_2 - \omega_3 x_2)\sin\theta,$$

$$x_2' = x_2 \cos\theta + \omega_2(\cos\theta - 1)[-(x_1\omega_1 + x_2\omega_2 + x_3\omega_3)] + (\omega_3 x_1 - \omega_1 x_3)\sin\theta,$$

$$x_3' = x_3 \cos\theta + \omega_3(\cos\theta - 1)[-(x_1\omega_1 + x_2\omega_2 + x_3\omega_3)] + (\omega_1 x_2 - \omega_2 x_1)\sin\theta.$$

Remark 1.60. For $\omega = e_1$ we get

$$\begin{aligned} x_1' &= x_1\cos\theta + x_1(1-\cos\theta), \\ x_2' &= x_2\cos\theta - x_3\sin\theta, \\ x_3' &= x_3\cos\theta + x_2\sin\theta. \end{aligned}$$

This is equivalent to

$$x' = \begin{pmatrix} 1 & 0 & 0 \\ 0 & \cos\theta & -\sin\theta \\ 0 & \sin\theta & \cos\theta \end{pmatrix} \begin{pmatrix} x_1 \\ x_2 \\ x_3 \end{pmatrix} = R_1(\theta)x.$$

Similarly, we achieve $x' = R_2(\theta)x$ for $\omega = e_2$ and $x' = R_3(\theta)x$ for $\omega = e_3$.

Definition 1.61. *Let* $u \in \mathbb{R}^3$. *Each vector* $x \in \mathbb{R}^3$ *permits the decomposition* $x = v + \lambda u$ *with* $v \in \{z : (z,u) = 0\} = u^\perp$. *Then the map*

$$R_u x = v - \lambda u$$

is called reflection along u, *i.e.,* R_u *is the identity onto* $u^\perp$ *and minus the identity on the line through* u .

Corollary 1.62. *Any reflection has the representation*

$$R_u x = x - 2\frac{(x,u)}{(u,u)}u.$$

Proof. Let $x = v + \lambda u$ then $R_u x = v - \lambda u$. Obviously, $\lambda = \frac{(x,u)}{(u,u)}$, such that $R_u x = x - 2\lambda u$. #

Theorem 1.63. (CARTAN–DIEUDONNE)
Each orthogonal transformation $R \in O(\mathbb{R}^3)$ *can be represented as a product of a finite number* s *of reflections along the vectors* ω_j, $(\omega_j, \omega_j) \neq 0$ $(j = 1, ..., s)$, *i.e.,* $R = R_{\omega_1} R_{\omega_2} ... R_{\omega_s}$.

Remark 1.64. Theorem 1.63. can be generalized to vector spaces with signature p, q and dimension n (cf. [121]).

1.3.2 Formula of Rodriguez

Just before W.R. HAMILTON discovered quaternions O. RODRIGUEZ obtained important theorems on 3-dimensional rotations. First, we make some preparations for the famous formula of RODRIGUEZ.

Proposition 1.65. *Let $x, y \in \operatorname{Vec} \mathbb{H}$, y invertible. Then the mapping*

$$\rho_y(x) = -yxy^{-1}$$

is a reflection at the plane orthogonal to the set $\{\lambda y, \lambda \in \mathbb{R}\}$.

Proof. Clearly, that $(yxy^{-1})^2 = yxy^{-1}yxy^{-1} = yx^2y^{-1} = x^2$. Furthermore, $\rho_y(x)$ is linear and we have $\rho_y(y) = -y$. Because all z with $(z, y) = 0$ satisfy the anti-commutativity relation with y (Proposition 1.46.) one gets $\rho_y(z) = -yzy^{-1} = zyy^{-1}$. #

Corollary 1.66. *Any rotation in $\mathbb{R}^3$ is of the form*

$$\rho_u(x) = uxu^{-1},$$

where $u \in \operatorname{Vec} \mathbb{H}$ and u^{-1} exists.

Remark 1.67. R. LIPSCHITZ was the first to obtain a representation formula of this type. He wrote contrary to our formula

$$\rho'_u(x) = u^{-1}xu.$$

Using Definition 1.52. we obtain immediately:

Lemma 1.68. *The mapping $x \longmapsto x' = e^{-\frac{\theta}{2}\omega}xe^{\frac{\theta}{2}\omega}$ with $\omega^2 = -1$ $(\theta \in \mathbb{R})$ describes a rotation of the point x around ω by the angle θ.*

Now we will substitute: $\theta\omega = ja$, $j = e_1e_2e_3$, and $a^2 = 1$. Then one writes

$$e^{ja} = \cos a + j \sin a.$$

The corresponding series expansion shows $\cos a = \cos |a|$, $\sin a = (a/|a|) \sin |a|$ and in this way $\tan a = (a/|a|) \tan |a|$.

To describe the product of two consecutive rotations we have only to composite two exponentials.

Lemma 1.69. *(formula of* RODRIGUEZ*). Let $x' = e^{-\frac{1}{2}ja}xe^{\frac{1}{2}ja}$ be a rotation of x around a by the angle $|a|$ and $x'' = e^{-\frac{1}{2}jb}x'e^{\frac{1}{2}jb}$ be a rotation of x' around b by the angle $|b|$. Then we get for the composition of both rotations $x'' = e^{-\frac{1}{2}jc}xe^{\frac{1}{2}jc}$ with $e^{\frac{1}{2}jc} = e^{\frac{1}{2}ja}e^{\frac{1}{2}jb}$, where*

$$\tan\frac{c}{2} = \frac{\tan\frac{a}{2} + \tan\frac{b}{2} - \tan\frac{a}{2} \times \tan\frac{b}{2}}{1 - (\tan\frac{a}{2}, \tan\frac{b}{2})}.$$

This describes a rotation around the axis c by the angle $|c|$.

Remark 1.70. (see [182]). Let A be an antisymmetric matrix of $\mathbb{R}^4$. There are two invariant rotation planes, which have only one common point. When the eigenvalues of A are $i\alpha$ and $i\beta$ then the rotation angles of e^A are α and β. It can be proved that e^A permits the representation

$$e^A = a_0 I + a_1 A + a_2 A^2 + a_3 A^3$$

with

$$a_0 = (\alpha^2 \cos\beta - \beta^2 \cos\alpha)(\alpha^2 - \beta^2)^{-1},\ a_1 = \left(\frac{\alpha^2}{\beta}\sin\beta - \frac{\beta^2}{\alpha}\sin\alpha\right)(\alpha^2 - \beta^2)^{-1}$$

$$a_2 = (\cos\beta - \cos\alpha)(\alpha^2 - \beta^2)^{-1},\ a_3 = \left(\frac{1}{\beta}\sin\alpha - \frac{1}{\alpha}\sin\alpha\right)(\alpha^2 - \beta^2)^{-1}$$

e^A is defined, e.g., in [10]. In addition this gives the formula:

$$\lim_{\beta\to\alpha} e^A = I\cos\alpha + \frac{A}{\alpha}\sin\alpha.$$

1.4 General Clifford algebras

Let us start with an introduction into one of the most exciting extensions of the real number system which incorporates the geometric concept of the direction, the so-called CLIFFORD *algebras*. This famous structure provides the key for consideration of a large number of geometrical ideas which we find in mathematics and physics. Also when we will later sometimes restrict our observations only to special types of CLIFFORD algebras, namely to the real algebras as well as the real and complex quaternions, we think that it is essential to discuss this notion to some extent.
HAMILTON's discovery of quaternions in 1843 [119] can be seen as the first step towards a geometric algebra. Maybe one should notice that the comprehensive use of the rotation group by O. RODRIGUES in 1840 [244] was from geometrical point of view a pre-discovery of the quaternions. In 1844 appears "Die Ausdehnungslehre" by the German mathematician H. GRASSMANN. The embedding of the quaternions into the GRASSMANN system by W.K. CLIFFORD in his paper "Applications of Grassmann's extensive algebra" [45] leads in 1878 to the birth of the CLIFFORD algebras. CLIFFORD himself called his algebra "Geometric algebra". Two years later in 1880 R. LIPSCHITZ reinvented these algebras during his investigations on the sums of squares. He was the first who gave geometrical applications also in higher dimensions. K.Th. VAHLEN introduced in 1902 [318] a multiplication rule between two basis elements of a CLIFFORD algebra. Using a relation with WALSH functions this rule could be improved by P.E. HAGMARK and P. LOUNESTO in 1986. E. CARTAN in 1908 [41] discovered the connections of general CLIFFORD algebras with some types of matrix algebras and proved a periodicity theorem. Following the book by B.L. VAN DER WAERDEN, "Spinor Analysis" [322], the term "spinor" was introduced by EHRENFEST in 1929. This name is attached to the discovery of the fundamental theories of W. PAULI and P. DIRAC (1928). Finally we mention here that M. RIESZ in 1947 used so-called primitive idempotents for his construction of spinor spaces as minimal left ideals in CLIFFORD algebras. The reader can find more on the historical facts in the very interesting article of P. LOUNESTO "Clifford algebras and Hestenes spinors" [179] dedicated to the 60th birthday of D. HESTENES.

1.4.1 Vector space with quadratic form

Let $(\mathbb{R}^n, Q) = V$ be a real vector space with the basis $\{e_1, ..., e_n\}$ and the non-degenerate quadratic form $Q : V \longmapsto \mathbb{R}$ with

$$Q(x) = x_1^2 + x_2^2 + ... + x_p^2 - x_{p+1}^2 - ... - x_{p+q}^2, \; p+q = n,$$

where $x = \sum_{i=1}^n x_i e_i$ represents an arbitrary element from $(\mathbb{R}^n, Q)$.
SYLVESTER's theorem guarantees that the natural numbers p, q do not depend of the choice of the basis $\{e_1, ...e_n\}$. Now, we will write for the pair $(\mathbb{R}^n, Q)$ the abbreviation $\mathbb{R}^{p,q}$. The tupel p, q is named *signature* of $\mathbb{R}^{p,q}$. The spaces $\mathbb{R}^{n,0}$ are called EUCLIDEAN *spaces* while all spaces of the type $\mathbb{R}^{0,n}$ are named *anti-Euclidean spaces.* In particular, the space $\mathbb{R}^{1,3}$ is called MINKOWSKI *space-time.*

1.4.2 A corresponding Clifford algebra

Let us now construct an associative algebra $\mathcal{A}$ with unit e_0 over $\mathbb{R}$ which contains copies of $\mathbb{R}$ and $\mathbb{R}^{p,q}$. Using the above-mentioned orthonormal basis we demand the following conditions:

(i) $e_i^2 = 1 \, (1 \leq i \leq p)$, $e_i^2 = -1 \, (p < i \leq n)$, $\quad e_i e_j + e_j e_i = 0 \; (i < j)$.

(ii) $\mathbb{R}^{p,q}$ generates $\mathcal{A}$ as an algebra over $\mathbb{R}$.

(iii) For $p - q = 1 \, mod \, 4$ shall be $e_N = e_1 e_2 ... e_n \neq \pm 1$.

The algebra $\mathcal{A}$ is called *universal* CLIFFORD *algebra* $Cl_{p,q}$. Sometimes we will also use the denotation $C\ell_{p,q}$.
If condition (iii) is not fulfilled, one speaks also of a CLIFFORD algebra, a so-called *non-universal* CLIFFORD *algebra.*

All elements $x \in Cl_{p,q}$ defined as a sum of terms of length k.

$$x = \sum_{1 \leq i_1 \leq ... \leq i_k \leq n} x_{i_1 ... i_k} e_{i_1} ... e_{i_k}$$

are said to be an element of *degree* k or k*-vector.* The subset of all 1-vectors we will denote by $\text{Vec}\, Cl_{p,q}$.

Lemma 1.71. *Let $p - q \neq 1 \, mod \, 4$, then the universal* CLIFFORD *algebra $Cl_{p,q}$ is simple; this means that there are no proper two-sided ideals. For $p - q = 1 \, mod \, 4$ the real* CLIFFORD *algebra $Cl_{p,q}$ is a direct sum of the simple ideals $\frac{1}{2}(1 + e_N)Cl_{p,q}$ and $\frac{1}{2}(1 - e_N)Cl_{p,q}$ with $e_N = e_1 ... e_n$. The centre Z of $Cl_{p,q}$ consists in the case of an even n only of scalar multiples of the unit, and in the case of odd dimensions it is ismorphic to $\mathbb{R} \bigotimes \mathbb{R} (e_N = 1, p - q = 1 mod 4)$ or to $\mathbb{C}$ $(e_N = -1, p - q = 3 \, mod \, 4)$.*

Furthermore, linear combinations of the degree $k = 2k'$ are called *even*, those with a degree $k = 2k' + 1$ are called *odd*. Taking the tensor product of the real algebras $Cl_{p,q}$ with the field of complex numbers one obtains the *complex* CLIFFORD *algebras* $\mathbb{C} \bigotimes Cl_{p,q}$. On account of its physical importance let us introduce the so-called *chirality operator* . It is nothing other than the multiplication by the element $\Theta = i^{\mu} e_N$, where $\mu = 0, 1, 2, 3$ and i is the imaginary unit. Elements of the CLIFFORD algebra $Cl_{n,0}$ are sometimes called *cliffors* [139].

Remark 1.72. Both relations (i), (iii) can be replaced equivalently by (i$'$) : $x^2 = Q(x)$ *for any vector* x *in* $\mathbb{R}^{p,q}$ and (iii$'$):$\mathcal{A}$ *is not generated by subspaces of* $\mathbb{R}^{p,q}$. In order to prove the first condition we have to set in the sequence for x the special elements e_i , e_j and $e_i + e_j$. A simple calculation leads to (i). The meaning of condition (iii) and of course also of condition (iii$'$) is to guarantee the uniquness of the Clifford algebra. When condition (iii) is fulfilled then $\dim Cl_{p,q} = 2^n$ and so universal Clifford algebras always have dimension 2^n. In the non-universal case the dimension is only 2^{n-1}.

Remark 1.73. P. LOUNESTO gave the following good example which shows that condition (iii) should not be omitted: *Consider the orthonormal basis* $\{e_1, e_2, e_3\}$ *in* $\mathbb{R}^3$. *Relations* $e_i^2 = -1,\ e_i e_j + e_j e_i = 0 \quad (i \neq j)$ *are said to be fulfilled. Furthermore, let* $e_3 = e_1 e_2$ *then we get the algebra of the real quaternions with the basis* $\{e_0, e_1, e_2, e_3\}$. *From the same vector space we can also obtain* $Cl_{0,3}$ *using the basis* $e_0, e_1, e_2, e_3 \neq e_1 e_2, e_1 e_2, e_1 e_3, e_2 e_3, e_1 e_2 e_3$ *which satisfy the relations* (i). *It can be seen that* $Cl_{0,3}$ *is isomorphic to* $\mathbb{H} \oplus \mathbb{H}$.

1.4.3 Low-dimensional Clifford algebras

A good possibility of representing CLIFFORD algebras, especially those of low dimensions, is to do it with the help of the PAULI *matrices* $\sigma_0, \sigma_1, \sigma_2, \sigma_3$ introduced in Section 1.1.3. In this way we will interchange the basic elements e_i $(i = 0, 1, 2, 3)$ by the PAULI matrices σ_i $(i = 0, 1, 2, 3)$. It is easy to see that $\sigma_i^2 = \sigma_0$ and that the elements σ_i also fulfil the anti-commutativity property. We will agree upon the use of the symbol " $=$ " for algebras. Let $M(d, F)$ denote the ring of $d \times d$ matrices over the field F. Let us now discuss the following examples:

$Cl_{0,0}$: In this case the basis consists only of the identity element σ_0. Therefore $Cl_{0,0} = \mathbb{R}$. An isomorphism is explicitly given by $x = x_0 \sigma_0$.

$Cl_{1,0}$: The basis in this algebra is given by σ_0 and σ_3. Obviously, $Cl_{1,0} = \mathbb{R} \oplus \mathbb{R}$. The isomorphism can be described by $x = x_0 \sigma_0 + x_1 \sigma_3$ where x is an element from $Cl_{1,0}$.

$Cl_{0,1}$: This algebra is generated by σ_0 and $i\sigma_2$ with $(i\sigma_2)^2 = -\sigma_0$, therefore $Cl_{0,1} = \mathbb{C}$. This is clear by the representation $x = x_0 \sigma_0 + x_1 i \sigma_2$.

$Cl_{2,0}$: The basis is given by $\sigma_0, \sigma_1, \sigma_3, \sigma_1\sigma_3$. Hence an arbitrary element x from $Cl_{2,0}$ can be represented by $x = x_0\sigma_0 + x_1\sigma_1 + x_2\sigma_3 + x_3\sigma_1\sigma_3$, so $Cl_{2,0} = M(2, \mathbb{R})$, where $M(2, \mathbb{R})$ is the algebra of real 2×2 matrices .

$Cl_{1,1}$: The matrices $\sigma_0, \sigma_1, i\sigma_2, \sigma_1 i\sigma_2$ form a basis in $Cl_{1,1}$. After a straightforward computation we obtain the relation $Cl_{1,1} = M(2, \mathbb{R})$.

$Cl_{0,2}$: The matrices $\sigma_0, -i\sigma_1, -i\sigma_2, \sigma_1\sigma_2\ (= -i\sigma_3)$ form a basis in $Cl_{0,2}$. For any element x of this algebra we obtain the representation

$$x = \begin{pmatrix} x_0 - ix_1 & -x_2 + ix_3 \\ x_2 + ix_3 & x_0 + ix_1 \end{pmatrix}$$

with $x_i \in \mathbb{R} \quad (i = 0, 1, 2, 3)$.

Later we will see that in this way can also be described a matrix realization of the *algebra of real quaternions* $\mathbb{H}$. Hence $Cl_{0,2} = \mathbb{H}$.

$Cl_{3,0}$: The matrices $\sigma_0, \sigma_1, \sigma_2, \sigma_3, \sigma_1\sigma_2, \sigma_2\sigma_3, \sigma_3\sigma_1, \sigma_1\sigma_2\sigma_3$ form a basis in the algebra $Cl_{3,0}$. We find that $\sigma_1\sigma_2 = -i\sigma_3, \sigma_2\sigma_3 = -i\sigma_1, \sigma_3\sigma_1 = -i\sigma_2$ and $\sigma_1\sigma_2\sigma_3 = -i\sigma_0$. An arbitrary element x of this algebra can be represented by

$$x = (x_0 - x_7 i)\sigma_0 + (x_1 - x_5 i)\sigma_1 + (x_2 - x_6 i)\sigma_2 + (x_3 - x_4 i)\sigma_3.$$

Then it is easy to get $Cl_{3,0} = \mathbb{C} \bigotimes \mathbb{H}$. This algebra coincides with the PAULI *algebra*. The denotation by P is made in honour of WOLFGANG PAULI.

$Cl_{1,3}$: In this algebra a basis can be constructed by the so-called *d-numbers* $\gamma_0, \gamma_1, \gamma_2, \gamma_3$ of DIRAC which are formal 2×2 matrices in which entries are the PAULI matrices. So we have

$$\gamma_0 = \begin{pmatrix} \sigma_0 & 0 \\ 0 & -\sigma_0 \end{pmatrix}, \gamma_i = \begin{pmatrix} 0 & -\sigma_i \\ \sigma_i & 0 \end{pmatrix}$$

for $i = 1, 2, 3$. Clearly, we have $\gamma_0^2 = \gamma_0$ and $\gamma_i^2 = -\gamma_0$. The basis in $Cl_{1,3}$ is now given by

$$\begin{aligned} &1, \\ &\gamma_0, \gamma_1, \gamma_2, \gamma_3, \\ &\gamma_1\gamma_0, \gamma_2\gamma_0, \gamma_3\gamma_0, \gamma_1\gamma_2, \gamma_2\gamma_3, \gamma_3\gamma_1, \\ &\gamma_0\gamma_1\gamma_2, \gamma_0\gamma_2\gamma_3, \gamma_0\gamma_3\gamma_1, \gamma_1\gamma_2\gamma_3, \\ &\gamma_0\gamma_1\gamma_2\gamma_3, \end{aligned}$$

where $1 : \mathbb{R}^{1,3} \longmapsto \mathbb{R}^{1,3}$ is the unit matrix. This algebra is called *space–time algebra*. One can show that this algebra is isomorphic to $M(4, \mathbb{R})$.

1.4.4 Table of matrix representations

CLIFFORD algebras are isomorphic as associative algebras with unit to corresponding matrix algebras. Setting now $s := (p+q)/2$ one can show that only eight different rings of matrices occur in the description of the structure of CLIFFORD algebras. The following table gives an overview on the possible structures.

$(p\text{-}q)$ mod 8	matrix ring
0	$M(2^{[s]}, \mathbb{R})$
1	$M(2^{[s]}, \mathbb{R}) \oplus M(2^{[s]}, \mathbb{R})$
2	$M(2^{[s]}, \mathbb{R})$
3	$M(2^{[\frac{s-1}{2}]}, \mathbb{C})$
4	$M(2^{[s-1]}, \mathbb{H})$
5	$M(2^{[\frac{s-3}{2}]}, \mathbb{H}) \oplus M(2^{[\frac{s-3}{2}]}, \mathbb{H})$
6	$M(2^{[s-1]}, \mathbb{H})$
7	$M(2^{[\frac{s-1}{2}]}, \mathbb{C})$

We mention that s depends only on the dimension of the algebra and not on its special structure. The denotation $[s]$ means the entire part of s.

1.4.5 Involutions

In a CLIFFORD algebra there are three involutions. Depending on the degree k of $x \in Cl_{p,q}$ we define the following:

(i) *grade involution* (or *main involution*) by the formula $\hat{x} := (-1)^k x$,

(ii) *reverse involution* (or *inversion*) by the relation $\tilde{x} := (-1)^{\frac{k(k-1)}{2}} x$,

(iii) *conjugation* by $\bar{x} := (-1)^{\frac{k(k+1)}{2}} x$.

Lemma 1.74. *These involutions satisfy the following isomorphic and anti-isomorphic conditions with respect to the multiplication of two elements*

$$(\widehat{xy}) = \hat{x}\hat{y}, \quad (\widetilde{xy}) = \tilde{y}\tilde{x}, \quad (\overline{xy}) = \bar{y}\bar{x}.$$

Reversion and conjugation can be extended complex-linear to complexified CLIFFORD algebras. In addition it is usual to introduce a *complex conjugation* by $x^* = x_1 - ix_2$ for $x = x_1 + ix_2$ $(x_k \in Cl_{p,q})$.

Example 1.75. Let be $x \in \mathbb{C} \bigotimes Cl_{3,0}$ the element

$$x = 1 + 3e_1 + 2e_2 + e_{21} + ie_{23} + e_{123},$$

where $e_{nm} := e_n e_m$ and $e_{knm} := e_k e_m e_n$, then we have for the above-mentioned involutions:

$$\begin{aligned}\hat{x} &= 1 - 3e_1 - 2e_2 + e_{21} + ie_{23} - e_{123},\\ \tilde{x} &= 1 + 3e_1 + 2e_2 - e_{21} - ie_{23} - e_{123},\\ \bar{x} &= 1 - 3e_1 - 2e_2 - e_{21} - ie_{23} + e_{123},\\ x^* &= 1 + 3e_1 + 2e_2 + e_{21} - ie_{23} + e_{123}.\end{aligned}$$

Definition 1.76. *Let $x \in Cl_{n,0} \cup Cl_{0,n}$ then $\sqrt{\mathrm{Sc}\,(\tilde{x}x)}$ is called absolute value of x. It is denoted by $|x|$.*

Remark 1.77. For $x \in Cl_{p,q}$ the product $\tilde{x}x$ is not necessarily real. P. LOUNESTO gave a counterexample: Let $p = 3, q = 1, x = (1 + e_1)(1 + e_{234}) \in Cl_{3,1}$ and $x\tilde{x} = 0 \in \mathbb{R}$ but $\tilde{x}x = 4e_{234} + 4e_{1234}$ obviously does not belong to $\mathbb{R}$. The so-called *square norm* which can be defined by $\bar{x}x = ||x||$ is often used. For more information see R. HARVEY (cf. [121]).

1.4.6 Spinor spaces

Without loss of generality one can introduce a *spinor space* as a minimal left ideal V of the real CLIFFORD algebra $Cl_{p,q}$. We mention that an ideal is a subalgebra which absorbs by multiplication with elements of $Cl_{p,q}$ the whole algebra. In this way we obtain for each spinor space the representation

$$V = Cl_{p,q}f,$$

where f is a primitive idempotent of the CLIFFORD algebra $Cl_{p,q}$. We have $\dim V = 2^r$ or $\dim V = 2^{r+1}$, where $r = [n/2]$ in dependence on $p - q = 0, 1, 2 \bmod 8$ or $p - q = 3, 4, 5, 6, 7 \bmod 8$. We will give here some examples of idempotent elements, and for detailed information we refer to corresponding papers and books (cf. [179], [235]).

Example 1.78. 1. Assume the orthogonal space $\mathbb{R}^{r,r}$ and let $\{e_1, ..., e_{2r}\}$ be a basis in it, then the primitive idempotent f can be constructed by

$$f = \left(\tfrac{1}{2}\right)^r (1 + e_1 e_{r+1}) \cdot ... \cdot (1 + e_r e_{2r}).$$

2. In $Cl_{1,3}$ with the orthonormal basis $\{e_0, e_1, e_2, e_3\}$ we find the idempotent element

$$f = \tfrac{1}{2}(1 + e_0).$$

Without loss of generality one can represent a *complex spinor space* always in the form

$$S_{p,q} = (\mathbb{C} \otimes Cl_{p,q})f,$$

where f is a primitive idempotent of the complexified algebra $\mathbb{C} \bigotimes Cl_{p,q}$. The elements in $S_{p,q}$ should be called CLIFFORD *algebraic spinors*. In lower dimensional cases the idempotents are known explicitly. Dependent on the signature we will give the idempotent elements for our examples by the following table:

signature	idempotent f
$0,2$	$\frac{1}{2}(1+ie_1)$
$1,1$	$\frac{1}{2}(1+e_1)$
$2,0$	$\frac{1}{2}(1+e_1)$
$0,3$	$\frac{1}{4}(1+ie_1)(1+ie_2e_3)$
$1,2$	$\frac{1}{4}(1+e_1)(1+ie_2e_3)$
$2,1$	$\frac{1}{4}(1+e_1)(1+e_2e_3)$
$3,0$	$\frac{1}{4}(1+e_1)(1+ie_2e_3)$
$1,3$	$\frac{1}{4}(1+e_1)(1+ie_2e_3)$

In the case of $p=0$ and $q=2m$ the space $S_{2m}=S_{0,2m}$ is called *space of* DIRAC *spinors*. Furthermore, the eigenspaces of the chirality operator $\theta \in Cl_{0,2m-1}$ with $\theta^2=1$ namely $S_{2m}^{+}=\{u \in S_{2m} : \theta u = u\}$ and $S_{2m}^{-}\{u \in S_{2m} : \theta u = -u\}$ are called *spaces of* WEYL *spinors*. The space S_{2m} splits into the subspaces $\mathrm{Re}\ S_{2m}$ and $\mathrm{Im}\ S_{2m}$, which are called *spaces of* MAJORANA *spinors*. For more details we refer to [63] and [179].

1.5 Tensors and cliffors

In 1902 the Italian mathematicians M.M.G. RICCI and T. LEVI-CIVITA introduced in their famous work "Méthodes de calcul differential absolu et leurs applications" [241] the so-called tensor algebra which is termed by some people as the mother of all algebras. Because the tensor calculus is a powerful tool for the treatment of mathematical problems in different fields of physics and engineering, we intend to work out the close connections of this calculus with CLIFFORD algebras. The advantages of working in CLIFFORD algebras instead of using tensors will be explained. C. CHEVALLEY in 1954 in his book "The Algebraic Theory of Spinors" [43] gave an introduction of a CLIFFORD algebra as the quotient algebra of the tensor algebra relative to a two-sided ideal. From this approach we will give a short survey. In order to obtain a better understanding of the relations between these important algebras we will start with a brief reminder about the construction of a tensor algebra.

1.5.1 Tensor algebra

Let $X = \mathbb{R}^{p,q}$ be a vector space of the signature p, q and the dimension n. Furthermore, let $Y := X \times X \times ... \times X$ (p times). We consider now the linear vector space F of all real-valued functions on Y which vanish with the exception of a finite number of points. Setting $y := (x_1, ..., x_p) \in Y$, then F consists of all expressions of the form

$$\sum_{i=1}^{m} \mu_i f_i(y)$$

with $y \in Y, m \in \mathbb{N}, \mu_i \in \mathbb{R}$. The functions $f_i(y) = 1$ for $y = y^i \in Y$ $(i = 1, .., m)$ and zero otherwise. For $y_u^k := (x_1, ..., x_{k-1}, u, x_{k+1}, ..., x_p) \in Y$ we define a vector subspace $F_0 \subset F$ by the generating expressions

$$f(y_{u+v}^k) = f(y_u^k) + f(y_v^k) \qquad f(y_{\lambda u}^k) = \lambda f(y_u^k)$$

with $\lambda \in \mathbb{R}$. The set of all classes of equivalence

$$\{f(y) + F_0\} =: y^\circ$$

again form a vector space which is denoted by $\bigotimes^p X$. Each element of $\bigotimes^p X$ we denote by $y^\circ := x_1 \otimes x_2 \otimes ... \otimes x_p$. Finally the sum

$$\sum_{p=0}^{\infty} \bigotimes^p X =: \bigotimes X$$

is just the *tensor algebra*.

This definition of the tensor algebra is doubtless correct, but nevertheless it is difficult to understand. Therefore we will describe the tensor product between two spaces by a set of linear mappings.

Let X_1 and X_2 be linear vector spaces of finite dimensions and their duals X_1^* and X_2^*. It is well known that X_2 is isomorpic to X_2^* (independent of the chosen basis!). In this way we identify an element $x_2 \in X_2$ with the element $f_2^* \in X_2^*$. Moreover, let $\Phi : L(X_1, X_2) \longmapsto X_1 \bigotimes X_2^*$ be a mapping which is defined by the relation

$$\Phi(\mathcal{L})(x_1, x_2) = f_2^*(\mathcal{L}(x_1))$$

with $\mathcal{L} \in L(X_1, X_2)$.

Lemma 1.79. [62] *$\Phi(\mathcal{L})$ is a bilinear mapping from $X_1 \times X_2^*$ onto $\mathbb{R}$.*

Theorem 1.80. *The vector spaces $L(X_1, X_2)$ and $X_1 \bigotimes X_2^*$ are isomorphic to each other. The isomorphism is given by the mapping Φ.*

Corollary 1.81. *We have the following isomorphisms:*

$$\begin{aligned} L(X_1, X_2^*) &= X_1 \bigotimes X_2 \\ L(X_1^*, X_2) &= X_1^* \bigotimes X_2^* \\ L(X_1, X_2) &= X_1 \bigotimes X_2^* \\ L(X_1^*, X_2^*) &= X_1^* \bigotimes X_2. \end{aligned}$$

Proof. If it is necessary, we have only to replace in Theorem 1.80. the spaces X_1 and X_2 by X_1^* and X_2^*, relatively. #

Example 1.82. (i) $\mathbb{R} \bigotimes \mathbb{R} = L(\mathbb{R}, \mathbb{R}^*) = L(\mathbb{R}, \mathbb{R}) = \mathbb{R}$

(ii) $\mathbb{R} \bigotimes \mathbb{C} = L(\mathbb{R}, \mathbb{C}^*) = L(\mathbb{R}, \mathbb{C}) = \mathbb{C}$

(iii) $\mathbb{R} \bigotimes \mathbb{H} = L(\mathbb{R}, \mathbb{H}^*) = L(\mathbb{R}, \mathbb{H}) = \mathbb{H}$

(iv) $\mathbb{C} \bigotimes \mathbb{C} = L(\mathbb{C}, \mathbb{C}^*) = L(\mathbb{C}, \mathbb{C}) = \mathbb{C} \bigoplus \mathbb{C}$

(v) $\mathbb{C} \bigotimes \mathbb{H} = L(\mathbb{C}, \mathbb{H}^*) = L(\mathbb{C}, \mathbb{H}) = M(2, \mathbb{C})$.

The identity is to understand as an isomorphism between $\mathbb{R}$-linear vector spaces.

Lemma 1.83. [235] *$\bigotimes X$ is a graded associative algebra with unit.*

1.5.2 Construction of a Clifford algebra

We will mention here only the general connections between tensors and cliffors.

Lemma 1.84. [43] *The set* $\{x \otimes x + Q(x)1 : x \in X\}$ *generates the two-sided ideal* $I(Q)$, *which is given by the formula*

$$I(Q) = \{y \otimes (x \otimes x + Q(x)) \otimes z : x \in X, y, z \in \bigotimes X\}.$$

Corollary 1.85. *Let* $B(x,\xi) = x_1\xi_1 + \ldots + x_p\xi_p - x_{p+1}\xi_{p+1} - \ldots - x_{p+q}\xi_{p+q}$, $x, \xi \in X$. *Then we have* $x \otimes \xi + \xi \otimes x + 2B(x,\xi) \in I(Q)$.

Proof. It is easy to see that

$$\begin{aligned}(x+\xi) &\otimes (x+\xi) - (x \otimes x) - (\xi \otimes \xi) + B(x+\xi, x+\xi) - B(x,x) - B(\xi,\xi) \\ &= x \otimes \xi + \xi \otimes x + 2B(x,\xi).\end{aligned}$$

Since the left-hand side belongs to $I(Q)$ the right-hand side also has to belong to $I(Q)$. #

Theorem 1.86. [88] *Let* X *be a linear vector space of signature* p, q. *Then the* CLIFFORD *algebra* $Cl_{p,q}$ *is just the quotient* $\bigotimes X/I(Q)$ *with the associated homomorphism* $\pi : \bigotimes X \longmapsto \bigotimes X/I(Q)$ *with* $\pi(1) = 1^\star$, *where* $1^\star$ *is the unit in the quotient algebra.*

Remark 1.87. Generally speaking one can identify in the above-mentioned sense anti-symmetric tensors with elements in a CLIFFORD algebra. If we evaluate this approach from the algebraic point of view then it is reasonable to do this exactly in this manner. But the advantages of a CLIFFORD algebra as geometric algebra can only be considered inadequate in the tensor algebra. According to D. HESTENES' criticism [128] we maintain that the elements of a CLIFFORD algebra which can be seen as generalized numbers characterized by the magnitude, direction, orientation and dimension represent in an optimal way the geometric properties of most of the physical terms. Anti-symmetric tensors should only be used if linear algebraic properties are the subject of investigation. It is well known that the properties of symmetric and anti-symmetric tensors are very different. Hence, one cannot expect that the tensor algebra can reflect such specific properties of anti-symmetric tensors as can be carried out in a CLIFFORD algebra.

1.6 Multivectors

The term "*vector*" has been introduced by W.R. HAMILTON and definitely regarded as a directed number. Firstly multivectors appear in H. GRASSMANN's paper of 1844. He invented his exterior algebra (now called GRASSMANN algebra). There is a close connection to a CLIFFORD algebra over an n-dimensional vector space. One can show that the GRASSMANN algebra and a corresponding CLIFFORD algebra are isomorphic in the sense of vector spaces.

1.6.1 Grassmann algebra

GRASSMANN's algebra is easy to construct from the tensor algebra by forming a proper quotient structure. For this reason we will consider all elements of the type $x \otimes x \in X \bigotimes X$, where $X = \mathbb{R}^{p,q}$.

Lemma 1.88. *The set* $J(X) = \{y \otimes (x \otimes x) \otimes z : x \in X, \quad y, z \subset \bigotimes X\}$ *form a two sided ideal in* $\bigotimes X$.

Definition 1.89. *The factor algebra* $\bigotimes X / J(X)$ *is called* GRASSMANN *algebra or exterior algebra over* X. *This algebra is denoted by* $\bigwedge X$ *and the elements in it (equivalence classes in* $\bigotimes X$!) *are written by 1 or* $x^{(p)} := x_1 x_2 \ldots x_p$, $x_i \in X$ $(p = 1, 2, \ldots)$. *The linear hull of all elements* $x^{(p)}$ *is denoted by* $\bigwedge^p X$. *In addition we set* $\bigwedge^0 X := \mathbb{R}$, $\bigwedge^1 X := X$. *The elements in* $\bigwedge^p X$ *are called p-vectors; p-vectors of the type* $x^{(p)}$ *are called simple. It is reasonable that 0-vectors are also called scalars and 1-vectors are also called vectors. For 2-vectors is usual to call these bivectors.*

Analogously to Section 1.5 we have

Lemma 1.90. $\bigwedge X$ *is a graded, associative and anticommutative algebra with unit.*

Proof. We will verify only the anticommutativity rule. The formula

$$(x + \xi) \otimes (x + \xi) - (x \otimes x) - (\xi \otimes \xi) = \xi \otimes x + x \otimes \xi.$$

is valid. Obviously, the left hand-side belongs to $J(X)$ and, therefore, we have $\xi \wedge x = -x \wedge \xi$ for $x, \xi \in X$. #

Remark 1.91. This approach shows that GRASSMANN's algebra can be identified with a subspace of $\bigotimes X$, the linear space of all skew tensors. In this context we obtain immediately from $x \wedge \xi = x \wedge \xi + \xi \wedge x + x \wedge \xi$ the representation

$$\frac{1}{2}(x \otimes \xi - \xi \otimes x) = x \wedge \xi.$$

Lemma 1.92. *Let $\{e_1, ..., e_n\}$ be a basis in X then*

$$1; e_1, ..., e_n; e_1 \wedge e_2, e_1 \wedge e_3, ..., e_{n-1} \wedge e_n; ...; e_1 \wedge ... \wedge e_n.$$

forms a basis in $\bigwedge X$ and $\dim \bigwedge X = 2^n$.

Corollary 1.93. *$\bigwedge X$ permits the representation $\bigwedge X = \sum_{p=1}^{n} \bigwedge^p X$, where especially $\bigwedge^0 X = \mathbb{R}$, $\bigwedge^1 X = X$, $\bigwedge^p X$ is for $p > 0$ the linear closure of the set $\{e_{i_1} \wedge ... \wedge e_{i_p}, 1 \leq i_1 < ... < i_p \leq n\}$.*

Proof. We have to show that $\bigwedge^p X = 0$ for $p > n$. In fact, in this case each exterior product $e_{i_1} \wedge ... \wedge e_{i_p}$ has at least two equal factors and therefore it is zero. #

Corollary 1.94. *Each p-vector is related to a skew symmetric tensor. The canonical mapping Alt: $\bigotimes X \longmapsto \bigwedge X$ is given by*

$$Alt(x_1 \otimes ... \otimes x_p) = \frac{1}{p!} \sum_{\sigma \in S_p} sgn(\sigma) x_{\sigma(1)} \otimes ... \otimes x_{\sigma(p)} = x_1 \wedge ... \wedge x_p,$$

where S_p is the special symmetric group, i.e., the group of all permutations of p elements.

1.6.2 Turn towards a Clifford algebra

Let us write for brevity $B(x, \xi) =: < x, \xi >$ (cf. Section 1.5, Corollary 1.8.3). This symmetric bilinear form can be extended to simple p-vectors in $\bigwedge^p X$ by the determination

$$< x_1 \wedge ... \wedge x_p, \xi_1 \wedge ... \wedge \xi_p >:= \det(< x_i, \xi_j >)_{i,j=1}^{p}.$$

Then we extend to all p-vectors by linearity and to all elements in $\bigwedge X$ by orthogonality.

Definition 1.95. [88] *Let $x^{(p)}$ be a simple p-vector and $x \in X$, then the operator of the exterior multiplication or creation operator E_x is defined by the relations*

$$E_x(1) = x \;,\; E_x(x^{(p)}) = x \wedge x^{(p)}$$

and by the linear extension onto $\bigwedge X$.

Definition 1.96. [88] *Let $x^{(p)}$ be a simple p-vector and $x \in X$. The operator of the interior multiplication or annihilation operator J_x is defined by*

$$J_x(1) = 0 \;,\; J_x(x^{(p)}) := \sum_{i=1}^{p} (-1)^{(i-1)} B(x, x_i) x_1 \wedge ... \wedge \hat{x}_i \wedge ... \wedge x_p$$

and by linear extension onto the whole $\bigwedge X$, where $\hat{x}_i$ means that this factor is omitted in this product.

Corollary 1.97. *Let $u \in \bigwedge^p X$ then $E_x(u) \in \bigwedge^{p+1} X$ and $J_x(u) \in \bigwedge^{p-1} X$. In addition, in the case of $\xi \in X$ holds, $J_x(\xi) =< x, \xi >= B(x, \xi)$.*

Lemma 1.98. *We have for $x \in X$, $u, v \in \bigwedge X$ the relation*

$$< E_x(u), v >=< u, J_x(v) >,$$

which means J_x is the adjoint operator to E_x.

Remark 1.99. In tensor calculus notation the interior multiplication is said to be *contraction*. In physical literature it is usually used for the denotation: $x\rfloor v, x \in X$, $v \in \bigwedge X$. In [179] this definition is generalized to $x \in \bigwedge X$.

Corollary 1.100. *Let u, v be simple p-vectors then we have the derivation property:*

$$J_x(u \wedge v) = J_x(u) \wedge v + (-1)^{deg\ u} u \wedge J_x(v).$$

Corollary 1.101. *We have $E_x^2 = J_x^2 = 0$.*

Proof. The first property is valid by reason of $x \wedge x = 0$. For the second one we consider

$$J_x^2(x^{(p)}) = \sum_{i=1}^{p} (-1)^{i-1} < x, x_i > J_x(x_1 \wedge \ldots \wedge \hat{x}_i \wedge \ldots \wedge x_p)$$

$$= \sum_{i \leq j < i \leq p} (-1)^{i+j-2} < x, x_i >< x, x_j > x_1 \wedge \ldots \wedge \hat{x}_j \wedge \ldots \wedge \hat{x}_i \wedge \ldots \wedge x_p$$

$$- \sum_{1 \leq i < j \leq p} (-1)^{i+j-2} < x, x_i >< x, x_j > x_1 \wedge \ldots \wedge \hat{x}_i \wedge \ldots \wedge \hat{x}_j \wedge \ldots \wedge x_p = 0.$$

Since both cases $i < j$ and $i > j$ are included in the sum, then $J_x^2(x^{(p)}) = 0$. #

Theorem 1.102. *Introducing the* CLIFFORD *product of $x \in X$ and $u \in \bigwedge X$ by*

$$xu := E_x(u) - J_x(u)$$

and extending this by linearity and associativity to all $\bigwedge X$, then $\bigwedge X$ is isomorphic to $Cl_{p,q}$ as a linear vector space.

1.7 Quaternionic matrices

In quantum mechanics (cf. [246]) and also in the treatment of multi-body problems it is advantageous to have a linear algebra of matrices with quaternionic entries. In this section we will deal with the main properties of quaternionic matrices and discuss applications. For a survey on quaternions and matrices with quaternionic entries we refer to the paper [82].
We have already seen in Section 1.1 that each real quaternion $q = \alpha + j\beta$, $\alpha, \beta \in \mathbb{C}$, $j^2 = -1$ can be identified with the matrix

$$M(q) = \begin{pmatrix} \alpha & -\overline{\beta} \\ \beta & \overline{\alpha} \end{pmatrix} \in M(2, \mathbb{C}).$$

Square matrices with n^2 complex entries are $\mathbb{C}$-linear mappings within the n-dimensional $\mathbb{C}$-linear vector space $\mathbb{C}^n$. Similarly, square matrices with n^2 quaternion entries can be considered as right module maps within $\mathbb{H}^n$ over the right module $\mathbb{H}$.

1.7.1 On the notion of a quaternionic matrix

It is quite natural that in non-commutativity of the entries causes the difficulties in the use of such matrices.

Remark 1.103. Call into mind the definition of a right module M over a ring R. For $m, m_1, m_2 \in M$ and $s, s_1, s_2 \in R$ it has to be $m_1 + m_2 \in M$, $ms \in M$. The following rules are required:

(i) $(m_1 + m_2)s = m_1 s + m_2 s$,

(ii) $(ms_1)s_2 = m(s_1 s_2)$,

(iii) $m(s_1 + s_2) = ms_1 + ms_2$.

Each quaternionic right-linear transformation L in $\mathbb{H}^n$ can be represented by a matrix of the $n \times n$-type

$$Q = \begin{pmatrix} q_{11} & \cdots & q_{1n} \\ \cdots & \cdots & \cdots \\ q_{n1} & \cdots & q_{nn} \end{pmatrix}, \qquad (q_{ij} \in \mathbb{H}),$$

where the multiplication is defined by the rule $L(x) = Qx$, where $(L(x))_i = \sum_{j=1}^{n} q_{ij} x_j$.

Proposition 1.104. *Each matrix $Q \in M(n, \mathbb{H})$ can be identified with a matrix*

$$\tilde{Q} = \begin{pmatrix} Q^1 & -\overline{Q}^2 \\ Q^2 & Q^1 \end{pmatrix} \in M(2n, \mathbb{C}),$$

where $Q = Q^1 + jQ^2$, $Q^i \in M(n, \mathbb{C})$.

Proof. The proof is an immediate consequence of the unique decomposition of an arbitrary element $x \in \mathbb{H}^n$ in the form $x = x^{(1)} + jx^{(2)}$, with $x^{(i)} \in \mathbb{C}^n$. #

Definition 1.105. *Let* $Q = (q_{ij}) \in M(n, \mathbb{H})$ *then* $Q^* = (\overline{q_{ji}}) \in M(n, \mathbb{H})$ *is called quaternionic conjugate matrix corresponding to* Q. *A matrix* Q *with* $Q^*Q = I$ *is called quaternionic unitary. If* $Q^* = Q$ *then* Q *is said to be quaternionic Hermitean. For* $x \in \mathbb{H}^n$ *the number* $||x|| = \sqrt{(x^*x)}$ *is called the Euclidean norm in* $\mathbb{H}^n$, *where* $x^* = (\overline{x}_1, ..., \overline{x}_n)$. *A matrix* $Q \in M(n, \mathbb{H})$ *is said to be non-singular if there exists a matrix* $P \in M(n, \mathbb{H})$ *with* $QP = PQ = I$.

Corollary 1.106. *Let* $U \in M(n, \mathbb{H})$ *be a unitary matrix and* $Q \in M(n, \mathbb{H})$ *then one has*

$$||UQ|| = ||QU|| = ||Q|| \quad \textit{with} \quad ||Q||^2 = \text{trace}(Q^*Q).$$

1.7.2 Eigenvalues and QR-factorization

In order to consider $\mathbb{H}$-right-linear systems of algebraic equations it is important to have a suitable formulation of an quaternionic eigenvalue problem for matrices of the set $M(n, \mathbb{H})$.

Definition 1.107 [38] *A complex number* $\lambda \in \mathbb{C}^+$ *(upper half plane) is called eigenvalue with an eigenvector* $x \in \mathbb{H}^n$ *of the quaternionic matrix* $Q \in M(n, \mathbb{H})$ *if the equation*

$$Q\,x = x\,\lambda$$

is satisfied.

Remark 1.108. The referred authors gain by the help of this definition that eigenvalues form a discrete finite set. For a quaternion $\omega \in \mathbb{H}$ with $|\omega| = 1$ one can prove the identity

$$Q(x\omega) = x\omega(\overline{\omega}\lambda\omega).$$

In the case of $\lambda \in \mathbb{R}$ the set $S = \{\overline{\omega}\lambda\omega : |\omega| = 1\}$ provides a continuum of eigenvectors, which correspond to λ. Reference [38] refers to a result of W. Gragg (1987). In order to justify our restrictive definition of an eigenvalue the following theorem was proved.

Theorem 1.109. [38] *For* $\lambda \in \mathbb{H}$ *one can find in* $S = \{\overline{\omega}\lambda\omega : |\omega| = 1\}$ *a unique number* $\lambda' \in \mathbb{C}^+ \subset \mathbb{H}$.

This theorem shows that always one element of S can be marked out for belonging to $\mathbb{C}^+$. Definition 1.105 seems to be very realistic and extremely useful for working with eigenvalues of matrices with quaternionic entries. From a practical point of view it is important to have some kind of a QR-factorization for quaternionic matrices.

Theorem 1.110. [38] *Any matrix $A \in M(n, \mathbb{H})$ admits a factorization*

$$A = QR,$$

where $Q \in M(n, \mathbb{H})$ with $||Q|| = 1$ and $R \in M(n, \mathbb{H})$ an upper triangular matrix. In case of the non-singularity of A this decomposition is true up to a unitary diagonal matrix, i.e. if $A = UT$ is another factorization of the QR-type with a unitary matrix U and an upper triangular matrix T then there exists a unitary diagonal matrix $D \in M(n, \mathbb{H})$ such that $Q = UD$ and $T = DR$. Uniqueness can be verified if the diagonal entries of R are real numbers.

In analogy to the SCHUR decomposition of complex matrices a similar result can also be proved in the quaternionic case.

Theorem 1.111. [38] *Let $A \in M(n, \mathbb{H})$ then there is a unitary matrix $U \in M(n, \mathbb{H})$ and an upper triangular matrix T such that*

$$A = UTU^*.$$

The diagonal entries of T can be represented by the eigenvalues of A.

Chapter 2

Clifford valued functions and forms

2.1 Classes of regular functions

2.1.1 Spaces of Clifford algebra valued functions

Let $G \subset \mathbb{R}^n$ be an open (nonempty) domain and $\Gamma = \partial G$ a piecewise smooth surface. We denote by $\mathcal{B}(\Omega)$ one of the following BANACH spaces of real or complex valued functions on Ω where Ω can be $G, \partial G = \Gamma$, or any suitable subset of $\overline{G}$:

(i) $C^{(k)}(\Omega)$ $(k = 1, 2, ...)$ is the space of k-times continuously differentiable functions in Ω.

(ii) $C^{(k,\alpha)}(\Omega)$ $(k = 0, 1, 2, ...;\ \alpha \in (0, 1])$ is the space of k-times differentiable functions, whose k-th derivative is HÖLDER continuous with the exponent α.

(iii) $L_p(\Omega)$, $(1 \leq p < \infty)$, is just the space of all functions, whose p-th power is LEBESGUE integrable in Ω.

(iv) $W_p^k(\Omega)$ is the space of k-times differentiable functions in SOBOLEV's sense, whose k-th derivative belongs to $L_p(\Omega)$.

(v) $\overset{\circ}{W}{}_p^k(G)$ $(1 \leq p < \infty;\ k = 1, 2, ...)$, is the space of k-times differentiable functions in SOBOLEV's sense which k-th derivative belongs to $L_p(G)$ and vanish on the boundary Γ.

Definition 2.1. *Let $\mathcal{E}$ be either $\mathbb{R}$ or $\mathbb{C}$. Then the space $\mathcal{B}(\Omega) \otimes_{\mathcal{E}} \mathcal{A}$, where $\mathcal{A}$ either $Cl_{p,q}$ or $\mathbb{C} \otimes Cl_{p,q}$ is called $\mathcal{E}$-linear space $\mathcal{B}(\Omega, \mathcal{A})$ of all $\mathcal{A}$-valued functions.*

Remark 2.2. For both $\mathcal{A} = Cl_{0,n}$ or $\mathcal{A} = Cl_{n,0}$ the space $\mathcal{B}(\Omega, \mathcal{A})$ is a BANACH space . This means that

$$\mathcal{B}(\Omega, \mathcal{A}) = \left\{ u = \sum_{i=0}^{\dim \mathcal{A}} u_i \varepsilon_i : u_i \in \mathcal{B} \right\}$$

is furnished with the norm $||u|| = \{\sum_{i=0}^{\dim \mathcal{A}} ||u_i||^2\}^{\frac{1}{2}}$. The elements ε_k are in general k-vectors such that $\{\varepsilon_0, ..., \varepsilon_{\dim \mathcal{A}}\}$ form a basis in $\mathcal{A}$.

Proposition 2.3. *Let u be given in a basis representation $u = \sum_{i=0}^{\dim \mathcal{A}} u_i(x)\varepsilon_i$, $x \in \Omega \subset \mathbb{R}^n$, where $\{\varepsilon_0, ..., \varepsilon_{\dim \mathcal{A}}\}$ is a basis in $\mathcal{A}$, then $u \in \mathcal{B}(\Omega, \mathcal{A})$ if and only if $u_i \in \mathcal{B}(\Omega)$.*

Corollary 2.4. *Topological properties as continuity, differentiability, and integrability of the coefficients $u_i(x), x \in \Omega$, in a suitable basis representation transfer to the $\mathcal{A}$-valued function $u = \sum_{i=0}^{\dim \mathcal{A}} u_i(x)\varepsilon_i$.*

2.1.2 Notion of Clifford regular functions

In general regular functions are those which take up the kernel of a proper differential operator.

Definition 2.5. *The operator*

$$D := \sum_{i=1}^{n} e_i \partial_i$$

which acts on the space $C^1(G, Cl_{p,q})$, where $G \subset \mathbb{R}^{p,q}$ is called DIRAC *operator and $\partial_i := \partial/\partial x_i$ is the i-th partial derivative and $x = \sum_{i=1}^{n} x_i e_i \in \mathbb{R}^{p,q}$.*

Definition 2.6. *The operator*

$$\partial := \partial_0 + D$$

which acts on $C^1(G, Cl_{p,q})$ with $G \subset \mathbb{R}^{p,q}$ is called CAUCHY FUETER *operator where $\partial_0 := \partial/\partial x_0$ and $x = x_0 + \sum_{i=1}^{n} x_i e_i \in \mathbb{R} \oplus \mathbb{R}^{p,q}$. (This denotation is not usual up to now. Often it is called generalized* CAUCHY–RIEMANN *operator. We think that our proposal is justified by the history.)*

Definition 2.7. *Let $G \subset \mathbb{R}^{p,q}$, $p + q = n$, be an n-dimensional vector space with signature p, q. A function $u \in C^1(G, Cl_{p,q})$ is called left* CLIFFORD *regular (right* CLIFFORD *regular) if and only if*

$$Du = 0 \; ((uD) = 0).$$

A function which is both left and right CLIFFORD *regular is called two-sided* CLIFFORD *regular. Functions which fulfil the condition*

$$\partial\, u = 0 \quad ((u\partial) = 0)$$

are called left (right) CLIFFORD *holomorphic. We agree that the denotations $Cl_{p,q}$- regular or $Cl_{p,q}$- holomorphic will sometimes be used if it is necessary to emphasize the special algebra.*

Remark 2.8. In $Cl_{0,n}$ left (right) Clifford regular functions are often called *left (right) monogenic* functions. In such algebras the DIRAC operator is sometimes called *standard* EUCLID*ean* DIRAC *operator* [88]. The set of all monogenic functions is denoted by $\mathcal{M}(G, Cl_{0,n})$. In [88] left (right) CLIFFORD regular functions are called *left(right)* CLIFFORD *analytic functions.*

Remark 2.9. Through the whole of our book we will view the action of D and ∂ as an action on the left. It is sufficiently known that there are corresponding results for action of these operators on the right. Sometimes we will simply say regular functions instead of left regular functions.

Remark 2.10. The CAUCHY–FUETER operator ∂ which acts on $C^1(G, Cl_{p,q})$ with $G \in \mathbb{R}^{p,q}$ can be considered as the DIRAC operator on $C^1(G, Cl_{p+1,q})$ with $G \in \mathbb{R}^{p+1,q}$. Further connections between both operators are discussed in [63] and [88].

Example 2.11. Let $G \subset \mathbb{R}^3$, $e_0 = (1,0,0,0)^T, e_1 = (0,1,0,0)^T, e_2 = (0,0,1,0)^T$, $e_3 = (0,0,0,1)^T$ form a basis in $\mathbb{R}^4$. Under the governing conditions

(i) $e_i e_j + e_j e_i = -2\delta_{ij}$ $(i,j = 1,2,3;\ i < j)$,

(ii) $e_0^2 = 1, e_i e_0 = e_0 e_i = e_i$,

(iii) $e_1 e_2 = e_3$,

$\{e_0, e_1, e_2, e_3\}$ also form a basis in the algebra of real quaternions $\mathbb{H}$. The DIRAC operator has the form

$$D := e_1\partial_1 + e_2\partial_2 + e_3\partial_3.$$

Together with $u = u_0 e_0 + \underline{u}$, $\underline{u} = \sum_{i=1}^{3} u_i e_i$, it follows that

$$Du = -\operatorname{div}\ \underline{u} + \operatorname{rot}\ \underline{u} + \operatorname{grad}\ u_0,$$

where

$$\operatorname{div}\ \underline{u} := \sum_{i=1}^{3} \partial_i u_i, \quad \operatorname{grad}\ u_0 := \sum_{i=1}^{3} e_i \partial_i u_0, \quad \operatorname{rot}\ \underline{u} := \begin{vmatrix} e_1 & e_2 & e_3 \\ \partial_1 & \partial_2 & \partial_3 \\ u_1 & u_2 & u_3 \end{vmatrix}.$$

For (uD) we get in an analogous way

$$(uD) = -\operatorname{div}\ \underline{u} - \operatorname{rot}\ \underline{u} + \operatorname{grad}\ u_0.$$

The CAUCHY–FUETER operator has the corresponding representation

$$\partial u = \partial_0 u_0 - \operatorname{div}\ \underline{u} + \partial_0 \underline{u} - \operatorname{rot}\ \underline{u} + \operatorname{grad}\ u_0.$$

Example 2.12. Let $G \in \mathbb{R}^3$, $\mathcal{A} = Cl_{3,0}$. The PAULI matrices $\sigma_0, \sigma_1, \sigma_2, \sigma_3$ form a basis in $Cl_{3,0}$. We get for the DIRAC operator the expression

$$D = \sigma_1\partial_1 + \sigma_2\partial_2 + \sigma_3\partial_3.$$

Together with $u = u_0\sigma_0 + \sum_{j=1}^{3} u_j\sigma_j$ it follows that

$$\begin{aligned} Du &= \sum_{i=1}^{3} \sigma_i \partial_i u_0 + \sum_{j=1}^{3}\sum_{i=1}^{3} \sigma_i\sigma_j \partial_i u_j = \operatorname{div} \underline{u} + \sum_{i<j} \sigma_i\sigma_j(\partial_i u_j - \partial_j u_i) \\ &= \operatorname{div} \underline{u} + \operatorname{grad}\ u_0 - i \operatorname{rot}\ \underline{u} \end{aligned}$$

where in this algebra

$$\operatorname{grad} u_0 := \sum_{i=1}^{3} \sigma_i \partial_i u_0 \quad \text{and} \quad \operatorname{rot} \underline{u} := \begin{vmatrix} \sigma_1 & \sigma_2 & \sigma_3 \\ \partial_1 & \partial_2 & \partial_3 \\ u_1 & u_2 & u_3 \end{vmatrix} .$$

div $\underline{u}$ is the same as in Example 2.11. For (uD) we get

$$(uD) = \operatorname{div} \underline{u} + \operatorname{grad} u_0 + i \operatorname{rot} \underline{u}.$$

Example 2.13. Let $G \in \mathbb{R}^4$, $\mathcal{A} = Cl_{1,3}$. A basis in $Cl_{1,3}$ is given by the aid of the d-numbers (cf. Section 1.4) $\gamma_0, \gamma_1, \gamma_2, \gamma_3$. We obtain now for the DIRAC operator

$$D := \gamma_0 \partial_0 + \gamma_1 \partial_1 + \gamma_2 \partial_2 + \gamma_3 \partial_3 .$$

For $u \in C^1(G, Cl_{1,3})$ we obtain

$$Du = [\partial_0 u_0 - \operatorname{div} \underline{u}] + \sum_{i<j} [\partial_i u_j - \partial_j u_i] \gamma_i \gamma_j .$$

We should mention that this DIRAC operator can also be considered as a CAUCHY–FUETER operator on $\mathbb{R} \oplus \mathbb{R}^3$.

Proposition 2.14. *Quaternionic regular functions can be also characterized by the following properties:*

(i) *quaternionic-regular functions are sourceless and in the case of* $u_0 =$ const *also rotationless vector fields.*

(ii) $Cl_{3,0}$*-regular functions are rotationless and sourceless vector fields with* $u_0 =$ const

Definition 2.15. *Let be* $\overline{\partial}$ *defined by*

$$\overline{\partial} := \partial_0 - D.$$

This operator is called adjoint CAUCHY–FUETER *operator.*
In $Cl_{p,q}$, $p + q = n$, *the operator* $\overline{D}$ *defined by*

$$\overline{D} = \sum_{i=1}^{n} \overline{e_i} \partial_i$$

is called adjoint DIRAC *operator.*

2.1.3 On quaternionic regular functions

In this section we will put together useful elementary properties of quaternionic regular functions. As we already know there are two essential quaternionic algebras, namely the algebras of real and complex quaternions. Analytic investigations reduce in most cases to the consideration of the DIRAC operator D and its kernel ker D. It makes sense to introduce the following denotations: $\mathcal{A}_{Cl}$ stands for both algebras either $Cl_{0,n}$ or $Cl_{m,0}$ and $\mathcal{A}_{\mathbb{H}}$ denotes either $\mathbb{H}$ or $Cl_{3,0}$. As previously mentioned, $\{e_0, e_1, e_2, e_3\}$ is a $\mathbb{R}$-linear basis in $\mathbb{H}$ and $\{\sigma_0, \sigma_1, \sigma_2, \sigma_3\}$ is a $\mathbb{C}$-linear basis in $Cl_{3,0}$. Assume now that G is a domain in $\mathbb{R}^3$, or sometimes in $\mathbb{R}^4$.

Proposition 2.16. *Let be $u \in C^2(G, Cl_{0,n})$ $(u \in C^2(G, Cl_{n,0}))$. Then we have*

$$\overline{D}D\,u = D\overline{D}\,u = \pm\Delta\,u,$$

where $\Delta u = \sum_{i=1}^{n} \partial_i^2\, u$.

Corollary 2.17. *Let* $u \in \ker D \cap C^2(G, \mathcal{A}_{Cl})$ *then* $u \in \ker\Delta \cap C^2(G, \mathcal{A}_{Cl})$, *i.e., CLIFFORD regular functions of this type are harmonic.*

Proposition 2.18 (generalized LEIBNIZ rule). *Let be* $u, v \in C^1(G, \mathcal{A}_{\mathbb{H}})$ *then*

$$D\,(uv) = (D\,u)v + \overline{u}\,D\,v + 2Sc(uD)v.$$

Proof. In [110] is contained the proof for the algebra of real quaternions. In the case of PAULI algebra we conclude similarly. #

Corollary 2.19 (product rules in vector analysis). *Let* $u, v \in C^1(G, \mathbb{H})$. *Then in accordance with the definition of the vector field operations (cf. Example 2.11)* grad, div, rot *we have the following relations:*

(i) $\operatorname{grad}(u_0 v_0) = (\operatorname{grad} u_0)\, v_0 + u_0 \operatorname{grad} v_0,$

(ii) $\operatorname{div}(u_0 \underline{v}) = (\operatorname{grad} u_0)\, \underline{v} + u_0 \operatorname{div} \underline{v},$

(iii) $\operatorname{rot}(u_0 \underline{v}) = \operatorname{grad} u_0 \times \underline{v} + u_0 \operatorname{rot} \underline{v},$

(iv) $\operatorname{grad}(\underline{u}, \underline{v}) = \underline{u} \times \operatorname{rot} \underline{v} + \underline{v} \times \operatorname{rot} \underline{u} + (\underline{u}, \operatorname{grad})\, \underline{v} + (\underline{v}, \operatorname{grad})\, \underline{u},$

(v) $\operatorname{rot}(\underline{u} \times \underline{v}) = \underline{u} \operatorname{div} \underline{v} - \underline{v} \operatorname{div} \underline{u} + (\underline{v}, \operatorname{grad})\, \underline{u} - (\underline{u}, \operatorname{grad})\, \underline{v}.$

Remark 2.20. It can be shown (cf. [97], [61]) that by using of so-called totally analytic variables the ordinary LEIBNIZ rule

$$D\,(uv) = (D\,u)v + u\,D\,v$$

holds. A variable $z(x)$ is said to be *totally analytic* if and only if $z(x)$ permits the representation $z(x) = \sum_{i=1}^{3} x_i d_i$ where $d_i \in \mathbb{H}$ with the commutativity property $d_i d_j = d_j d_i$. For more information we refer to [110].

Proposition 2.21 (multiplier problem). *Let $u \in C^1(G, \mathbb{H}) \cap \ker D$ be an arbitrarily given function. If for all $v \in \ker D \cap C^1(G, \mathbb{H})$ also $vu \in \ker D \cap C^1(G, \mathbb{H})$ then $u = const.$*

Proof. A straightforward calculation leads to the result. #

Proposition 2.22. *The class of $\mathcal{A}_{\mathbf{H}}$-regular functions does not contain the squares of each of its elements.*

Proof. Let $x = x_1\sigma_1 - x_2\sigma_2 \in \ker D \cap C^1(G, Cl_{3,0})$. Then we have

$x^2 = (x_1\sigma_1 - x_2\sigma_2)(x_1\sigma_1 - x_2\sigma_2) = x_1^2 + x_2^2.$

Hence $D\, x^2 = 2(x_1\sigma_1 + x_2\sigma_2) \neq 0.$ #

Remark 2.23. For a totally analytic variable z with $z(x) \in \ker D \cap C^1(\mathbb{R}^4, \mathbb{H})$ follows with necessity $z^k \in \ker D \cap C^1(R^4, \mathbb{H})$.

Remark 2.24. We studied in [110] a generalization of an interpolation formula of LAGRANGE's type with quaternionic regular functions. The result is as follows: let $z(x) := \sum_{i=1}^{3} x_i d_i$, $z(a_i{}^{(j)}) := \sum_{i=1}^{3} a_i{}^{(j)} d_i$, $a_i^{(j)} \in \mathbb{H}$. We write in abbreviated form:

$z_j(x) := z(x) - z(a^{(j)}),\ z_{kj} := z(a^{(k)}) - z(a^{(j)})$

and demand $z_{kj} \neq 0$ for $k \neq j$. A quaternionic regular interpolation function is then given by the following formula:

$$(L_n\, u)(x) = \sum_{k=1}^{n} \left[\frac{z_1(x)}{z_{k1}} \cdot \ldots \cdot \frac{z_{k-1}(x)}{z_{k(k-1)}} \frac{z_{k+1}(x)}{z_{k(k+1)}} \cdot \ldots \cdot \frac{z_n(x)}{z_{kn}} \right] u_k$$

with

(i) $(L_n u)^k \in \ker D \cap C^1(\mathbb{R}^4, \mathbb{H})$ $(k = 1, \ldots, n)$,

(ii) $(L_n u)(a^{(j)}) = u_j$ $(j = 1, \ldots, n)$.

Proposition 2.25. *Let $u \in \ker D \cap C^1(G, \mathcal{A})$, $u(x) \neq 0$ then $u^{-1} \in \ker D \cap C^1(G, \mathcal{A})$ not in general.*

Proof. It is sufficient to give a counterexample . Taking again (cf. Proposition 2.22) $u(x) = x_1\sigma_1 - x_2\sigma_2 \in \ker D \cap C^1(G, Cl_{3,0})$ then for $0 \notin G$ we obtain

$$u^{-1} = \frac{x_2\sigma_2 - x_1\sigma_1}{x_1^2 + x_2^2}$$

and one can easily verify that

$$D\,u^{-1} = \frac{2(x_1^2 - x_2^2)}{(x_1^2 + x_2^2)^2}\sigma_0 + i\,\frac{4x_1x_2}{(x_1^2 + x_2^2)^2}\sigma_3 \neq 0.$$

The same example can be taken in ker $D \cap C^1(G, \mathbb{H})$. #

Omission of a LEIBNIZ rule in the classical sense leads to considerable peculiarities in the construction of systems of regular functions. We will formulate some easy principles to produce quaternionic regular functions.

2.1.4 Quaternionic regular polynomials

Let $\{e_0, e_1, e_2, e_3\}$ be a basis in $\mathbb{H}$ and be $x = \sum_{i=0}^{3} x_i e_i$. We assume as before that $e_0^2 = e_0,\ e_ie_j + e_je_i = -2\delta_{ij}$. We consider the new variables

$$z_k := x_0e_k - x_ke_0 \ (k = 1, 2, 3).$$

For $\partial = \sum_{i=0}^{3} e_i\partial_i$ one has $\partial z_k = 0$. Unfortunately, the product of two of such variables is not quaternionic regular, i.e. $\partial\, z_k z_l \neq 0$ for any k and l. Nevertheless, one can find homogeneous quaternionic regular polynomials of any degree by symmetrization. These polynomials are given by the formula

$$P_m(x) = \frac{1}{m!}\sum_{(\mu_1,\dots,\mu_m)} z_{\mu_1}\cdot\ldots\cdot z_{\mu_m},$$

where $(\mu_1, ..., \mu_m)$ covers all distinguishable permutations of $\{1, ..., m\}$ and $(\mu_1, ..., \mu_m)$ is an arbitrary combination of m elements out of repetitions of the set $\{1, 2, 3\}$. For $m = 0$ we fix $P_0(x) = e_0$. This construction was first made by R. DELANGHE [61] and is also valid in $\mathcal{A}_{Cl}$. Comprehensive results on CLIFFORD regular polynoms can be found in [48] and also in [63].

2.1.5 Conjugate harmonic functions

Let G be an open set in $\mathbb{R}^n$ and star-shaped with respect to the origin and $u_0 : G \longmapsto \mathbb{R}$ is harmonic in G then

$$u(x) = u_0(x) - \mathrm{Vec}\int_0^1 t^{n-2}(D\,u_0)(tx)x\,dt$$

is $Cl_{0,n}$- regular in G. In [35] this formula was modified into a new one using the so-called *spherical* DIRAC *operator*. With the denotations

$$x = r\,\omega,\ x = \sum_{i=1}^{n} e_ix_i,\ r = |\,x\,|,\ \omega = \sum_{i=1}^{n} e_i\omega_i \in S^{n-1}$$

we obtain

$$u(x) = u(r\omega) - \frac{1}{r^{n-2}} \int_0^r \rho^{n-3} \Gamma u(\rho\omega) d\rho,$$

where

$$\Gamma = -\omega \wedge \partial_\omega \text{ with } \partial_\omega = \sum_{i=1}^{n} e_i \partial_{\omega_i}.$$

2.1.6 Extensions of complex holomorphic functions

Let $t = \xi + i\eta \in \mathbb{C}$, $x = x_0 + \underline{x} \in \mathbb{H}$, and $\omega = \underline{x}/|\underline{x}|$. Now define maps $g : \mathbb{C} \longmapsto \mathbb{H}$ by

$$g(t) = \xi + \omega\eta \quad \text{and} \quad h : \mathbb{H} \longmapsto \mathbb{C}^+ \quad \text{with} \quad h(x) = x_0 + i|\underline{x}|,$$

where $\mathbb{C}^+$ denotes the upper half plane of $\mathbb{C}$. Now we consider an open subset $U \subseteq \mathbb{C}^+$ and a complex holomorphic function $u : U \longmapsto \mathbb{C}$ then

$$v(x) := \Delta\, g(u(h(x)))$$

is a quaternionic regular function on $h^{-1}(U) \subset \mathbb{H}$. In this way the operator

$$V : Cl_{0,1} \cap \ker \partial_{\bar{z}} \longmapsto \mathbb{H} \cap \ker D$$

is defined with $v = V(u)(x) = \Delta g(u(h(x)))$ and $\partial_{\bar{z}} = 1/2(\partial/\partial\xi + i\,\partial/\partial\eta)$.

Proposition 2.26. [78] *Let $u_i : \mathbb{C}^+ \supseteq U \longmapsto \mathbb{C}$ $(i = 1, 2)$ complex holomorphic functions then*

$$V(u_1 + u_2) = V(u_1) + V(u_2),$$
$$V(\lambda u) = \lambda V(u) \qquad (\lambda \in \mathbb{R}),$$
$$V(u_1 u_2) = V(u_1) V(u_2).$$

Proposition 2.27. [190] *Let $G_k, (k = 1, 2, 3)$, open sets in $\mathbb{C}$ and let $h_k : G_k \longmapsto \mathbb{C}$, complex holomorphic functions. Furthermore let $h_k = \operatorname{Re} h_k + i \operatorname{Im} h_k$. Further let*

$$G := \{x = x_0 + e_1 x_1 + e_2 x_2 + e_3 x_3 \in \mathbb{H} : x_0 + e_k x_k \in G_k\}$$

then the function $H : G \longmapsto \mathbb{H}$ given by

$$H(x) = \sum_{k=1}^{3} \operatorname{Re} h_k(x_0, x_k) + \sum_{k=1}^{3} e_k \operatorname{Im} h_k(x_0, x_k)$$

is quaternionic regular in G.

Proposition 2.28. [225] *Let G be an open domain in $\mathbb{H}$ and $u : \mathbb{H} \longmapsto \mathbb{C}$ be an $\mathbb{R}$-differentiable function. Then u is quaternionic regular if and only if u is complex holomorphic.*

Remark 2.29. For more interesting properties of this type we refer to the papers of [190], [85] and [226]. In [225] Proposition 2.28 is generalized to $\mathbb{H}^n$.

2.1.7 Cauchy kernel functions

We will mention here only an important type of quaternionic regular functions which we use in collocation methods for boundary value problems of partial differential equations. Let $\underline{x}^{(k)}$ be a system of points outside a given domain $G \subset \mathbb{R}^3$. Then there exists a system of quaternionic regular functions $\{\phi_k\}$ which are defined by

$$\phi_k(\underline{x}) = \frac{\underline{x} - \underline{x}^{(k)}}{|\underline{x} - \underline{x}^{(k)}|^3}.$$

Under some conditions completeness and advantageously numerical properties are possible to obtain. Further, we will use this system of singularity functions for calculations.

2.2 Different approaches to regular functions

Complex function theory is the result of a long development involving the contributions of many mathematicians. Famous names are involved in the history of this fundamental mathematical field. Doubtless the power of the classical complex function theory comes from the equivalence of the different approaches of RIEMANN, CAUCHY and WEIERSTRASS, which can be seen as the pillars of this theory. As is known from the history, CAUCHY's name is connected with the notion of complex differentiability, the WEIERSTRASS approach bases on convergent power series and RIEMANN introduced the so-called CAUCHY–RIEMANN equations. Hence complex function theory is a unifying generalization of three different theories. This means that one can start equivalently with power series, CAUCHY–RIEMANN systems or with functions which have a complex derivative. In all cases the whole theory comes out.

2.2.1 A non-optimistic result

In Section 2.1 we followed the ideas of B. RIEMANN and investigated, in some sense, multidimensional generalizations of complex differential operators. Now we will look for possibilities in order to deduce almost the same state as in the function theory of one complex variable. First steps towards a similar CAUCHY theory in CLIFFORD algebras were not optimistic. The Russian mathematician A.S. MEJLIHZON [198] achieved the following result in 1948:

Lemma 2.30. *Let $G \subseteq \mathbb{H}$; $x, \Delta x \in \mathbb{H}$ such that x, $x + \Delta x \in G$ for all sufficiently small Δx and u be a $\mathbb{R}$-differentiable $\mathbb{H}$-valued function. If the limit*

$$\lim_{\Delta x \to 0} \frac{u(x + \Delta x) - u(x)}{\Delta x}$$

does exist, then u must have the representation

$$u(x) = a + xb,$$

where $a, b \in \mathbb{H}$ are constant quaternions.

Remark 2.31. Thus the class of all differentiable functions defined similarly to the complex case is too small and, therefore, out of interest. Clearly, also the WEIERSTRASS approach does not work. Taking as regular functions only the quaternionic power series on each point in a open domain of $\mathbb{R}^4$, then one gets a class of "regular" functions which is too large for obtaining reasonable results.

2.2.2 Hypercomplex unified theory

Up to ten years ago it seemed to be relatively hopeless to find a way for the development of a unified hypercomplex analysis. Using well-defined hypercomplex variables the paravector space $P\mathrm{Vec}\, Cl_{0,n}$ transforms into the space $\mathcal{H}^n$ which has similarities with $\mathbb{C}^n$. The space $\mathcal{H}^n$ is completely determined as the image of the mapping $h : P\mathrm{Vec}\, Cl_{0,n} \longmapsto \mathcal{H}^n \subset P\mathrm{Vec}\, Cl_{0,n}$ where

$$h(x) = z = (z_1, ..., z_n) \text{ with } z_k = x_k - x_0 e_k.$$

Obviously, h has an inverse map and the following formula holds:

$$x = h^{-1}(z) = \mathcal{N} \sum_{k=1}^{n} z_k e_k \text{ with } \mathcal{N} u = \frac{1}{n} u_0 + \underline{u}.$$

In this section we are following the work of H.R. MALONEK [186], [187], [188] and F. SOMMEN [281], [282] as well as the explanations in [36].

Definition 2.32. *Let $u : P\mathrm{Vec}\, Cl_{0,n} \longmapsto Cl_{0,n}$ be an $\mathbb{R}$-differentiable function and $u(x) = u(h^{-1}(z)) =: u^*(z)$. The function u is called left (right)* CLIFFORD *differentiable if there exists a left (right) $Cl_{0,n}$−linear mapping A with $A \in L(\mathcal{H}^n, P\mathrm{Vec}\, Cl_{0,n})$ such that*

$$\lim_{\Delta z \to 0} \frac{|u^*(z + \Delta z) - u^*(z) - A\Delta z|}{|\Delta z|} = 0,$$

where $A = (A_1, ..., A_n)$ and $\Delta z = (\Delta z_1, ..., \Delta z_n)$.

Proposition 2.33. *A function u is left (right)* CLIFFORD *differentiable if and only if u is left (right)* CLIFFORD *regular.*

Remark 2.34. In [282] is given the definition of left (right) CLIFFORD regular functions in terms of differentials. It is said that a function $u : P\mathrm{Vec}\, Cl_{0,n} \longmapsto Cl_{0,n}$ is left (right) CLIFFORD regular if and only if the differential du has the representation $du = \sum_{i=1}^{n} dz_i a_i$ with $a_i : P\mathrm{Vec}\, Cl_{0,n} \longmapsto Cl_{0,n}$ $(i = 1, ...n)$. Further generalizations can be found in [116].

Corollary 2.35. *The* CAUCHY *approach and the* RIEMANN *approach are equivalent in the sense described above.*

We already know that FUETER's polynomials (cf. Section 2.1) (sometimes also named inner spherical monogenics) are CLIFFORD regular. With the help of such polynomials one can deduce the WEIERSTRASS approach, which is based only on such polynomials.

Definition 2.36. *Let $a_k \in Cl_{0,n}$ $(k = 1, ..., n)$. Then the product*

$$\frac{1}{n!} \sum_{\sigma \in S_n} a_{\sigma(1)} a_{\sigma(2)} \cdot ... \cdot a_{\sigma(n)}$$

is called permutational or symmetric product. It will be denoted by $a_1 \times a_2 \times \ldots \times a_n$.

Proposition 2.37. *The permutational product permits the following properties:*

(i) *Permutations of factors do not change the value.*

(ii) *Operation "$\times$" is distributive with respect to addition.*

(iii) *This product is not associative in general.*

(iv) *The following formula is valid:*

$$\begin{aligned} a_1 \times \ldots \times a_n &= \frac{1}{n}\sum_{k=1}^{n} a_k(a_1 \times \ldots \times \hat{a}_k \times \ldots \times a_n) \\ &= \frac{1}{n}\sum_{k=1}^{n}(a_1 \times \ldots \times \hat{a}_k \times \ldots \times a_n)a_k, \end{aligned}$$

where in the permutational product $a_1 \times \ldots \times \hat{a}_k \times \ldots \times a_n$ *the factor* a_k *is missing.*

(v) *Let* $\nu = (\nu_1, \ldots, \nu_n)$ *and* $a_i \neq a_j$, $(i \neq j)$, *then*

$$a_1^{\nu_1} \times \ldots \times a_n^{\nu_n} = \frac{\nu!}{|\nu|!}\sum_{\sigma \in S_n} a_{\sigma(1)} \cdot \ldots \cdot a_{\sigma(|\nu|)},$$

where $\nu! = \nu_1! \ldots \nu_n!$ *and* $|\nu| = \nu_1 + \ldots. + \nu_n$.

Corollary 2.38. FUETER *polynomials* $P_n(x)$ *can be written in the form*

$$\begin{aligned} P_\nu(x) &= (-1)^{|\nu|}(x \times e_1)^{\nu_1} \times \ldots \times (x \times e_n)^{\nu_n} \\ &= (-1)^{|\nu|} z_1^{\nu_1} \times \ldots \times z_n^{\nu_n} =: z^\nu. \end{aligned}$$

Proof. The proof is straightforward. We compute only

$$\begin{aligned} &2(x \times e_k) = xe_k + e_k x = x_0 e_k + e_k x_0 + x_k e_k^2 + e_k^2 x_k \\ &+\sum_{i \neq k}[x_i e_i e_k + e_k e_i x_i] = 2x_0 e_k - 2x_k = -2z_k. \qquad \# \end{aligned}$$

Definition 2.39. *Let* $\nu = (\nu_1, \ldots, \nu_n)$, *then*

$$P(x) = \sum_{k=0}^{\infty}\sum_{|\nu|=k} z^\nu a_\nu \qquad \left(P(x) = \sum_{k=0}^{\infty}\sum_{|\nu|=k} a_\nu z^\nu\right)$$

is called left (right) multiple permutational power series, where $a_\nu \in Cl_{0,n}$.

Remark 2.40. In [186] a formula is given for the radius of convergence r of the series considered above. It reads very similar to the well-known CAUCHY–HADAMARD formula

$$r = (\overline{\lim_{k\to\infty}}(\max_{|\nu|=k}|a_\nu|)^{\frac{1}{k}})^{-1}.$$

Theorem 2.41. *Let* $P(x) = \sum_{k=0}^{\infty}\sum_{|\nu|=k} z^\nu a_\nu$. *If* $P(x)$ *converges in the ball* $B_r(0)$ *with the convergence radius* r, *then* $P(x)$ *is* CLIFFORD *differentiable and* $\partial P/\partial x_i = A_i$ *(cf. Definition 2.32)* $(i = 1, ..., n)$, *and, therefore,* CLIFFORD *regular.*

In generalization of Definition 2.32 we obtain

Definition 2.42. *Let* $u : P\mathrm{Vec}\, Cl_{0,n} \longmapsto Cl_{0,n}$ *with* $u(x) = u^*(z)$ *an* $\mathbb{R}$-*differentiable function defined in a neighbourhood* $B_r(x)$ *and*

$$Cl_{0,n}^m = L_{l,(r)}(\mathcal{H}^n, Cl_{0,n}^{m-1}) \quad \text{with} \quad Cl_{0,n}^1 = L_{l,(r)}(\mathcal{H}^n, Cl_{0,n}).$$

Then the function u *is called* k*-fold* CLIFFORD *differentiable in* x *if there exists a* $C\ell_{0,n}$*-linear mapping* $A^{(k)} \in L_{l,(r)}(\mathcal{H}^n, Cl_{0,n}^m)$ *such that*

$$\lim_{\Delta z\to 0}\frac{|u_{l(r)}^{*(k-1)}(z+\Delta z) - u_{l(r)}^{*(k-1)}(z) - A^{(k)}\Delta z|}{|\Delta z|} = 0.$$

Corollary 2.43. *Each multiple permutational power series of a function* u *is infinitely* CLIFFORD *differentiable inside the convergence domain and represents a* TAYLOR *series expansion for* u *where*

$$u(x) = \sum_{k=0}^{\infty}\sum_{|\nu|=k}\frac{(-1)^{|\nu|}}{k!}\frac{\partial^\nu u}{\partial x^\nu}(0)(x\times e_1)^{\nu_1}\times \ldots \times (x\times e_n)^{\nu_n}.$$

Conversely, if $u \in \ker\partial \cap C^1(B_r(0), Cl_{0,n})$ *then* u *permits the representation*

$$u(x) = \sum_{k=0}^{\infty}\sum_{|\nu|=k} a_\nu P_\nu(x)$$

with

$$a_\nu = \tfrac{1}{\nu!}(\partial^\nu u)(0)$$

$$\partial^\nu := \frac{\partial^{|\nu|}}{\partial x_1^{\nu_1}\cdot\ldots\cdot\partial x_n^{\nu_n}}.$$

Remark 2.44. In this way we found that WEIERSTRASS' approach in the sense considered above is equivalent to the concept of CLIFFORD regularity and therefore the whole unified approach of the complex function theory in the plane is realized on the basis of fundamental results of R. DELANGHE, H.R. MALONEK, and F. SOMMEN in $Cl_{0,n}$-analysis. Recently in [155] the main results of BERS' approach (cf. [23]) can be also transferred to CLIFFORD analysis.

2.3 Paravector valued elementary functions

Let us start our explanations with the introduction of an exponential function. For this section we assume that x is a paravector in $Cl_{0,n}$, i.e. $x \in PVecCl_{0,n} \subset Cl_{0,n}$. It permits the representation $x = x_0 + \underline{x}$ with $\underline{x} = \sum_{i=1}^{n} x_i e_i$. Furthermore, we denote by $\omega(x) := \underline{x}/|\underline{x}|$. Clearly, $\omega(x)$ belongs to the unit sphere in $\mathbb{R}^n$.

Definition 2.45. *Let x be a paravector in $Cl_{0,n}$. The exponential function e^x is defined by the potential series*

$$\sum_{k=0}^{\infty} \frac{x^k}{k!}. \tag{2.1}$$

Remark 2.46. The series (2.1) is normally convergent. It can be easily shown that for an arbitrary $\varepsilon > 0$ it is always possible to find a sufficiently large number N such that for any $r, s > N$ holds

$$\left| \sum_{k=r}^{s} \frac{x^k}{k!} \right| \leq \sum_{k=r}^{s} \frac{K^k |x|^k}{k!} < \varepsilon,$$

where K is a constant which only depends on n and satisfies the inequality $|xy| \leq K^2 |x||y|$. Note that $|x|^2 = x\overline{x} = x_0^2 + x_1^2 + ... + x_n^2$.

Proposition 2.47. *The exponential function permits the representation*

$e^x = e^{x_0}(\cos|\underline{x}| + \omega(x)\sin|\underline{x}|)$.

Proof. Using the *Cauchy* product of two potential series we get

$$\sum_{k=0}^{\infty} \frac{x^k}{k!} = \sum_{k=0}^{\infty} \frac{(x_0 + \underline{x})^k}{k!} = \sum_{k=0}^{\infty} \frac{1}{k!} \sum_{\ell=0}^{k} \binom{k}{\ell} x_0^{\ell} \underline{x}^{k-\ell}$$

$$= \sum_{k=0}^{\infty} \sum_{\ell=0}^{k} \frac{x_0^{\ell} \underline{x}^{k-\ell}}{\ell!(k-\ell)!} = \sum_{k=0}^{\infty} \frac{x_0^{\ell}}{\ell!} \sum_{m=0}^{\infty} \frac{\underline{x}^m}{m!} = e^{x_0} e^{\underline{x}}.$$

It remains to consider $e^{\underline{x}}$. We have

$$\sum_{k=0}^{\infty} \frac{\underline{x}^k}{k!} = \sum_{\ell=0}^{\infty} \frac{\underline{x}^{2\ell}}{(2\ell)!} + \sum_{\ell=0}^{\infty} \frac{\underline{x}^{2\ell+1}}{(2\ell+1)!}$$

$$= \sum_{\ell=0}^{\infty} (-1)^{\ell} \frac{|\underline{x}|^{2\ell}}{(2\ell)!} + \omega(x) \sum_{\ell=0}^{\infty} (-1)^{\ell} \frac{|\underline{x}|^{2\ell+1}}{(2\ell+1)!} = \cos|\underline{x}| + \omega(x)\sin|\underline{x}|. \qquad \#$$

Proposition 2.48. *Let x, y be paravectors in $Cl_{0,n}$. If $xy = yx$ we get*

$$e^{x+y} = e^x e^y. \tag{2.2}$$

Proof. Following Definition 2.45 we have

$$e^{x+y} = \sum_{k=0}^{\infty} \frac{(x+y)^k}{k!} = \sum_{k=0}^{\infty}\sum_{\ell=0}^{k} \frac{x^\ell y^{k-\ell}}{\ell!(k-\ell)!} = \sum_{\ell=0}^{\infty} \frac{x^\ell}{\ell!} \sum_{m=0}^{\infty} \frac{y^m}{m!} = e^x e^y. \qquad \#$$

Remark 2.49. The commutativity is not necessary for the validity of rule (2.2). Take for instance $x = 3\pi e_1$, $y = 4\pi e_2$ then $xy = 12\pi^2 e_1 e_2 = -12\pi^2 e_2 e_1 = -yx$. On the other hand we have

$$|x+y| = 5\pi, \quad |x| = 3\pi \quad \text{and} \quad |y| = 4\pi.$$

Furthermore,

$$e^{x+y} = \cos\, 5\pi + \frac{3\pi e_1 + 4\pi e_2}{5\pi} \sin\, 5\pi = -1.$$

In the same way we also get

$$e^x = \cos\, 3\pi = -1 \quad \text{and} \quad e^y = \cos\, 4\pi = 1.$$

Thus we find $e^{x+y} = e^x e^y$.

Corollary 2.50. *For any paravector $x \in Cl_{0,n}$ we have*

(i) $e^x \neq 0$,

(ii) $e^{-x} e^x = 1$,

(iii) $e^{\omega(x)\pi} = -1$,

(iv) $e^{k\underline{x}} = (e^{\underline{x}})^k \quad (k \in \mathbb{N})$ *(*MOIVRE*'s formula)*,

(v) $|e^x| = e^{x_0}$.

Proof. Owing to Proposition 2.47 we obtain

$$|e^x| = e^{x_0}|(\cos|\underline{x}| + \omega(x)\sin|\underline{x}|)| = e^{x_0}\sqrt{(\cos^2|\underline{x}| + \sin^2|\underline{x}|)} = e^{x_0} > 0.$$

Because of $\omega(x)^2 = -1$ then it holds that

$$e^x e^{-x} = e^{x_0}(\cos|\underline{x}| + \omega(x)\sin|\underline{x}|)e^{-x_0}(\cos|\underline{x}| - \omega(x)\sin|\underline{x}|) = 1.$$

Property (iii) follows from

$$\begin{aligned} e^{\omega(x)\pi} &= \cos|\omega(x)\pi| + \omega(\omega(x)\pi)\sin|\omega(x)\pi| \\ &= \cos\, \pi + \omega(\omega(x)\pi)\sin\, \pi = -1. \end{aligned}$$

Especially, we have

$$e^{2\underline{x}} = (e^{\underline{x}})^2.$$

Assume now the validity of $e^{k\underline{x}} = (e^{\underline{x}})^k$ for some k. Then one can easily obtain this identity for the $k+1$. #

Proposition 2.51. *Let x be a paravector in $Cl_{0,n}$. Then the exponential function may be described by the limit*

$$e^x = \lim_{m \mapsto \infty} \left(1 + \frac{x}{m}\right)^m .$$

Proof. Consider the difference

$$e^x - \left(1 + \frac{x}{m}\right)^m = \sum_{k=0}^{\infty} \left(\frac{1}{k!} - \frac{\binom{k}{m}}{m^k} \right) x^k .$$

Since $\left(1 + \frac{x}{m}\right)^m$ is a polynomial this series is normally convergent. The same is valid for the series e^x. The coefficients in front of x^k are positive. Now we obtain

$$\left| e^x - \left(1 + \frac{x}{m}\right)^m \right| \le \sum_{k=0}^{\infty} \left(\frac{1}{k!} - \frac{\binom{k}{m}}{m^k} \right) |x|^k = e^{|x|} - \left(1 + \frac{|x|}{m}\right)^m ,$$

which tends to zero for m to infinity. #

Definition 2.52. *Let x be a paravector in $Cl_{0,n}$. Then hyperbolic and trigonometric functions are defined by the following formulae:*

$$\sinh x \;:=\; \frac{e^x - e^{-x}}{2}, \quad \cosh x := \frac{e^x + e^{-x}}{2}.$$

For $|\underline{x}| \neq 0$ we define:

$$\sin x \;:=\; \frac{e^{x\omega(x)} - e^{-x\omega(x)}}{2}\omega(x), \quad \cos x := \frac{e^{x\omega(x)} + e^{-x\omega(x)}}{2}.$$

It holds that

$$\sin x = -\sin(-x) \quad \textit{and} \quad \cos x = \cos(-x).$$

Proposition 2.53. *The hyperbolic functions* $\sinh x$ *and* $\cosh x$ *permit the following representations:*

$$\cosh x = \omega(x)\frac{e^{x_0} - e^{-x_0}}{2}\sin|\underline{x}| + \frac{e^{x_0} + e^{-x_0}}{2}\cos|\underline{x}|,$$
$$\sinh x = \omega(x)\frac{e^{x_0} + e^{-x_0}}{2}\sin|\underline{x}| + \frac{e^{x_0} - e^{-x_0}}{2}\cos|\underline{x}|.$$

Proof. A straightforward calculation provides the result. #

Proposition 2.54. *The trigonometric functions* sin *and* cos *permit the representations:*

$$\sin x = \omega(x)\frac{e^{|\underline{x}|} - e^{-|\underline{x}|}}{2} \cos\, x_0 + \frac{e^{|\underline{x}|} + c^{-|\underline{x}|}}{2} \sin\, x_0, \tag{2.3}$$

$$\cos x = \omega(x)\frac{e^{-|\underline{x}|} - e^{|\underline{x}|}}{2} \sin\, x_0 + \frac{e^{-|\underline{x}|} + e^{|\underline{x}|}}{2} \cos\, x_0. \tag{2.4}$$

The right-hand sides of these representations can be used for the definition of sin *and* cos *in the case of* $|\underline{x}| = 0$.

Proof. This result is an immediate consequence of Definition 2.52. #

Remark 2.55. Proposition 2.53 and Proposition 2.54 can be seen as analogues to corresponding formulae of the complex plane, i.e. in $Cl_{0,1} = \mathbb{C}$. For $z = x + iy \in \mathbb{C}$ one has for instance the relations

$$\sin\, z = \sinh\, y \cos\, x + \cosh\, y \sin\, x$$
$$\cos\, z = \sinh\, y \sin\, x + \cosh\, y \cos\, x.$$

Corollary 2.56. *The following relations are valid*

(i) $\cosh\, x = \cos(\omega(x)x)$,

(ii) $\sinh\, x = \omega(x) \sin(\omega(x)x)$,

(iii) $\cosh\, \underline{x} = \cos|\underline{x}|$,

(iv) $\sinh\, \underline{x} = -\omega(x) \sin|\underline{x}|$.

Proof. First of all we state

$$\omega(x)x = \frac{\underline{x}x_0}{|\underline{x}|} - |\underline{x}|.$$

Replacing in formulae (2.3) as well as (2.4) x by $\omega(x)x$ we get

$$\cos\omega(x)x = \omega\left(\frac{\underline{x}x_0}{|\underline{x}|}\right)\frac{e^{-|x_0|} - e^{|x_0|}}{2}\sin(-|\underline{x}|) + \frac{e^{|x_0|} + e^{-|x_0|}}{2}\cos|\underline{x}|,$$

$$\sin\omega(x)x = \omega\left(\frac{\underline{x}x_0}{|\underline{x}|}\right)\frac{e^{|x_0|} - e^{-|x_0|}}{2}\cos|\underline{x}| + \frac{e^{|x_0|} + e^{-|x_0|}}{2}\sin(-|\underline{x}|).$$

Due to $\omega(\frac{\underline{x}x_0}{|\underline{x}|}) = \frac{\underline{x}}{|\underline{x}|}\frac{x_0}{|x_0|}$ we finally obtain

$$\begin{aligned}
\cos\omega(x)x &= \omega(x)\frac{e^{x_0} - e^{-x_0}}{2}\sin|\underline{x}| + \frac{e^{x_0} - e^{-x_0}}{2}\cos|\underline{x}| = \cosh\, x \\
-\omega(x)\sin\omega(x)x &= \omega(x)\frac{e^{x_0} - e^{-x_0}}{2}\cos|\underline{x}| - \frac{e^{x_0} + e^{-x_0}}{2}\sinh\, |\underline{x}|)(-\omega(x)) \\
&= \sinh\, x.
\end{aligned}$$

#

Remark 2.57. We note the relations

$$|\sin\, x|^2 = \sin^2 x_0 + \sinh^2 |\underline{x}|,$$
$$|\cos\, x|^2 = \cos^2 x_0 + \cosh^2 |\underline{x}| - 1.$$

More effort requires ultimately a straightforward computation of the relations

$$\sin(x+y) = \sin x \cos y + \cos x \sin y,$$
$$\cos(x+y) = \cos x \cos y - \sin x \sin y,$$

where x, y are paravectors in $Cl_{0,n}$. Moreover, one can easily see the formulae

$$\sin(k\pi + \underline{x}) = \omega(x)(-1)^k \sinh \underline{x} \quad (k \in \mathbb{N}),$$
$$\cos(k\pi + \underline{x}) = (-1)^k \cosh\, \underline{x}.$$

Furthermore, it follows that $\sin x$ and $\cos x$ have no other zeros than the well known real ones. Now it is quite clear that in this natural way CLIFFORD valued tangent- and cotangent functions can be introduced and all canonical theorems can be verified.

Definition 2.58. *Let $x \in Cl_{0,n}$ be a paravector, i.e. $x = x_0 + \underline{x}$, $x \neq x_0 < 0$. Then a* CLIFFORD *valued logarithmic function* $\log\, x$ *is defined by*

$$\log\, x := \ln|x| + \omega(x)\text{arc}\cos\frac{x_0}{|x|}, \quad |\underline{x}| \neq 0 \quad or\ |\underline{x}| = 0, x_0 > 0.$$

Corollary 2.59. *The logarithmic function $\log x$ has the following properties:*

(i) $\log\, 1 = 0$ *and* $\log\, e_i = \frac{\pi}{2}\ (i = 1, ..., n)$,

(ii) $1 - \dfrac{1}{|x|} - \text{arc}\tan\dfrac{|\underline{x}|}{|x_0|} \leq |\log\, x| \leq |x| - 1 + \text{arc}\tan\dfrac{|\underline{x}|}{|x_0|}.$

Proposition 2.60. *The function* $\log\, x$ *is the inverse to the exponential function e^x introduced above, which means*

$$e^{\log\, x} = x \quad \text{and} \quad \log\, e^x = x.$$

Proof. Applying Proposition 2.48 we have

$$e^{\log\, x} = e^{\ln|x|}e^{\omega(x)\text{arc}\cos\frac{x_0}{|x|}}.$$

Thus we get

$$e^{\log\, x} = |x|\left(\cos \text{arc}\cos\frac{x_0}{|x|} + \omega(x)\sin \text{arc}\cos\frac{x_0}{|x|}\right).$$

We find immediately

$$\cos \operatorname{arc}\cos \frac{x_0}{|x|} + \omega(x) \sin \operatorname{arc}\cos \frac{x_0}{|x|} = \frac{x_0}{|x|} + \omega(x)\frac{|\underline{x}|}{|x|} = \frac{x}{|x|}.$$

The other relation is verified by

$$\begin{aligned} \log\, e^{x} &= \log\, e^{x_0}(\cos|\underline{x}| + \omega(x)\sin|\underline{x}|) \\ &= \ln\, e^{x_0} + \omega(x)\operatorname{arc}\cos\left(\frac{e^{x_0}\cos|\underline{x}|}{e^{x_0}}\right) = x_0 + \frac{\underline{x}}{|\underline{x}|}|\underline{x}| = x. \qquad \# \end{aligned}$$

Proposition 2.61. *Let $xy = yx$ then the well-known logarithmic rule*

$$\log\,(xy) = \log\, x + \log\, y$$

is valid.

Definition 2.62. *Let α be a real number. The general potential function x^α is given by*

$$x^\alpha := e^{\alpha \log\, x}.$$

One example will confirm this definition.

Example 2.63. Let $x = \underline{x}$ and $\alpha = \frac{1}{2}$. We have to line out the following calculation:

$$\begin{aligned} \underline{x}^{\frac{1}{2}} &= e^{\frac{1}{2}(\ln|\underline{x}| + \omega(x)\operatorname{arc}\cos\frac{x_0}{|x|} + \omega(x)2k\pi)} \\ &= \sqrt{|\underline{x}|}\left[\cos(\frac{1}{2}\operatorname{arc}\cos 0 + k\pi)\omega(x)\sin(\frac{1}{2}\operatorname{arc}\sin 0 + k\pi)\right] \\ &= \sqrt{|\underline{x}|}\left[\cos(\frac{\pi}{4} + k\pi) + \frac{\underline{x}}{|\underline{x}|}\sin(\frac{\pi}{4} + k\pi)\right] \end{aligned}$$

for $k = 0, 1$.

It would be desirable that these types of elementary functions are also regular in some sense. But calculations with the CAUCHY–FUETER operator

$$D = e_1\partial_1 + \ldots + e_n\partial_n$$

were not successful. Therefore our elementary functions will not be CLIFFORD analytic. By the use of spherical coordinates in $\mathbb{R}^n$ this operator D obtains the following representation:

$$D = \partial_0 + \omega(x)\partial_{|\underline{x}|} + \omega(x)\frac{1}{|\underline{x}|}\Gamma,$$

where Γ is just the operator

$$\Gamma = \sum_{i<j} e_i e_j (x_i\partial_j - x_j\partial_i).$$

The main problem is caused by the latter operator. It provides to multiply $\omega(x)$ by $(n-1)$. We obtain $(n-1)\omega(x)$. This expression is disturbing the structure of our elementary functions. Besides, we note that scalar and vector fields which are only depending on $|\underline{x}|$ are lying in the kernel of Γ.

Definition 2.64. *The differential operator*

$$D_{ra} := \frac{1}{2}(\partial_0 - \omega(x)\partial_{|\underline{x}|})$$

is called radial CAUCHY–FUETER *operator.*

Note the denotation $D_{ra}u := u'$ for a paravector $u \in Cl_{0,n}$.

Let us now investigate in which manner this operator is acting on the elementary functions.

Proposition 2.65. *We list here the following differential actions:*

(i) $\partial_{|\underline{x}|}\omega(x) = 0$,

(ii) $\partial_{|\underline{x}|}|x| = \frac{|\underline{x}|}{|x|}$,

(iii) $\partial_{|\underline{x}|}\underline{x} = \omega(x)$.

Proof. For $u \in C^2(\mathbb{R}^n, Cl_{0,n})$ we have for instance

$$\partial_{|\underline{x}|}\omega_i(x) = \partial_{|\underline{x}|}\partial_i|\underline{x}| = \partial_i\partial_{|\underline{x}|}|\underline{x}| = \partial_i 1 = 0,$$

where $\omega_i(x) = x_i/|\underline{x}|$. We note only for (iii) that $\underline{x} = (\underline{x}/|\underline{x}|)|\underline{x}|$. Property (ii) is easy to verify. #

Proposition 2.66. *Let u, v be paravector valued functions. If $\omega(x)u = u\omega(x)$ then for D_{ra} the* LEIBNIZ *rule holds.*

Proof. We have

$$2(uv)' = \partial_0(uv) - \omega(x)\partial_{|\underline{x}|}(uv) = 2u'v + u\partial_0 v - \omega(x)u\partial_{|\underline{x}|}v = 2(u'v + v'u).\#$$

Theorem 2.67. *The radial differential operator D_{ra} generates in the set of elementary functions the following rules:*

(i) $(e^x)' = e^x$,

(ii) $(\sin x)' = \cos x, \quad (\cos x)' = -\sin x$,

(iii) $(\sinh x)' = \cosh x, \quad (\cosh x)' = \sinh x$,

(iv) $(\log x)' = \frac{1}{x}$,

(v) $(x^\alpha)' = \alpha x^{\alpha-1}$ *where* $\alpha \in \mathbb{R}$.

Proof. Using Proposition 2.65 the properties (i), (ii), and (iii) follow immediately. Therefore, we will only show the differentiability rules (iv) and (v). We have

$$(\partial_0 - \omega(x)\partial_{|\underline{x}|})\log x = \partial_0 \log x - \omega(\underline{x})\partial_{|\underline{x}|}\log x.$$

Let us consider the first term:

$$\partial_0 \ln|x| + \partial_0\omega(x)\text{arc}\cos\frac{x_0}{|x|} = \frac{x_0}{|x|^2} + \omega(\underline{x})\frac{-1}{\sqrt{1-\left(\frac{x_0}{|x|}\right)^2}}\partial_0\frac{x_0}{|x|}$$

$$= \frac{x_0}{|x|^2} - \frac{\underline{x}}{|\underline{x}|}\frac{|x|}{|\underline{x}|}\frac{|x| - \frac{x_0^2}{|x|}}{|x|^2} = \frac{x_0 - \underline{x}}{|x|^2} = \frac{\overline{x}}{|x|^2} = \frac{1}{x}.$$

The second term provides

$$-\omega(x)\partial_{|\underline{x}|}\log x = -\omega(x)(\partial_{|\underline{x}|}(\ln|x| + \partial_{|\underline{x}|}\text{arc}\cos\frac{x_0}{|x|})$$

$$= -\omega(x)\left(\frac{1}{|x|}\partial_{|\underline{x}|}|x| + \frac{(-1)}{\sqrt{1-\left(\frac{x_0}{|x|}\right)^2}}\frac{(-x_0)}{|x|^2}\partial_{|\underline{x}|}|x|\right)$$

$$= \frac{x_0}{|x|^2} - \frac{\underline{x}}{|x|^2} = \frac{\overline{x}}{|x|^2} = \frac{1}{x}.$$

These calculations show (iv). To verify (v) we have to proceed as follows:

$$(x^\alpha)' = (e^{\alpha\log x})' = (e^{\alpha\ln|x| + \alpha\omega(x)\text{arc}\cos\frac{x_0}{|x|}})'.$$

Using Proposition 2.47 and the LEIBNIZ rule we obtain

$$\alpha|x|^{\alpha-1}\left(\cos\alpha\text{arc}\cos\frac{x_0}{|x|} + \omega(x)\sin\alpha\text{arc}\cos\frac{x_0}{|x|}\right)$$
$$+|x|^\alpha\left(\cos\alpha\text{arc}\cos\frac{x_0}{|x|} + \omega(\underline{x})\sin\alpha\text{arc}\cos\frac{x_0}{|x|}\right)'.$$

Relationship (v) arises now by straightforward but mostly real calculations.
#

Definition 2.68. *Let $f : P\text{Vec}\, Cl_{0,1} \mapsto P\text{Vec}\, Cl_{0,1}$ be a paravector valued function in $Cl_{0,n}$ of the type $f = f_0 + \omega(x)f_1$ with $f_0, f_1 \in \mathbb{R}$ and $h = h_0 + \omega(x)|\underline{h}|$. Such a function is called radially differentiable or radially* CLIFFORD *regular if and only if*

$$\lim_{h\mapsto 0}\frac{(f(x+h) - f(x)\overline{h}}{|h|^2}$$

exists. In the case of existence the radial derivative is just $f'(x) = D_{ra}$.

Theorem 2.69. *Let* $f_i = f_i(x_0, |\underline{x}|)$, $(i = 0, 1)$ *real-valued functions and* $f = f_0 + \omega(x) f_1$. *Further, there exist continuous partial derivatives* $\partial_0 f_i$ *and* $\partial_{|\underline{x}|} f_i\ (i = 0, 1)$. *Then the function* f *is radially differentiable if and only if the relations*

$$\partial_0 f_0 = \partial_{|\underline{x}|} f_1, \qquad \partial_0 f_1 = -\partial_{|\underline{x}|} f_0$$

are valid.

Corollary 2.70. *All above defined elementary functions are radially* CLIFFORD *regular.*

Remark 2.71. The conception of radially CLIFFORD regular functions goes back to R. FUETER's first papers. Later these ideas appear in papers of different authors. In this context should be mentioned M. SCE [264], A. SUDBURY [313], F. SOMMEN [286], [280], H. LEUTWILER [168], [169], [170] K. NONO [216], and M. MARINOV [193]. Contributions in [36] on this subject should also be mentioned.

2.4 Regular functions in symmetric domains

Holomorphic functions of a single complex variable are closely connected with CLIFFORD regular functions in domains with axial symmetry. Here will be discussed generalized HERMITE polynomials as well as an axial exponential function. It is also possible to describe classes of regular polynomials which are complex potentials of axial regular vector fields. Our explanations start with results of the paper [181] and continue with investigations outlined in [51], [52] and [140]. In some sense this section can be seen as a continuation of Section 2.3.

Let $x_0 \in \mathbb{R}, x \in \operatorname{Vec} C\ell_{n,0}\ (e_i^2 = 1\,; i = 1, ..., n), a \in S^n$. Furthermore, let x be given in cylindric coordinates: $x = \underline{x} + x_0 a = \sum_{i=1}^n (x_i e_i + x_0 a_i)$.

Proposition 2.72. *For*

$$y = \frac{1}{2}(nax + (n-2)xa) \text{ follows } Dy = 0.$$

Proof. Clearly $D(xa) = na$ and $D(ax) = -(n-2)a$, then this statement follows immediately. #

Remark 2.73. Let $s(x) = \frac{x}{|x|^n}$. Then

$$q(x) = D(a, s) = \frac{1}{|x|^n}\left(a - n(a, x)\frac{x}{|x|^2}\right)$$

can be seen as the field of the *n-dimensional dipole*. With $w = -s(x)q(x^{-1})$ is given the *complex potential* of the vector field $v = -(n-1)a$.

Proposition 2.74. *The polynomials*

$$p_k(x) = \frac{(-1)^k}{k!}\, s(x)q_k(x) \quad \text{with} \quad q_k(x) = (\partial_{x_n}^k s)(x)$$

are axially symmetric CLIFFORD *regular and homogenous of degree* $-(k+n-1)$.

Proof. We note only that for any CLIFFORD regular function $u = u(x)$ the function $-s(x)u(x^{-1})$ is also CLIFFORD regular for $(|x| \neq 0)$. #

Remark 2.75. Polynomial $q_k(x)$ describes an axial multipole.

Theorem 2.76. [181] *The polynomials* $p_k(x)$ *permit the representation*

$$p_k(x) = \sum_{j=1}^{k} p_{kj}(ax)^{k-j}(xa)^j,$$

where the coefficients can be obtained by the recurrence relation

$$p_{kj} := \frac{1}{k}\left(\frac{n-2}{2} + (k-j)\right)(p_{k-1,j} + p_{k-1,k-j}).$$

Proof. The proof is given in [181]. #

Remark 2.77. For $n = 2$, Sc $p_k(x)/|x|^k$ coincides with the CHEBYCHEV polynomial $T_k(\cos\theta) = \cos n\theta$. For $n = 3$ we have Sc $p_k(x)/|x|^k$ coincides with the well-known LEGENDRE polynomial $P_k(\cos\theta)$ and for $n = 4$

$$\text{Sc } p_k(x) = |x|^k \left(\frac{k}{2} + 1\right) \frac{\sin(k+1)\theta}{\sin\theta}.$$

Here $|\underline{x}| = |x| \sin\theta$ and $x_n = |x| \cos\theta$.

Now we will assume the CLIFFORD algebra $Cl_{0,n}$ ($e_i^2 = -1; i = 1, ..., n$). Further, let $x = x_0 e_0 + |\underline{x}|\omega(x)$, where $\omega(x) = \underline{x}/|\underline{x}|$.

Proposition 2.78. [286] *The functions*

$$u_i(x) = [A_i(x_0, |\underline{x}|) + \omega(x)\, B_i(x_0, |\underline{x}|)] P_i(\omega(x))$$

are CLIFFORD *regular iff the following system of the* VEKUA *type*

$$\begin{aligned} \partial_{x_0} A_i - \partial_{|\underline{x}|} B_i &= \frac{i+m-1}{|\underline{x}|} \\ \partial_{x_0} B_i + \partial_{|\underline{x}|} A_i &= \frac{i}{|\underline{x}|} A_i \end{aligned}$$

is fulfilled. Hereby P_i is an inner spherical CLIFFORD *regular function of order i, i.e. $P_i(\lambda \underline{y}) = \lambda^i P_i(\underline{y})$ ($\lambda \in \mathbb{R}$; $\underline{y} \in \text{Vec}\, Cl_{0,n}$).*

Remark 2.79. In the case of $i = 0$ we get functions of the type $p_k(x)$ (cf. Theorem 2.76.).

Theorem 2.80. [51] *Let u be a holomorphic function of the complex variable z on the set $\{z = x_0 + i|\underline{x}|t;\ -1 \le t \le 1\}$ for all $x = x_0 + |\underline{x}|\omega(x)$ in a suitable domain $G \subset \mathbb{R} \oplus \mathbb{R}^n$. Setting $\nu = k + n/2$, then*

$$u_k(x) = \frac{\Gamma(\nu)}{\Gamma(\nu - 1/2)\Gamma(1/2)} \int_{-1}^{1} (1-t^2)^{\nu-3/2} u(x_0 + i|\underline{x}|t)(1 - it\omega(x))dt\, P_k(\underline{x})$$

is (axial) CLIFFORD *regular of degree k for all $x \in G$.*

Corollary 2.81. *The axial exponential function [286]*

$$E_{k,n}(x_0, |\underline{x}|) := 2^{\nu-1}\Gamma(\nu) e^{x_0} |\underline{x}|^{1-n/2} (J_{\nu-1}(|\underline{x}|) + \omega(x) J_\nu(|\underline{x}|))$$

with $\lim_{|\underline{x}|\to 0} |\underline{x}|^{-k} E_{k,n}(x_0, |\underline{x}|) = e^{x_0}$ *permits the integral representation*

$$E_{k,n}(x_0, |\underline{x}|) = \frac{\Gamma(\nu)|\underline{x}|^k}{\Gamma(\nu - 1/2)\Gamma(1/2)} \int_{-1}^{1} (1-t^2)^{\nu-3/2} e^{x_0 + i|\underline{x}|t}(1 - it\omega(x))dt,$$

i.e., $E_{n,k}(x_0, |\underline{x}|)$ *is the image of* e^x *of this transformation.*

Proof. We refer only to the standard integral representation of the BESSEL function (cf. [248], p. 431).

$$J_v(|\underline{x}|) = \frac{(|\underline{x}|/2)^\nu}{\Gamma(\nu + 1/2)\Gamma(1/2)} \int_{-1}^{1} (1-t^2)^{\nu-1/2} e^{i|\underline{x}|t} dt. \quad \#$$

Here we will demonstrate exemplified for other generalizations, how to obtain CLIFFORD regular HERMITE polynomials which form a basis in $L_2(\mathbb{R}^n, \exp(-|\underline{x}|^2/2; Cl_{0,n})$. F. SOMMEN in [286] could develop a method which is quite similar to the complex case. Let $z = x + iy \in \mathbb{C}$ then the standard HERMITE polynomials H_n can be generated by the TAYLOR series expansion

$$e^{z^2/2} = \sum_{n=0}^{\infty} e^{-y^2/2} \frac{H_n(y)}{n!}(ix)^n.$$

In this way HERMITE polynomials arise by holomorphic extension of the function

$$e^{(iy)^2/2}.$$

In the same manner generalized HERMITE polynomials may be obtained by axial CLIFFORD regular extension of

$$e^{-|\underline{x}|^2/2} P_k(\underline{x})$$

from $x_0 = 0$ to $x_0 \neq 0$, where P_k is an inner spherical polynomial of the degree k.

Let $\partial_x := \sum_{i=1}^n e_i \partial_i$.

Definition 2.82. *The axial monogenic extension of*

$$u_k(x) = |\underline{x}|^k e^{-|\underline{x}|^2/2} P_k(\omega(x))$$

to $x_0 \neq 0$ *is given by*

$$u_k(x) = e^{-|\underline{x}|^2/2} \sum_{\ell=0}^{\infty} \frac{H_{\ell,n,k}(\underline{x}) x_0^\ell}{\ell!} P_k(\underline{x}),$$

where

$$H_{\ell,n,k}(\underline{x}) P_k(\underline{x}) := e^{\frac{|\underline{x}|^2}{2}} (-1)^\ell \partial_{\underline{x}}^\ell \left(e^{-\frac{|\underline{x}|^2}{2}} P_k(\underline{x}) \right).$$

Then the functions $H_{\ell,n,i}(\underline{x})$ *are called generalized* HERMITE *polynomials. Note that*

$$H_{\ell,1,0}(\underline{x}) = \omega(x)^{\ell} H_{\ell}(|\underline{x}|),$$

where $H_{\ell}(|\underline{x}|)$ *are the classical* HERMITE *polynomials.*

Remark 2.83. In [51] there are given relations which express the close connection of generalized HERMITE polynomials to LAGUERRE functions. Further, generalized HERMITE polynomials could be explicitly calculated. Similarly, generalized GEGENBAUER polynomials may be obtained, too.

Remark 2.84. In addition to the explanations in Section 2.1 we note a further principle in order to transform holomorphic functions into CLIFFORD regular ones, which is very closely connected with the so-called "plane wave decompositions" (cf. [288]). It reads as follows:

Let $u = u(z)$ *be a holomorphic function in some domain in* $\mathbb{C}$ *then*

$$(1 - i\underline{v})u(x_0 + i(\underline{x}, \underline{v}))$$

is always axial CLIFFORD *regular in* $\mathbb{R} \oplus \mathbb{R}^n$, *where* $\underline{v} = \sum\limits_{k=1}^{n} v_k e_k \in S^n$.

Remark 2.85. Most of these ideas have been transferred to domains with biaxial symmetry. We refer to the papers [140] and [52].

2.5 Uniqueness of Clifford regular functions

In this very short section we will present so-called uniqueness sets for CLIFFORD regular functions. Sometimes it will work similarly to the complex case. But we have to look carefully for the appropriate analogue in $\mathbb{R}^n$. In order to prove a uniqueness result for holomorphic functions u in a domain G of the complex plane, it is sufficient that u is given on a set which has an accumulation point in G. In this formulation the corresponding statement is not true. Other results, as for instance LUSIN's theorem, can be proved similarly.

Let us start with the following counterexample.

Example 2.86. Let $n = 2, Cl_{0,n} = \mathbb{H}, G = B_1(0)$. In addition let $x = x_1e_1 + x_2e_2 + x_3e_3$. By $x_3 = 0$ the x_1, x_2-plane is defined. Furthermore, the function $u(x) = x_2e_1 - x_1e_2 \in \ker D$. On the x_3-axis we have obviously $u(x) = 0$ and each point in $\{x : x_1 = x_2 = 0\}$ is an accumulation point, but u is not identical zero on $B_1(0)$.

Remark 2.87. This example shows also that not all zeros of an $\mathbb{H}$-regular function are isolated.

Theorem 2.88. *Let $u \in \ker D \cap C^1(G, Cl_{0,n})$. We fix a point $a \in G$ and consider the sets*

$$l_\omega(a) := \{x \in G : \ x = a + t\omega, \ t > 0, \ \omega \in S^n\}.$$

Assume E as being such a set that the point a is an accumulation point of each of the sets $E \cap l_\omega(a)$. If $u = 0$ on E then $u = 0$ on G.

Proof. Let ω be an arbitrary but fixed direction and $B_\varepsilon(a)$ a ball around a with radius ε. There exists a sequence $(x_n(\omega))$ with $x_n(\omega) \to a$ such that $u(x_n(\omega)) = 0$ $(n = 1, 2, ...)$. For each ω there is a natural number $N(\omega)$ such that $x_n(\omega) \in B_\varepsilon(a)$ for $n \geq N(\omega)$. As S^n is compact in $\mathbb{R}^n$ we find the largest number N_0. Now we obtain that for all $\omega \in S^n$ and $n \geq N_0$ is $x_n(\omega) \in B_\varepsilon(a)$. Let n tend to infinity and we have from TAYLOR's formula

$$0 = u(x_n(\omega)) = \sum_{k=0}^{\infty} a_k(a, \omega)|x_n(\omega) - a|^k$$

$a_0(a, \omega) = 0$. Division by $|x_n(\omega) - a|$ yields

$$0 = \sum_{k=1}^{\infty} a_k(a, \omega)|x_n(\omega) - a|^{k-1}$$

with the consequence $a_1(a, \omega) = 0$. Repeating this procedure again and again we finally obtain

$$a_k(a, \omega) = 0 \quad (k = 0, 1, ..., n, ...).$$

Hence, $u \equiv 0$ in $B_\varepsilon(a)$ and it immediately follows that $u \equiv 0$ in the whole domain G. #

Remark 2.89. A point a which is an accumulation point in each of the sets $E \cap l_\omega(a)$ is sometimes also called *spatial accumulation point.*

In order to prove LUSIN's theorem we need the following extension result.

Lemma 2.90. *Let G_i (i=1,2) be bounded domains in $\mathbb{R}^n, G_1 \cap G_2 = \emptyset, \partial G_i = \Gamma_i$, $\Gamma_1 \cap \Gamma_2 = \Gamma_0$ $(i = 1,2)$. The boundaries Γ_i have to be piecewise smooth* LIAPUNOV *surfaces. Furthermore, let u_i belong to $(ker\, D)(G_i) \cap C(\overline{G}_i, Cl_{0,n})$ $(i = 1,2)$ and coincide on Γ_0. Then the $Cl_{0,n}$-valued function*

$$u(x) = \begin{cases} u_1(x) & (x \in G_1) \\ u_2(x) & (x \in G_2) \\ u_1(x) = u_2(x) & (x \in \Gamma_0) \end{cases}$$

is CLIFFORD *regular in $G_1 \cup G_2 \cup \Gamma_0$.*

Proof. We have to verify that u in Γ_0 is CLIFFORD regular. Let $B_\varepsilon(x)$ be again a ball around $x \in \Gamma_0$ of radius ε, where $\varepsilon > 0$ is chosen so small that $B_\varepsilon(x) \subset G_1 \cup G_2 \cup \Gamma_0$. Define in $B_\varepsilon(x)$ by $u^*(x) := (F_{S_\varepsilon} u)(x)$ with $\partial B_\varepsilon(x) = S_\varepsilon$ a $Cl_{0,n}$-regular function. Under consideration of the orientation of $\Gamma_i = (\partial B_\varepsilon \cap G_i) \cup \Gamma_0$ $(i = 1,2)$ we obtain

$$u^*(x) = (F_{\Gamma_1} u)(x) + (F_{\Gamma_2} u)(x).$$

For $x \in G_1 \cup \Gamma_i$ we have $u^*(x) = u_i(x)$ $(i = 1,2)$. In particular, there is in Γ_0 $u^*(x) = u(x)$ and, therefore, the function $u = u(x)$ is CLIFFORD regular in Γ_0. #

Theorem 2.91. (LUSIN's theorem) *Let γ be a connected submanifold contained in $\Gamma = \partial G$. If the $Cl_{0,n}$-valued function*

$$u \in (ker\, D)(G) \cap C(\overline{G} \cup \gamma, Cl_{0,n})$$

vanishes on γ then the function $u = u(x)$ is in G identical zero.

Proof. Making use of Lemma 2.90 we have to extend u throughout the submanifold γ into some domain G' by zero. The function

$$u'(x) = \begin{cases} u(x) & x \in G \\ 0 & x \in G' \setminus \overline{G} \\ 0 & x \in \gamma \end{cases}$$

satisfies the assumptions of Lemma 2.90. The function u' is $Cl_{0,n}$-regular in $G \cup G' \cup \gamma$. For the uniqueness of a $Cl_{0,n}$-regular function Theorem 2.88 yields $u' \equiv 0$. #

2.6 Hypercomplex exterior calculus

K. HABETHA introduced in 1976 in his paper [116] special hypercomplex differential forms of first order. In 1978 A. SUDBERY defines in his famous paper "Quaternionic analysis" [313] the module $\mathcal{E}^r(G,\mathbb{H})$ of $\mathbb{H}$-valued differentiable r-forms. These considerations were based on earlier results of H.G. HAEFELI (1947) ([118]). By the aid of the quaternionic exterior calculus he succeeded in establishing in a fashionable manner fundamental theorems of a quaternionic function theory. J. RYAN gave in [251] interesting generalizations. Later on V. SOUČEK in his thesis [294] and also in [295] continued together with R. DELANGHE and F. SOMMEN with these considerations up to a generalized residue theory on manifolds (cf. [291], [292], [64]). Comprehensive investigations are made by F. SOMMEN [281] and H.R. MALONEK [186] (cf. Section 3.7). While H.R. MALONEK studied in a very clearly manner a calculus of hypercomplex alternative differential forms as the basis of CAUCHY'S theorem, F. SOMMEN combined monogenic forms and homology theory. More recently we should mention the papers [62], [85], and [87] with different topics. In the latter the HODGE–DE RHAM system on r-forms is solved. The special meaning of RIESZ transform is analysed.

Let us start with the consideration of $\mathbb{H}$-valued forms.

Definition 2.92. *Let F_0, F_1, F_2, F_3 be real-valued r-forms with C^∞-coefficients defined in $G \subset \mathbb{R}^4$. Then an $\mathbb{H}$-valued r-form is given by*

$$F = F_0 e_0 + F_1 e_1 + F_2 e_2 + F_3 e_3.$$

Corollary 2.93. *An arbitrary r-form F permits the representation*

$$F = \sum_{0 \le i_1 < \ldots < i_r \le 3} f_{i_1 \ldots i_r} dx_{i_1} \wedge \ldots \wedge dx_{i_r},$$

where $f_{i_0 \ldots i_r}$ are suitable $\mathbb{H}$-valued $C^\infty(G)$-functions.

Corollary 2.94. *Let F_1, F_2 be $\mathbb{H}$-valued r-forms. We define a sum and a multiplication by $\lambda \in \mathbb{H}$ in the following way:*

(i) $$F_1 + F_2 := \sum_{0 \le i_1 < \ldots < i_r \le 3} \left(f^{(1)}_{i_1 \ldots i_r} + f^{(2)}_{i_1 \ldots i_r} \right) dx_{i_1} \wedge \ldots \wedge dx_{i_r},$$

(ii) $$\lambda F := \sum_{0 \le i_1 < \ldots < i_r \le 3} \lambda f_{i_1 \ldots i_r} dx_{i_1} \wedge \ldots \wedge dx_{i_r},$$

the space of all $\mathbb{H}$-valued r-forms becomes a $\mathbb{H}$-left module. It is denoted by $\mathcal{E}^r(G,\mathbb{H})$. By definition $\mathcal{E}^0(G,\mathbb{H})$ coincides with $C^\infty(G,\mathbb{H})$. Sometimes we will use for an r-form F the denotation $F^{(r)}$.

Definition 2.95. *The exterior product $\wedge$ of an r-form $F^{(r)}$ and a s-form $F^{(s)}$ is defined by*

$$(F^{(r)} \wedge F^{(s)})(f_1, ..., f_{r+s}) = \frac{1}{r!s!} \sum_p \operatorname{sgn} p \, F^{(r)}(f_{p_1}, ..., f_{p_r}) F^{(s)}(f_{p_{r+1}}, ..., f_{p_{r+s}})$$

where p runs over all permutations of $r + s$ objects.

Definition 2.96. *The operator $d : \mathcal{E}^r(G, \mathbb{H}) \to \mathcal{E}^{r+1}(G, \mathbb{H})$ defined by*

(i) $d^2 = 0$,

(ii) $d(F^{(r)} \wedge F^{(s)}) = dF^{(r)} \wedge F^{(s)} + (-1)^r F^{(r)} \wedge dF^{(s)}$,

is called DE RHAM *operator.*

Lemma 2.97. [118] *Let $F \in C^1(G, \mathbb{H})$, then there exist $\mathbb{H}$-valued functions f_i $(i = 0, 1, 2, 3)$ such that*

$$dF = \sum_{i=0}^{3} f_i \, \overline{dx} \, e_i, \tag{2.5}$$

where $\overline{dx} = dx_0 e_0 - dx_1 e_1 - dx_2 e_2 - dx_3 e_3$.

Lemma 2.98. [313] *Let $F \in C^1(G, \mathbb{H})$ be (right) quaternionic regular with the representation (2.5). Then kdF with*

$k = (\overline{f}_1 f_2 \overline{f}_3 - \overline{f}_3 f_2 \overline{f}_1)^{-1}$

is quaternionic regular on both sides.

Theorem 2.99. (cf. [85]) *Let $q = x_0 e_0 + x_1 e_1 + x_2 e_2 + x_3 e_3$. Then the following $\mathbb{H}$-valued 1-forms*

$$\begin{aligned} d\overline{q}_0 = d\overline{q} &= dx_0 - dx_1 e_1 - dx_2 e_2 - dx_3 e_3 \\ d\overline{q}_1 &= dx_0 - dx_1 e_1 + dx_2 e_2 + dx_3 e_3 \\ d\overline{q}_2 &= dx_0 + dx_1 e_1 - dx_2 e_2 + dx_3 e_3 \\ d\overline{q}_3 &= dx_0 + dx_1 e_1 + dx_2 e_2 - dx_3 e_3 \end{aligned}$$

generate $\mathcal{E}^1(G, \mathbb{H})$ and an arbitrary quaternionic 1-form defined on G can be written as

$$dF = \sum_{i=0}^{3} (d\overline{q}_i) \frac{\partial}{\partial \overline{q}_i} F,$$

where

$$\frac{\partial F}{\partial \overline{q}_i} = \sum_{j=0}^{3} e_j \frac{\partial F}{\partial x_j}.$$

Corollary 2.100. DE RHAM*'s operator* $d : \mathcal{E}^{(i)}(G, \mathbb{H}) \to \mathcal{E}^{(i+1)}(G, \mathbb{H})$ $(i = 0, 1)$ *permits the decomposition*

$$\begin{aligned} d &= \overline{\partial}_0 + \overline{\partial}_1 + \overline{\partial}_2 + \overline{\partial}_3, \\ \textit{where} \qquad \overline{\partial}_j(F) &= d\overline{q}_j \frac{\partial F}{\partial \overline{q}_j} \quad (j = 0, 1, 2, 3) \quad (i = 0) \end{aligned}$$

and

$$\overline{\partial}_j \left(d\overline{q}_k F \right) = \left(d\overline{q}_k \wedge d\overline{q}_j \frac{\partial F}{\partial \overline{q}_j} \right) - \frac{1}{2}(d\overline{q}_k \wedge d\overline{q}_j + d\overline{q}_j \wedge d\overline{q}_k) \frac{\partial F}{\partial \overline{q}_k} \quad (i = 1).$$

Furthermore, it holds that

$$\overline{\partial}_j \overline{\partial}_j F = 0 \qquad (j = 0, 1, 2, 3).$$

Lemma 2.101. [295] *The* DE RHAM *operator* d *allows the decomposition*

$$du = Du + \vec{D}u,$$

where

$$\begin{aligned} Du &= dx \partial u, \quad \vec{D}u = dx_1 \partial^1 u + dx_2 \partial^2 u + dx_3 \partial^3 u \\ \partial u &:= \frac{1}{4}(\partial_0 - e_1\partial_1 - e_2\partial_2 - e_3\partial_3), \quad \partial^i := -e_i \partial u e_i \qquad (i = 1, 2, 3). \end{aligned}$$

Remark 2.102. Sometimes D is identified with the DIRAC operator and $\vec{D}$ with the so-called *twister operator*, which plays a role in quantum physics. (cf. [13]). For generalizations to higher dimensions we will introduce the following denotations. At first we follow the consideration in [186].

Let $x = \sum_{i=0}^{n} x_i e_i$ be an arbitrary paravector in $Cl_{0,n}$. Then

$$dz_0 := e_0 dx_0 \quad dz_k := -e_k dx_0 + e_0 dx_k \quad (k = 1, ..., n) .$$

This is a differential form with coefficients of $P\mathrm{Vec}\, Cl_{0,n}$. Because of the properties $e_i e_j + e_j e_i = 0$ $(i \neq j)$ $e_0^2 = 1$, $e_i^2 = -1$ $(i = 1, ..., n)$, we get immediately

$$dx_0 = e_0 dz_0 \; , \; dx_k = e_k dz_0 + e_0 dz_k \; .$$

Let now $p_k : G \to \mathbb{R}$ be real-valued functions. Note that $f_k(x) = \tilde{f}_k(z)$, where $z = (z_0, ..., z_n)$ with $z_i = x_i e_0 - x_0 e_i$ $(i = 1, ..., n)$ and $z_0 = e_0$.

Definition 2.103. *The special 1-forms* dz_k $(k = 0, ..., n)$ *are called fundamental 1-forms.*

Theorem 2.104. *An arbitrary differential form $F \in$ PVec $Cl_{0,n}$ has the unique representation*

$$F = \sum_{k=0}^{n} f_k(z) dz_k.$$

Definition 2.105. *The exterior product $\wedge$ of two fundamental forms dz_k and dz_l is defined by*

$$\begin{aligned} dz_k \wedge dz_l &:= e_0 dx_k \wedge dx_l - e_k dx_0 \wedge dx_l - e_l dx_k \wedge dx_0 \quad (k, l = 1, ..., n) \\ dz_k \wedge dz_0 &:= dx_k \wedge dx_0. \end{aligned}$$

The exterior product of different fundamental 1-forms is called fundamental 2-form

Corollary 2.106. *The exterior product defined in Definition 2.105 is an alternating differential form.*

Corollary 2.107. *Let $A = (\alpha_0, ..., \alpha_{r-1})$ an ordered r-tupel of numbers from $\{0, 1, ..., r-1\}$ with $0 < \alpha_0 < \alpha_1 < ... < \alpha_{r-1} < n$. Then each fundamental r-form defined by $dz_A := dz_{\alpha_0} \wedge ... \wedge dz_{\alpha_{r-1}}$ permits the representation*

$$F_A = \sum_A f_A dz_A \quad \text{where} \quad f_A = (f_{A_1}, ..., f_{A_{\binom{n+1}{r}}}).$$

Definition 2.108. *The exterior product of two r-forms F_A, F_B is defined by*

$$F_A \wedge F_B := \sum_{A,B} f_A f_B dz_A \wedge dz_B,$$

where F_B has the same structure as F_A.

Lemma 2.109. *The fundamental $(n+1)$-form is real and we have*

$$dz_0 \wedge ... \wedge dz_n = dx_0 \wedge ... \wedge dx_n \, .$$

Theorem 2.110. [187] *Let $F \in C^1(G, Cl_{0,n})$. Then dF allows the decomposition*

$$dF = D_R F \, dz_0 + \sum_{k=1}^{n} \partial_k F dz_k,$$

where $D_R := \sum_{i=0}^{n} e_i \partial_i$ is the CAUCHY–FUETER *operator which is acting from the right.*

Remark 2.111. The reader can find the same formula in [282].

Definition 2.112. *Let* $H_A = \sum f_A dz_A g_A$*, where* $f_A, g_A \in C^1(G, \mathbb{R}^{\binom{n+1}{r}})$*. Then the action of* DE RHAM*'s operator is defined as follows:*

$$dH_A := \sum_A df_A \wedge dz_A g_A + (-1)^r \sum_A f_A dz_A \wedge dg_A.$$

A continuous differential form F *is called closed, if* $dF = 0$*, and exact if there exists a continuously differentiable function* F' *on* G *such that* $dF = F'$.

Remark 2.113. Now one can obtain all statements on closed and exact differential forms which are based on SCHWARZ's theorem. Furthermore, CAUCHY's integral theorem and its consequences can be easily obtained.

It is not difficult to check that an arbitrary r-form F on G can be represented by

$$F = \sum_{i_1 \dots i_r} f_{i_1 \dots i_r} e_{i_1} \wedge \dots \wedge e_{i_r},$$

where $f_{i_1 \dots i_r}$ is a real-valued function on G and $e_{i_k} \subset \{e_0, e_1, \dots, e_{n-1}\} \subset \mathbb{R} \oplus \mathbb{R}^{n-1}$.

Setting in Definition 1.94 $E_{e_i} =: \mu_i$ (creation operator) and in Definition 1.95 $J_{e_i} = \mu_i^*$ (annihilation operator) one can show the identities

Lemma 2.114. [87] *We have*

(i) $\mu_i \mu_j + \mu_j \mu_i = 0,$

(ii) $\mu_i^* \mu_j^* + \mu_j^* u_i^* = 0,$

(iii) $\mu_i^* \mu_j + \mu_i \mu_j^* = \delta_{ij}\, I.$

Theorem 2.115. [87] *The* HODGE–DE RHAM (d, d^*)*-operators*

$$d = \sum_{i=1}^{n-1} \mu_i \partial_i + \mu_0 \partial_0, \quad d^* = \sum_{i=1}^{n-1} \mu_i^* \partial_i + \mu_0^* \partial_0$$

fulfil the equality

$$dd^* + d^* d = \Delta.$$

Corollary 2.116. *Any solution of the first order system*

$$\begin{aligned} dF^{(r)} &= 0, \\ d^* F^{(r)} &= 0 \end{aligned}$$

will have harmonic components. Locally the solution is given by $F^{(r)} = d\phi^{(r-1)}$*, where* $\phi^{(r-1)}$ *is an (r-1)-form with harmonic components such that* $d^* \phi^{(r-1)} = 0$.

Remark 2.117. Corollary 2.116 means that the system $dF = 0, \quad d^*F = 0$ is just a generalized CAUCHY–RIEMANN system in the sense of STEIN–WEISS [307].

Definition 2.118. *Let R_i be the usual* RIESZ-*transform* $(i = 1, ..., n-1)$. *Then the operator $R = \sum_{i=1}^{n-1} \mu_i R_i$ is called exterior* RIESZ *transform and the operator*

$$R^* = \sum_{i=1}^{n-1} \mu_i^* R_i$$

is called interior RIESZ *transform. A differential form $F : \mathbb{R}^n \to \Lambda(\mathbb{R}^n)$ is called massless if and only if $R^*F = 0$.*

Remark 2.119. If F is an 1-form, then massless means divergence free. For more information in this topic we refer to the excellent work edited by J. RYAN [259].

2.7 Wirtinger Calculus

It is well known that in the GAUSSian plane the application of the partial differentiations

$$\frac{\partial}{\partial z} := \frac{1}{2}\Big(\frac{\partial}{\partial x} - i\frac{\partial}{\partial y}\Big) \qquad \frac{\partial}{\partial \overline{z}} := \frac{1}{2}\Big(\frac{\partial}{\partial x} + i\frac{\partial}{\partial y}\Big)$$

leads to a very appropriate representation of the classical results of the function theory of one complex variable. The operators $\partial/\partial z$ and $\partial/\partial\overline{z}$ has been introduced by H. POINCARÉ. In the papers of W. WIRTINGER it was based on a corresponding calculus. H. POINCARÉ used the denotations ∂ and $\overline{\partial}$, respectively. In the function theory of several variables the use of these complex partial derivatives seems to be necessary. Obviously, we have the identities

$$\partial_x = \partial + \overline{\partial}\,,\ \partial_y = i(\partial - \overline{\partial}).$$

Holomorphic functions are exactly those which depend only on $z = x + iy$ (and do not depend on $\overline{z} = x - iy$). The aim of this section is to transfer this idea to the CLIFFORD calculus. We follow here investigations made in [186].

First we have to study suitable hypercomplex variables.

Proposition 2.120. *Suppose $x_0, x_1, ..., x_n$ $Cl_{0,n}$-linearly independent. Let $z_0 = x_0e_0$, $z_k := x_ke_0 - x_0e_k$ $(k = 1, ..., n)$. Then the system $z_0, z_1, ..., z_n$ is $Cl_{0,n}$-(left and right)-linearly independent.*

Proof. Let $\lambda_k \in Cl_{0,n}$ $(k = 0, 1, ..., n)$. We consider the equation

$$\lambda_0 z_0 + \lambda_1 z_1 + ... + \lambda_n z_n = 0$$

and obtain

$$(\lambda_1 e_0 - \lambda_0 e_1)x_1 + ... + (\lambda_n e_0 - \lambda_0 e_n)x_n + (\lambda_0 e_0)x_0 = 0.$$

Since $x_0, x_1, ..., x_n$ are $Cl_{0,n}$-linearly independent follows also $\lambda_k e_0 = \lambda_0 e_k = 0$ $(k = 1, ..., n)$ and $\lambda_0 e_0 = 0$. #

Lemma 2.121. *The matrix*

$$E := \begin{pmatrix} e_0 & & 0 \\ -e_1 & \ddots & \\ -e_n & & e_0 \end{pmatrix}$$

transforms the $(n+1)$*-tupel* $(x_0, .., x_n)^T$ *into the* $(n+1)$*-tupel* $(z_0, ..., z_n)$. *Conversely, the* $(n+1)$*-tupel* $(x_0, ..., x_n)^T$ *can be uniquely expressed for right linear combinations by*

$$\begin{pmatrix} x_0 \\ \vdots \\ x_n \end{pmatrix} = \begin{pmatrix} e_0 & & 0 \\ -e_1 & \ddots & \\ -e_n & & e_0 \end{pmatrix} \begin{pmatrix} z_0 \\ \vdots \\ z_n \end{pmatrix}$$

Denotation 2.122. We will use in the following the denotations:

(i) $z := z_0 + \underline{z}$, with $\underline{z} = (z_1, ..., z_n)$

(ii) $\overline{z}^H := z_0 - \underline{z}$

(iii) $\overline{z}^{Cl} := z_0 + \overline{\underline{z}}^{Cl} = (x_0 e_0, ..., x_k e_0 + x_0 e_k, ...)$.

We agree with the identification $\overline{z} := \overline{z}^{Cl}$.

Lemma 2.123. *The variable* $\underline{z}$ *permits the representation*

$$\underline{z} = \underline{x} e_0 + J x_0, \quad \underline{x} := (x_1, ..., x_n)$$

with $J := (-e_1, ..., -e_n)$.

Proof. We have only to mention that

$$\underline{z} = (x_1 e_0 - x_0 e_1, ..., x_n e_0 - x_0 e_n) = (x_1, ..., x_n) e_0 + (-e_1, ..., -e_n) x_0 . \qquad \#$$

Remark 2.124. In this way H.R. MALONEK proposed to consider J as a new imaginary unit; $\underline{x}$ is then the "*real part*" of $\underline{z}$ and x_0 is the "*imaginary part*" of $\underline{z}$. Sometimes the notation $\underline{x} =: \mathrm{Re}\underline{z}$ and $x_0 =: \mathrm{Im}\underline{z}$ is used.

Lemma 2.125. *Let* $H^n = \{\underline{z}\}$. H^n *can be given the structure of a* HILBERT *space.*

Proof. At first we define an addition and a multiplication with elements of $P\mathrm{Vec}\, Cl_{0,n}$. Setting now:

$$\begin{aligned} z' + z &:= (z_1' + z_1, ..., z_n' + z_n) \\ \lambda z &:= (\lambda z_1, \lambda z_2, ..., \lambda z_n) \qquad (\lambda \in P\mathrm{Vec} Cl_{0,n}) \end{aligned}$$

with $z' = (z_1', ..., z_n')$ and $z_k' = x_k' e_0 - x_0' e_k \quad (k = 1, ..., n)$. An zero element is given by $0 := (0, ..., 0)$. With these determinations H^n has the structure of a $Cl_{0,n}$-(left)-module. We introduce the "scalar" product

$$z \cdot z' = \sum_{k=1}^{n} z_k \overline{z}_k'$$

with $\overline{z}'_k := x'_k e_0 - x'_0 \overline{e}_k$. It is easy to verify all properties of a real scalar product. A detailed calculation delivers

$$\begin{aligned} z \cdot \overline{z}' &= \sum_{k=1}^{n} (x_k e_0 - x_0 e_k)(x'_k e_0 - x'_0 e_k) = \\ &= \sum_{k=1}^{n} (x_k x'_k + x_0 x'_0) e_0 + (x_k x'_0 - x_0 x'_k) e_k \in P\mathrm{Vec}\, Cl_{0,n}. \end{aligned}$$

For $z = z'$ we get the norm

$$|z|_H^2 = \sum_{k=1}^{n} x_k^2 + n x_0^2 = \sqrt{z \cdot \overline{z}} \ . \qquad \#$$

Proposition 2.126. *The following formulae are true:*

(i) $z + \overline{z}^H = 2z_0$ (iii) $z - \overline{z}^H = 2\underline{z}$

(ii) $z + \overline{z}^{Cl} = 2x$ (iv) $z - \overline{z}^{Cl} = 2Jx_0$.

Proof. It remains to prove (ii) and (iv). We have

$$z = (x_0 e_0, ..., x_k e_0 - x_0 e_k, ...) \quad \text{and} \quad \overline{z}^{Cl} = (x_0 e_0, ..., x_k e_0 + x_0 e_k, ...).$$

Then

$$z + \overline{z}^{Cl} = 2(x_0, \underline{x}) = 2x.$$

On the other hand we have

$$z - \overline{z}^{Cl} = 2(0, -x_0 e_1, ..., -x_0 e_n) = 2Jx_0. \qquad \#$$

Now we have to define suitable multidimensional analogues to the basic operators $\partial/\partial z$ and $\partial/\partial \overline{z}$ of WIRTINGER's calculus. Recalling the complex representation of DE RHAM's operator d, which can be written in the form

$$d = \frac{\partial}{\partial z} dz + \frac{\partial}{\partial \overline{z}} d\overline{z}$$

we look for a multidimensional version of this representation.

Theorem 2.127. *Let $f \in C^1(G, Cl_{0,n})$. Then df permits the representation*

$$df = Df\, dz_0 + \nabla f \cdot \overline{dz} \tag{2.6}$$

where $\nabla f = (\partial_1 f, ..., \partial_n f)$.

Proof. Suppose $f = f(z)$. Then for $f \in C^1(G, Cl_{0,n})$

$$\Delta f(z) = \partial_0 f \Delta x_0 + ... + \partial_n f \Delta x_n + o(|\Delta x|),$$

where $\lim\limits_{\Delta x \to 0} \frac{0(|\Delta x|)}{|\Delta x|} = 0$. Now we transform into $z_0, ..., z_n$ variables:

$$\Delta f(z) = \partial_0 f \Delta x_0 + \sum_{k=1}^{n} \partial_k f(-e_k \Delta x_0 + \Delta x_k e_0) + \sum_{k=1}^{n} \partial_k f e_k \Delta x_0 + o(|\Delta x|)$$

$$= (Df)\Delta z_0 + \sum_{k=1}^{n} \partial_k f \cdot \Delta z_k + o(|\Delta x|).$$

Setting now

$$df := Df \Delta z_0 + \sum_{k=1}^{n} \partial_k f \cdot \Delta z_k,$$

and substituting for f subsequently $x_0, x_1, ..., x_n$ we get $\Delta z_k = dz_k \quad (k = 0, 1, ..., n)$. Hence

$$df = Df\, dz_0 + \sum_{k=1}^{n} \partial_k f dz_k.$$

With the denotations

$$\nabla f = (\partial_1 f, ..., \partial_k f) \quad \text{and} \quad dz = (dz_1, ..., dz_n)$$

we achieve the result. #

Corollary 2.128. *For the basic operators in* WIRTINGER *calculus we have only to put*

$$\frac{\partial}{\partial z_0} := D \qquad \text{and} \qquad \frac{\partial}{\partial \overline{z}} := \nabla.$$

Furthermore, it follows that f is CLIFFORD *analytic iff $Df = 0$.*

Remark 2.129. In formula (2.6) correspond $Df, \nabla f$ to $\partial/\partial\overline{z}, \partial/\partial z$ in the complex plane and also $dz_0, \overline{dz}$ correspond to $d\overline{z}, dz$, respectively.

Remark 2.130. In [186] are also studied other systems $z_0, z_1, ..., z_n$. Replacing $z_0 = x_0 e_0 = \mathrm{Sc}\, x$ by $z_0 := \overline{x} = x_0 e_0 - \underline{x}$, one gets a further basis of operators in WIRTINGER calculus.

Chapter 3

Clifford operator calculus

3.1 Teodorescu transform

This section is dedicated to the consideration of a very special weakly singular integral operator, which will play a key role in the following.

Let G be a domain in $\mathbb{R}^n$ with a piecewise smooth boundary $\partial G = \Gamma$. We will use here the following norms:

$$||u||_\infty := ||u||_{L_\infty(G,C\ell_{0,n})} = \text{vrai}\max_{x\in G} |u(x)|,$$

where $|\cdot|$ is just the CLIFFORD operator norm (cf. Section 1.4). Furthermore, for $1 \leq p < \infty$ we define

$$||u||_p := ||u||_{L_p(G,C\ell_{0,n})} = \left(\int_G |u|^p dx\right)^{1/p}$$

$$||u||_{L_p^\alpha(G,C\ell_{0,n})} := \left(\int_G [|u||x|^\alpha]^p dx\right)^{1/p}$$

We abbreviate

$$||u||_{W_p^1(G,C\ell_{0,n})} =: ||u||_{p,1} = \left(\int_G |u|^p + \sum_{i=0}^{3} |\frac{\partial}{\partial x_i} u|^p dx\right)^{1/p}.$$

By $|G|$ we denote the volume of G.

Definition 3.1. *Let $u \in C(G, C\ell_{0,n})$. Then the linear integral operator*

$$(T_G u)(x) = -\int_G e(x-y)u(y)dy \tag{3.1}$$

with

$$e(x) = \frac{1}{\sigma_n}\frac{\overline{\omega}(x)}{|x|^{n-1}},\ \omega(x) = \frac{x}{|x|},\ x = \underline{x} = \sum_{i=1}^{n} x_i e_i$$

is called TEODORESCU *transform over G.*

We remark that T_G corresponds to the known T-operator from the one variable complex analysis. In the case $n = 2$ it coincides with the complex T-operator up to a constant factor. First in this section we consider the existence of this integral and some elementary properties.

Lemma 3.2. *Let $u \in C(\overline{G}, C\ell_{0,n})$. Then the integral (3.1) exists for all $x \in \mathbb{R}^n$ and we have*

$$||T_G u||_C \le \frac{1}{\sigma_n} \max_{x \in \overline{G}} \int_G \frac{1}{|x-y|^{n-1}} dy ||u||_C .$$

Proof. Because of the the uniform convergence of $\int_G 1/(|x-y|^{n-1}) dy$ it is continuous according to y and, therefore, we can take the maximum over $\overline{G}$. #

This means that T_G is a continuous operator in $C(\overline{G})$.

Remark 3.3. L. HEDBERG proved in [123] that there exists a constant $C > 0$, that for $n > \alpha > 0$

$$\int_G \frac{1}{|x-y|^{n-\alpha}} dy \le C\,(\mathrm{diam}\, Q)^\alpha$$

holds, where Q is the smallest cube which contains G.

Proposition 3.4. *Let $u \in L_1(G, C\ell_{0,n})$. Then the integral*

$$-\frac{1}{\sigma_n} \int_G e(x-y) u(y) dy$$

exists for all $x \in \mathbb{R}^n \setminus \overline{G}$ and will also be denoted by $(T_G u)(x)$. Moreover, the inequality

$$|(T_G u)(x)| \le \frac{1}{\sigma_n} [dist(x, G)]^{1-n} ||u||_1 \;, x \in \mathbb{R}^n \setminus \overline{G}$$

holds.

Proof. We get

$$\begin{aligned} |(T_G u)(x)| &\le \frac{1}{\sigma_n} \left| \int_g e(x-y) u(y) dy \right| \le \frac{1}{\sigma_n} \int_G |e(x-y)|\, |u(y)| dy \\ &\le \frac{1}{\sigma_n} [dist(x, G)]^{1-n} ||u||_1 . \qquad \# \end{aligned}$$

Corollary 3.5. *For $|x| \to \infty$ it follows that $|(T_G u)(x)| \to 0$. Furthermore, we have $(T_G u)(x) \in C^\infty(\mathbb{R}^n \setminus \overline{G}, C\ell_{0,n})$.*

This property will be essential for the investigation of boundary value problems.

Lemma 3.6. *Let $u \in L_1(G, C\ell_{0,n})$. Then $(T_G u)(x)$ exists almost everywhere on $\mathbb{R}^n$ and belongs to $L_q(G', C\ell_{0,n})$ for q with $1 < q < n/(n-1)$ and any domain $G' \subset \mathbb{R}^n$.*

Proof. Let $v \in L_p(G, C\ell_{0,n}), p > n$. Then we get by HÖLDER's inequality

$$V(x) = \int_G \frac{|v(y)|}{|x-y|^{n-1}} dy \leq \int_G \frac{1}{|x-y|^{(n-1)q}} dy \|v\|_p.$$

Clearly, $(n-1)q < n$. Thus $V(x)$ is defined by a uniformly convergent integral and therefore continuously. Then $V(x)u(x) \in L_1(G, C\ell_{0,n})$. Using FUBINI's theorem we obtain the identity

$$\int_G |u(y)| V(y) dy = \int_G |v(y)| U(y) dy \tag{3.2}$$

with

$$U(x) = \int_G \frac{|u(y)|}{|x-y|^{n-1}} dy.$$

Hence we conclude $U(x) \in L_q(G, C\ell_{0,n})$, where $1/p + 1/q = 1$. In this way we get $(T_G u)(x) \in L_q(G, C\ell_{0,n})$. Together with Proposition 3.4. the statement follows. #

Now, we will prove continuity results of T_G in the scale of SOBOLEV spaces.

Lemma 3.7. *Let $u \in C_0(\mathbb{R}^n, C\ell_{0,n})$ then*

$$\partial_k (T_G u)(x) = -\frac{1}{\sigma_n} \int_G \partial_{k,x} e(x-y) u(y) dy + \bar{e}_k \frac{u(x)}{n}.$$

Proof. The kernel of the TEODORESCU transform is weakly singular, thus we are not able to obtain the derivative in a straightforward manner. We write

$$(T_G u)(x) = -\frac{1}{\sigma_n} \lim_{\varepsilon \to 0} \int_{G_\varepsilon} e(x-y) u(y) dy,$$

where $G_\varepsilon = \{y \in G : |x-y| \geq \varepsilon\}$ with arbitrarily small $\varepsilon > 0$. Then

$$\partial_k \int_{G_\varepsilon} -\frac{1}{\sigma_n} e(x-y) u(y) dy$$

$$= -\frac{1}{\sigma_n} \int_{G_\varepsilon} \partial_{k,x} e(x-y) u(y) dy + \frac{1}{\sigma_n} \int_{r=\varepsilon} \frac{1}{\varepsilon^{n-1}} \sum_{i=1}^{n} \frac{(x_i - y_i) e_i}{|x-y|} \cos(r, y_k) u(y) dS_\varepsilon$$

$$= -\frac{1}{\sigma_n} \int_{G_\varepsilon} \partial_{k,x} e(x-y) u(y) dy + \frac{1}{\sigma_n} \int_{S_1} \sum_{i=1}^{n} (x_i - y_i) e_i \cos(r, y_k) u(x + \varepsilon \Theta) dS_1,$$

where

$$\Theta = \frac{y-x}{|y-x|} \quad \text{and} \quad \cos(r, y_k) = \alpha_k = \frac{y_k - x_k}{|x-y|}.$$

We remember that $\alpha = \sum_{i=1}^{n} \alpha_i e_i$ is the outward pointing normal vector at the point x. If $\varepsilon \to 0$ then the second integral converges uniformly to

$$\frac{1}{\sigma_n} \int_{S_1} \sum_{i=1}^{n} (x_i - y_i) e_i \cos(r, x_k) dS_1 u(x).$$

Because of

$$\frac{1}{\sigma_n} \int_{S_1} \sum_{i=1}^{n} (x_i - y_i) e_i \cos(r, y_k) dS_1 =$$

$$= \frac{1}{\sigma_n} \sum_{i=1}^{n} e_i \int_{r<1} \frac{\partial (x_i - y_i)}{\partial y_k} dy = \frac{\overline{e}_k}{\sigma_n} \int_{S_1} \Big\{ \int_0^1 r^{n-1} dr \Big\} dS_1 = \frac{\overline{e}_k}{n}$$

we get

$$\frac{1}{\sigma_n} \int_{S_1} \sum_{i=1}^{n} (x_i - y_i) e_i \cos(r, y_k) dS_1 \; u(x) = \overline{e}_k \frac{u(x)}{n}.$$

On the other hand,

$$-\partial_{k,x} e(x-y) = \partial_{k,x} \sum_{i=1}^{n} \frac{(x_i - y_i)\overline{e}_i}{|x-y|^n} =$$

$$= \Big(\sum_{i \neq k} -n \frac{(x_k - y_k)(x_i - y_i)\overline{e}_i}{|x-y|^{n+2}} \Big) + \frac{\overline{e}_k}{|x-y|^n} - \frac{n(x_k - y_k)^2 \overline{e}_k}{|x-y|^{n+2}}$$

$$= \frac{1}{|x-y|^n} \Big\{ -n \sum_{i=1}^{n} \frac{(x_k - y_k)(x_i - y_i)\overline{e}_i}{|x-y|^2} + \sum_{j=1}^{n} \frac{(x_j - y_j)^2}{|x-y|^2} \overline{e}_k \Big\}.$$

Because of

$$\int_{S_1} \Big\{ -n \sum_{i=1}^{n} \frac{(x_k - y_k)(x_i - y_i)\overline{e}_i}{|x-y|^2} + \sum_{j=1}^{n} \frac{(x_j - y_j)^2}{|x-y|^2} \overline{e}_k \Big\} dS_1 =$$

$$= \int_{S_1} \Big\{ -n \sum_{i=1}^{n} \frac{(x_i - y_i)}{|x-y|} \overline{e}_i \cos(r, y_k) + \sum_{j=1}^{n} \frac{(x_j - y_j)}{|x-y|} \overline{e}_k \cos(r, y_j) \Big\} dS_1$$

$$= \Big\{ -n \frac{\sigma_n}{n} \overline{e}_k + \sum_{j=1}^{n} \frac{\sigma_n}{n} \overline{e}_k \Big\} = 0,$$

the integral

$$-\frac{1}{\sigma_n}\int\limits_{G_\varepsilon}\partial_{k,x}e(x-y)u(y)dy$$

converges uniformly in x to the singular integral

$$-\frac{1}{\sigma_n}\int\limits_{G}\partial_{k,x}e(x-y)u(y)dy$$

for $\varepsilon \to 0$.
Summing up

$$\begin{aligned}\partial_k(T_G u)(x) &= \partial_k\Big\{\lim_{\varepsilon\to 0}\int\limits_{G_\varepsilon} -\frac{1}{\sigma_n}e(x-y)u(y)dy\Big\}\\ &= -\frac{1}{\sigma_n}\lim_{\varepsilon\to 0}\partial_k\int\limits_{G_\varepsilon} e(x-y)u(y)dy\\ &= -\frac{1}{\sigma_n}\int\limits_{G}\partial_{k,x}e(x-y)u(y)dy+\overline{e}_k\frac{u(x)}{n} \quad . \qquad \#\end{aligned}$$

Theorem 3.8. *The operator*

$$\partial_k T : L_p(\mathbb{R}^n, C\ell_{0,n}) \to L_p(R^n, C\ell_{0,n})$$

is continuous and

$$\|\partial_k T_G\|_{L(L_p(R^n,C\ell_{0,n}))} \le \Big(C\sigma_n^{-\frac{1}{p}}+\frac{1}{n}\Big) \quad (1<p<\infty).$$

Proof. We obtain from Lemma 3.7.:

$$\partial_k T_G u(x) = -\frac{1}{\sigma_n}\int\limits_{G}\partial_{k,x}e(x-y)u(y)dy+\overline{e}_k\frac{u(x)}{n}, \text{ where } u\in C_0^1(\mathbb{R}^n, C\ell_{0,n}).$$

Using the Theorem of CALDERON–ZYGMUND (cf. [202] §3, XI) it can be proved that

$$\Big\|\frac{1}{\sigma_n}\int\limits_{\mathbf{R}^n} e(x-y)u(y)dy\Big\|_p \le C\Big(\int\limits_{S^1}\Big(\frac{1}{\sigma_n}\frac{(x-y)}{|x-y|}\Big)^q dS_1\Big)^{\frac{1}{q}}\|u\|_p,$$

$$\Big(\int\limits_{S_1}\Big(\frac{1}{\sigma_n}\frac{(x-y)}{|x-y|}\Big)^q dS_1\Big)^{\frac{1}{q}} = \frac{1}{\sigma_n}\Big(\int\limits_{S_1}dS_1\Big)^{\frac{1}{q}} = \sigma_n^{\frac{1}{q}-1} = \sigma_n^{-\frac{1}{p}},$$

where $1/p+1/q=1$. It is complicated to give an explicit expression for the constant C. Since $C_0^1(\mathbb{R}^n, C\ell_{0,n})$ is dense in $L_p(\mathbb{R}^n, C\ell_{0,n})$ and thus the inequality can be extended to $L_p(\mathbb{R}^n, C\ell_{0,n})$, i.e.

$$\Big\|\frac{1}{\sigma_n}\int \partial_{x,k}e(x-y)u(y)dy\Big\|_p \le C\sigma_n^{-\frac{1}{p}}\|u\|_p.$$

On the other hand,

$$\|\bar{e}_k \frac{u}{n}\|_p \leq \frac{1}{n}\|u\|_p.$$

Collecting all estimates we get

$$\left\|\partial_k T_G u\right\|_p \leq \left(C \cdot \sigma_n^{-\frac{1}{p}} + \frac{1}{n}\right)\|u\|_p. \qquad \#$$

Theorem 3.9. *Let G be a bounded domain. Then*

$$T_G : L_p(G, C\ell_{0,n}) \to W_p^1(G, C\ell_{0,n})$$

is continuous.

Proof. We will extend the function u defined on G continuously to $\mathbb{R}^n$:

$$u = \begin{cases} u & \text{in} \quad G, \\ 0 & \text{in} \quad \mathbb{R}^n \setminus G. \end{cases}$$

Then

$$\|T_G u\|_{p,1} = \left(\|T_G u\|_p + \sum_{k=1}^{n} \|\partial_k T_G u\|_p\right)^{1/p} \leq \|T_G u\|_p + \sum_{k=1}^{n} \|\partial_k T_G u\|_p$$

$$\leq \operatorname{diam} G\, \|u\|_p + \sum_{k=1}^{n} (C\sigma_n^{-\frac{1}{p}} + \frac{1}{n})\|u\|_p \leq (\operatorname{diam} G + nC\sigma_n^{-\frac{1}{p}} + 1)\|u\|_p. \qquad \#$$

By the help of Sobolev embedding theorems it is easy to obtain the following results on the continuity of T_G between other important spaces.

Corollary 3.10.

(i) $T_G : [L_p(G, C\ell_{0,n}), C^{0,\frac{p-n}{p}}(G, C\ell_{0,n})]$ *is continuous for* $p > n$.

(ii) *[274] Let $u \in L_p(G, C\ell_{0,n}), 1 < p < n$. Then for all $r < \frac{np}{n-q}$ we have that*

$T_G : L_p(G, C\ell_{0,n}) \to L_r(G, C\ell_{0,n})$ is compact.

For all $\varepsilon > 0$ there exists $\delta > 0$ with

$$\|T_G u(x) - T_G u(x')\|_r \leq \varepsilon \|u\|_p$$

for $|x - x'| < \delta$. Especially, we have that T_G is compact in $L_2(G, C\ell_{0,n})$.

Remark 3.11. Also in the case of unbounded domains, continuity properties of the operator T_G can be proved. Unfortunately, the proofs require additional expertise. We will only formulate some results and refer to the literature. The Teodorescu transform T_G is in the case of $p < n$ a continuous mapping from $L_p(G, C\ell_{0,n})$ into $L_p^{-1}(G, C\ell_{0,n})$. For

$$1 - \frac{n}{p} < \alpha < n\left(1 - \frac{1}{p}\right)$$

the transform T_G is continuous from $L_p^\alpha(G, C\ell_{0,n})$ into $L_p^{\alpha-1}(G, C\ell_{0,n})$. In order to prove this we begin with the inequality

$$\left|\int_G e(x-y)u(y)dy\right| \leq \int_G |e(x-y)||u(y)|dy \leq \frac{1}{\sigma_n}\int_G \frac{|u(y)|}{|x-y|^{n-1}}dy.$$

The integral on the right-hand side is a real-valued weakly singular integral in $\mathbb{R}^n$. The crucial point for proof of the mapping properties is to consider the integral

$$\int_{\mathbf{R}^n}\int_{\mathbf{R}^n} \frac{u(y)v(y)}{|x|^\beta |x-y|^{n-1}|y|^\alpha}dydx$$

with $\alpha + \beta = 1$. Note that the kernel function is homogeneous of degree $-n$ (because $\alpha + \beta = 1$). Such integrals are considered for instance in [202] and [306].

Remark 3.12. Corresponding considerations for different kinds of unbounded domains can be found in [19], [20] and [106].

For our theoretical investigations of boundary value problems this knowledge about the continuity of T_G is sufficient. But for more practical use, for instance in numerical methods, we need something more. It is necessary to fix more exactly the norm of T_G in the considered spaces. This number will be included below in some parameters of numerical methods. In general we apply the principle "The better the estimate of $||T_G||$ the better the method". To prove norm estimates for the TEODORESCU transform we can use all the classical proofs – most of them based on an application of HÖLDER's inequality. We will cite only some results to demonstrate the ideas of such proofs. Those interested in the mapping properties alone may skip the rest of this section.

Proposition 3.13. *Let $1 \leq q < n/(n-1)$. Then*

$$\int_G |e(x-y)|^q dx \leq \sigma_n^{1-q}\frac{(\text{diam } G)^{n-q(n-1)}}{n-q(n-1)}.$$

Proof. We have

$$|e(x-y)| = \left|\frac{1}{\sigma_n}\frac{y-x}{|x-y|^n}\right| = \frac{1}{\sigma_n}\frac{1}{|x-y|^{n-1}}$$

and thus

$$\begin{aligned}\int_G |e(x-y)|^q dy &= \frac{1}{\sigma_n^q}\int_G \frac{1}{|x-y|^{q(n-1)}}dy \\ &\leq \frac{1}{\sigma_n^q}\int_{|y|<\text{diam } G} \frac{1}{|y|^{q(n-1)}}dy = \sigma_n^{1-q}\frac{(\text{diam } G)^{n-q(n-1)}}{n-q(n-1)}.\end{aligned}$$ #

Theorem 3.14. *Let G be a bounded domain in $\mathbf{R}^n$, $u \in L_p(G, C\ell_{0,n})$. Then*

(i) $T_G : L_p(G, C\ell_{0,n}) \to L_\infty(G, C\ell_{0,n})$ *for $n < p < \infty$*

with

$$||T_G u||_\infty \leq \frac{1}{\sigma_n^{1/p}}\left(\frac{(\text{diam } G)^{n-p(\frac{n-1}{p-1})}}{n-p(\frac{n-1}{p-1})}\right)^{\frac{p-1}{p}}||u||_p,$$

(ii) $||T_G u||_p \leq \text{diam } G\ ||u||_p$ *for $1 < p < \infty$.*

Proof. At first we have

$$\begin{aligned}\left|\int_G e(x-y)u(y)dy\right| &\leq \int_G |e(x-y)||u(y)|dy \leq \left(\int_G |e(x-y)|^q dy\right)^{\frac{1}{q}} \left(\int_G |u(y)|^p dy\right)^{\frac{1}{p}} \\ &\leq \sigma_n^{-1/p}\left(\frac{(\text{diam } G)^{n-q(n-1)}}{n-q(n-1)}\right)^{1/q} \|u\|_p.\end{aligned}$$

This proves (i). In order to verify (ii) we have to estimate the following:

$$\begin{aligned}\left|\int_G e(x-y)u(y)dy\right| &\leq \int_G |e(x-y)||u(y)|dy = \int_G |e(x-y)|^{1-\frac{1}{p}}|e(x-y)|^{1/p}|u(y)|dy \\ &\leq \left(\int_G |e(x-y)|dy\right)^{1/q}\left(\int_G |e(x-y)||u(y)|^p dy\right)^{1/p},\end{aligned}$$

where $1/p + 1/q = 1$. Using now Proposition 3.13. we get

$$\left|\int_G e(x-y)u(y)dy\right| \leq (\text{diam } G)^{1/q}\left(\int_G |e(x-y)||u(y)|^p dy\right)^{1/p}$$

Thus

$$\begin{aligned}&\int_G \left|\int_G e(x-y)u(y)dy\right|^p dx \leq (\text{diam } G)^{p/q}\int_G\int_G |e(x-y)||u(y)|^p dy dx \\ &= (\text{diam } G)^{p/q}\int_G\int_G |e(x-y)|dx|u(y)|^p dy \leq (\text{diam } G)^{\frac{p}{q}+1}\int_G |u(y)|^p dy.\end{aligned}$$

Consequently, we obtain $\|T_G u\|_p \leq \text{diam } G\, \|u\|_p$.

Theorem 3.15. *Let $u \in L_p(G, C\ell_{0,n}), p > n$, then it holds that*

$$\|T_G u\|_p \leq \frac{1}{\sigma_n}\max_{x\in\overline{G}}\left(\int_G \frac{1}{|x-y|^{(n-1)}}dy\right)^{p/q}|G|^{1/p}\|u\|_p$$

Proof. One has only to accept the inequality

$$\left(\int_G\left(\int_G \frac{1}{|x-y|^{(n-1)q}}dy\right)^{p/q}dx\right)^{1/p} \leq \max_{x\in\overline{G}}\left(\int_G \frac{1}{|x-y|^{(n-1)q}}dy\right)^{p/q}|G|^{1/p}.$$

Lemma 3.16. *Let $u \in L_p(G, C\ell_{0,n}), p > n$, then we have*

$$|(T_G u)(x)| \leq \frac{1}{\sigma_n}\left(\int_G \frac{1}{|x-y|^{(n-1)q}}dy\right)^{1/q}\|u\|_p\ ,\ \textit{where}\ \frac{1}{p}+\frac{1}{q}=1.$$

Proof. This follows immediately by HÖLDER's inequality, where the occurring integral is, under the above conditions, weak singular. #

Corollary 3.17. *Let $u \in L_p(G, C\ell_{0,n}), p > n, G$ is a domain with a finite volume. Then the following estimate is valid:*

$$||T_G||_{L(L_p,C)} \leq C(n,p)|G|^{\frac{p-n}{p}} \quad with \quad C(n,p) = \sigma_n^{\left(\frac{p-n}{p}+\frac{p-1}{p}\right)} \cdot \left(\frac{1}{n\left(\frac{p-n}{p-1}\right)}\right)^{\frac{p-1}{p}}.$$

For $p \to \infty$ we get $||T_G||_{L(L^\infty,C)} \leq \frac{\sigma_n^2}{n}|G|$.

Proof. Using Lemma 3.16. we have only to estimate the integral

$$I^q = \int_G \frac{1}{|x-y|^{(n-1)q}}dy \quad \left(\frac{1}{p}+\frac{1}{q}=1\right).$$

Similarly to the action of SCHMIDT's inequality in the analysis of one complex variable we replace the domain G by a ball $B_R(x)$ of the same volume. Obviously, the radius R has to be $R = (|G|\sigma_n)^{1/n}$. Then it holds that

$$I^q \leq \int_{B_R(x)} \frac{1}{|x-y|^{(n-1)q}}dy \leq \int_0^R r^{(n-1)(1-q)}dr\ \sigma_n$$

with $r = |x-y|$. Further we obtain

$$\begin{aligned} I^q &\leq \frac{R^{(n-1)(1-q)+1}\sigma_n}{(n-1)(1-q)+1} = \frac{R^{n+q-nq}}{n+q-nq}\sigma_n \\ &= \frac{|G|^{1-q+\frac{q}{n}}\sigma_n^{2-q+\frac{q}{n}}}{n+q-nq} \leq \frac{|G|^{\frac{p-n}{p-1}}\sigma_n^{\frac{p-n}{p-1}+1}}{n\left(\frac{p-n}{p-1}\right)}. \end{aligned}$$

Consequently, it follows that $I \leq C(p,n)|G|^{\frac{p-n}{p}}$

with

$$C(p,n) = \left(\frac{1}{n\left(\frac{p-n}{p-1}\right)}\right)^{\frac{p-1}{p}} \sigma_n^{\left(\frac{p-n}{p}+\frac{p-1}{p}\right)},$$

which can be obtained by the help of the the uniform convergence of the integral

$$\int_G \frac{1}{|x-y|^{(n-1)q}}dy. \qquad \#$$

For all that follows we will omit the index G in T_G if the domain G is clear from the context.

Remark 3.18. In Appendix A we investigate the range of the TEODORESCU transform. For the practical use of this transform it is necessary to compute the images of certain given functions. To avoid numerical integration we try to find the images as explicit expressions. In particular, in the case of a ball we are able to describe the images exactly for sufficiently large classes of functions. This possibility justifies the denotation "transform".

3.2 Around Cauchy's integral formula

In this section we will concern ourselves with $C\ell_{0,n}$-versions of classical theorems of the function theory of one complex variable. In the centre of the considerations will be the generalization of BOREL–POMPEIU's formula. Let $G \subset \mathbb{C}$ be a domain which is bounded by a LIAPUNOV curve Γ. Furthermore, let $\overline{\partial} = (1/2)(\partial/\partial\xi + i\partial/\partial\eta)$, where $\zeta = \xi + i\eta$ be the CAUCHY–RIEMANN operator. Then BOREL-POMPEIU's formula reads as follows:

$$\frac{1}{2\pi i}\int\limits_{\Gamma}\frac{u(\zeta)}{\zeta - z}d\Gamma_\zeta - \frac{1}{2\pi i}\int\limits_{G}\frac{1}{\zeta - z}(\overline{\partial}u)(\zeta)d\zeta = \begin{cases} u(z) & , \quad z \in G \\ 0 & , \quad z \in \mathbb{C}\setminus\overline{G}. \end{cases}$$

Proposition 3.19. *Let $u \in C^1(G, C\ell_{0,n}) \cap C(\overline{G}, C\ell_{0,n})$. Then*

$$\int\limits_G (Du)(y)dy = \int\limits_\Gamma \alpha(y)u(y)d\Gamma_y,$$

where $\alpha(y) = \sum_{i=1}^n \alpha_i(y)e_i$ is the outward pointing normal unit vector at y.

Proof. Using GAUSS' formula we obtain

$$\begin{aligned}\int\limits_G (Du)(y)dy &= \sum_{i=1}^n\sum_A e_ie_A\int\limits_G \partial_i u_A(y)dy \\ &= \sum_{i=1}^n\sum_A e_ie_A\int\limits_\Gamma \alpha_i(y)u_A(y)d\Gamma_y \\ &= \int\limits_\Gamma \alpha(y)u(y)d\Gamma_y. \qquad \#\end{aligned}$$

Corollary 3.20. (CAUCHY's integral theorem.)
Let $u \in C^1(G, C\ell_{0,n}) \cap C(\overline{G}, C\ell_{0,n})$ and $u \in \ker D$ $(Du = 0!)$ then

$$\int\limits_\gamma \alpha(y)u(y)d\Gamma_y = 0,$$

where γ is an arbitrary surface with $\gamma \subset G$.

Let

$$E(x) = \frac{1}{\sigma_n}\frac{1}{(2-n)}|x|^{-(n-2)} \qquad (n > 2) \quad \text{with} \quad \sigma_n = \frac{2\pi^{n/2}}{\Gamma(\frac{n}{2})}$$

then we obtain $\Delta E(x) = 0$ $(x \neq 0)$ (σ_n is again the surface area of the unit sphere in $\mathbb{R}^n$). It can be easily seen that the DIRAC operator $D = \sum_{i=1}^n e_i \partial_i$ yields

$$e(x) = \overline{D}E(x) = \frac{-x}{\sigma_n |x|^n}.$$

Proposition 3.21. *Let $u \in C^2(G, C\ell_{0,n}) \cap C(\overline{G}, C\ell_{0,n})$. Then*

$$(DT_G u)(x) = \begin{cases} u(x) & \text{in } G, \\ 0 & \text{in } \mathbb{R}^n \setminus \overline{G}. \end{cases}$$

Proof. Firstly, we note that in case of $\partial_{ij} = \partial_{ji}$

$$\begin{aligned} D\overline{D} &= \sum_{i=1}^n \sum_{j=1}^n e_i \overline{e}_j \partial_{ij} = -\sum_{i=1}^n \sum_{j=1}^n e_i e_j \partial_{ij} \\ &= -\sum_{i=j} e_i e_j \partial_{ij} - \sum_{i>j} (e_i e_j - e_j e_i) \partial_{ij} = \Delta. \end{aligned}$$

Furthermore, we have for $x \in G$

$$\begin{aligned} (DT_G u)(x) &= D\Big(-\int_G e(x-y) u(y) dy \Big) = D\Big(-\int_G \overline{D}E(x-y) u(y) dy \Big) \\ &= \Delta\Big(-\int_G \frac{1}{\sigma_n (2-n)} \frac{1}{|x-y|^{n-2}} u(y) dy \Big). \end{aligned}$$

The latter integral is just the volume potential which solves the POISSON equation. Hence $(DT_G u)(x) = u(x)$.

For $\Delta E(x) = 0$ in $\mathbb{R}^n \setminus \overline{G}$ it follows our assertion. #

Proposition 3.22. *Let $u, v \in C^1(G, C\ell_{0,n}) \cap C(\overline{G}, C\ell_{0,n})$. Then*

$$\int_G [(uD)v + u(Dv)] dy = \int_\Gamma u(y) \alpha(y) v(y) d\Gamma_y,$$

where $(uD) = \sum_{i=1}^n \partial_i u e_i$, $Dv = \sum_{i=1}^n e_i \partial_i v$, and $\alpha(y)$ denotes the the outward pointing normal unit vector at y.

Proof. This immediately follows from GAUSS' theorem. We have

$$\int_G [(uD)v + uDv] dy = \int_G \sum_{i=1}^n [\partial_i u e_i v + u e_i \partial_i v] dy$$

$$= \sum_{i=1}^n \int_G \partial_i (u e_i v) dy = \int_\Gamma u(y) \alpha(y) v(y) d\Gamma_y. \qquad \#$$

Theorem 3.23. *Let $G \subset \mathbb{R}^n$ be a domain which is bounded by a piecewise* LIAPUNOV *surface Γ. Then for each $u \in C^1(G, C\ell_{0,n}) \cap C(\overline{G}, C\ell_{0,n})$*

$$\int_\Gamma e(x-y)\alpha(y)u(y)d\Gamma_y - \int_G e(x-y)(Du)(y)dy = \begin{cases} u(x) & , \quad x \in G \\ 0 & , \quad x \notin \overline{G} \end{cases} , \tag{3.3}$$

where $\alpha(y)$ denotes again the outward pointing normal unit vector at y.

Remark 3.24. Using STOKES' Theorem 3.23. can be proved on compact n-dimensional, oriented C^∞-manifolds which are lying in a domain of $\mathbb{R}^n$ cf. [88], [36].

Definition 3.25. *Let $u \in C^1(G, C\ell_{0,n}) \cap C(\overline{G}, C\ell_{0,n})$. The operator F_Γ defined by*

$$(F_\Gamma u)(x) := \int_\Gamma e(x-y)\alpha(y)u(y)d\Gamma_y$$

is called CAUCHY-BITSADZE *operator.*

Remark 3.26. In this new notation formula (3.3) has the representation

$$(F_\Gamma u)(x) + (T_G Du)(x) = \begin{cases} u(x) & , \quad x \in G \\ 0 & , \quad x \in \mathbb{R}^n \setminus \overline{G}. \end{cases}$$

Proof. (Theorem 3.23.) Let x be an interior point of G, $B_\varepsilon(x)$ a ball centred at x with the radius ε $(\varepsilon > 0)$. Further, let $G_\varepsilon := G \setminus B_\varepsilon(x)$. Applying Proposition 3.21. and choosing for $u = e(x)$ then it follows that

$$\begin{aligned} &\int_{G_\varepsilon} e(x-y)(Dv)(y)dy = \\ &= \int_\Gamma e(x-y)\alpha(y)v(y)d\Gamma_y - \int_{S_\varepsilon} e(x-y)\alpha(y)v(y)dS_{\varepsilon_y}. \end{aligned} \tag{3.4}$$

On S_ε $(y \in S_\varepsilon)$ holds

$$e(x-y)\alpha(y) = \frac{1}{\sigma_n}\frac{\overline{x}-\overline{y}}{|x-y|^n}\frac{y-x}{|x-y|} = \frac{1}{\sigma_n}\frac{1}{|x-y|^{n-1}}.$$

It remains to show that the latter integral converges to $v(x)$ if ε tends to zero. We have

$$\lim_{\varepsilon \to 0} \int_{S_\varepsilon} e(x-y)\alpha(y)v(y)dy = \lim_{\varepsilon \to 0}\left(\frac{1}{\sigma_n \varepsilon^{n-1}} \int_{S_\varepsilon(x)} v(y)dy\right) = v(x).$$

For $\varepsilon \to 0$ the first integral in (3.4) exists and BOREL-POMPEIU's formula for interior points is shown. For the second part we have to repeat the proof without excluding a neighbourhood of x. #

Remark 3.27. BOREL–POMPEIU's formula is also valid in the case of multiply connected domains. Suppose G_i $(i = 0, 1, ..., n)$ be simply connected bounded domains in $\mathbb{R}^n$ with smooth boundaries Γ_i $(i = 0, 1, ..., n)$ which fulfil the relations

(i) $G_0 \supset \bigcup_{i=1}^{n} \overline{G}_i = K$ and (ii) $\overline{G}_i \cap \overline{G}_j = \emptyset \quad (i \neq j)$. Then

$$(F_{\Gamma_0}u)(x) - \sum_{i=1}^{n}(F_{\Gamma_i}u)(x) + (T_{G_0\setminus K}Du)(x) = \begin{cases} u(x) & \text{in} \quad G_0 \setminus K \\ 0 & \text{in} \quad \mathbb{R}^n \setminus [G_0 \setminus K] \end{cases}$$

for every function in $C^1(G_0 \setminus K, C\ell_{0,n}) \cap C(\overline{G_0 \setminus K}, C\ell_{0,n})$.

Now it is easy to transfer most of the results of the theory of one complex variable which are connected with CAUCHY's integral theorem.

Theorem 3.28. (CAUCHY's integral formula). *Let $G \subset \mathbb{R}^n$ be a domain with a piecewise* LIAPUNOV *boundary. Further let $u \in \ker D$, then*

$$(F_\Gamma u)(x) = \begin{cases} u(x) & \text{in} \quad G \\ 0 & \text{in} \quad \mathbb{R}^n \setminus \overline{G} \end{cases} \tag{3.5}$$

holds.

Proof. Because of $Du = 0$ formula (3.5) holds. #

Corollary 3.29. *Let $u \in \ker D$. Then u has partial derivatives $\partial_{i_1 i_2 \ldots i_k}$ $(i_l \in \{1, ..., n\}; l \in \{1, ..., k\})$ of any order.*

Proof. The kernel function $e(x - y)$ has for $x \neq y$ partial derivatives of any order. #

Corollary 3.30. (Mean-value theorem.) *Let $B_r(x) \subset \mathbb{R}^n$ be a ball centred at x of radius r. For any $u \in \ker D$*

$$u(x) = \frac{n}{\sigma_n r^n} \int_{B_r(x)} u(y)dy \ .$$

Note that $\sigma_n r^n/n$ is just the volume of the ball $B_r(x)$.

Proof. CAUCHY's integral formula yields

$$\begin{aligned} u(x) &= \int_{S_r(x)} e(x-y)\alpha(y)u(y)dS_{r_y} = \frac{1}{\sigma_n} \int_{S_r(x)} \frac{\overline{y} - \overline{x}}{|x-y|^n}\alpha(y)u(y)dS_{r_y} \\ &= \frac{1}{\sigma_n r^n} \int_{S_r(x)} (\overline{y} - \overline{x})\alpha(y)u(y)dS_{r_y}. \end{aligned}$$

By the aid of Proposition 3.22. we get, consequently,

$$\begin{aligned} u(x) &= \frac{1}{\sigma_n r^n} \int\limits_{B_r(x)} [(\overline{y}-\overline{x})D]u(y)dy + \frac{1}{\sigma_n r^n} \int\limits_{B_r(x)} (\overline{y}-\overline{x})Du(y)dy \\ &= \frac{n}{\sigma_n r^n} \int\limits_{B_r(x)} u(y)dy. \qquad \# \end{aligned}$$

Corollary 3.31. (Mean-value formula.) *Let $B_r(x)$ again be a ball centred at x of radius r. Then*

$$u(x) = \frac{1}{\sigma_n r^{n-1}} \int\limits_{S_r(x)} u(y)dS_{r_y},$$

where $u \in (\ker D)(B_r(x)) \cap C(\overline{B}_r(x))$.

Proof. Applying Cauchy's integral formula then

$$u(x) = \frac{1}{\sigma_n} \int\limits_{S_r(x)} \frac{\overline{y}-\overline{x}}{|y-x|^n} \frac{y-x}{|y-x|} u(y)dS_{r_y} = \frac{1}{\sigma_n r^{n-1}} \int\limits_{S_r(x)} u(y)dS_{r_y}. \qquad \#$$

Theorem 3.32. (Maximum modulus theorem). *Let $G \subset \mathbb{R}^n$ be a connected bounded domain and $u \in (\ker D)(G) \cap C(\overline{G})$ with $\overline{G} = G \cup \Gamma$, $\Gamma = \partial G$. If there exists some $z \in G$ with*

$$|u(x)| \leq |u(z)| \qquad (\forall\, x \in G)$$

it is necessary and sufficient that u is constant.

Proof. For $z \in G$ there exists a ball $B_\varepsilon(z) \subset G$. Applying the mean value formula we obtain

$$u(z) = \frac{1}{\sigma_n \varepsilon^{n-1}} \int\limits_{S_\varepsilon} u(x)dS_{\varepsilon,x}. \tag{3.6}$$

Assume now that $|u(x)| \leq |u(z)|$ and u is not constant in $B_\varepsilon(z)$. Then there exists a decomposition of S_ε into two sets

$$S'_\varepsilon = \{x \in S_\varepsilon : |u(x)| = |u(z)|\} \quad S''_\varepsilon = \{x \in S_\varepsilon : |u(x)| < |u(z)|\}$$

Formula (3.6) yields

$$|u(z)| < \frac{1}{\sigma_n \varepsilon^{n-1}} \left[\int\limits_{S'_\varepsilon} |u(x)| dS'_{\varepsilon,x} + \int\limits_{S''_\varepsilon} |u(z)| dS''_{\varepsilon,x} \right] = \frac{|S'_\varepsilon| + |S''_\varepsilon|}{\sigma_n \varepsilon^{n-1}} |u(z)| = |u(z)|.$$

This is a contradiction. Hence on S_ε we have $|u(x)| = |u(z)|$.

Choosing now smaller balls $B_{\varepsilon'}(z)$ $(\varepsilon' < \varepsilon)$ we obtain the same. Therefore, $|u(x)| = |u(z)|$ in the whole ball $B_\varepsilon(z)$. Since G is connected we get immediately $|u(x)| = \text{const.}$ on G and for the continuity of $|u(x)|$ such also on $\overline{G}$. It remains to show that $u(x)$ is also a constant. From $|u|^2 = \text{const.}$ we conclude

$$0 = D|u|^2 = D(\sum_A u_A^2) = 2\sum_A \sum_{j=1}^n (\partial_j u_A)u_A e_j$$

and also

$$0 = \Delta|u|^2 = \overline{D}D|u|^2 = 2\sum_A \sum_{i=1}^n \sum_{j=1}^n \left[(\partial_{ij}u_A)u_A e_i e_j + (\partial_j u_A)(\partial_i u_A)e_i e_j\right].$$

Since the components of a CLIFFORD left regular function are harmonic one gets

$$0 = \sum_A \sum_{i=1}^n (\partial_i u_A)^2$$

and consequently $\partial_i u_A = 0$ $(i = 1, ..., n)$. #

Corollary 3.33. *Let $G \subset \mathbb{R}^n$ be a connected and bounded domain and $u \in (\ker D)(G) \cap C(\overline{G})$ then*

$$\sup_{x\in\overline{G}} |u(x)| = \sup_{x\in\partial G} |u(x)|.$$

Theorem 3.34. (LIOUVILLE's theorem.) *Let $u \in (\ker D)(\mathbb{R}^n)$. If $|u(x)| \leq M$ $(x \in \mathbb{R}^n)$ then u is a constant.*

Proof. Let $B_R(x)$ a ball centred at x of radius R. CAUCHY's integral formula leads to

$$(\partial_j u)(x) = \int_{S_R} \partial_j e(x-y)\alpha(y)u(y)dS_{R,y}.$$

We obtain for $j = 1, ..., n$:

$$\begin{aligned}
\partial_j e(x-y) &= \sum_{i=1}^n \frac{\partial_j(y_i - x_i)|x-y|^n - (y_i - x_i)n|x-y|^{n-2}(y_j - x_j)}{|x-y|^{2n}} e_i \\
&= \sum_{i=1}^n \frac{-\delta_{ij}|x-y|^n - n(y_i - x_i)n\,|x-y|^{n-2}(y_j - x_j)}{|x-y|^{2n}} e_i.
\end{aligned}$$

Hence, it follows that

$$|\partial_j e(x-y)| \leq \frac{(n+1)n}{|x-y|^n} \qquad (j = 1, ..., n).$$

Therefore, for $R \to \infty$

$$|\partial_j u(x)| \leq \frac{(n+1)n}{R^n \sigma_n} M \sigma_n R^{n-1} = \frac{(n+1)n}{R} M \to 0.$$

In this way $\partial_j u = 0$ $(j = 1, ..., n)$ in $\mathbb{R}^n$ and $u = \text{const.} \in C\ell_{0,n}$. #

Theorem 3.35. (MORERA's theorem.)
Let $u \in C^1(G, C\ell_{0,n}), Du \in L_r(G, C\ell_{0,n}), r > n$. If for all balls $B_r(x)$ $(r > 0$, $x \in G)$

$$\int_{B_r(x)} \alpha(y) u(y) d\gamma(y) = 0 ,$$

then $Du = 0$ in G.

Proof. Let $x \in G$ be an arbitrary point (B_{r_k}) $k \in N$ a regular sequence of balls which is contracting to $x \in G$. We have $x \in \bigcap_k B_{r_k}$. For the validity of LEBESGUE's theorem we obtain for each $v \in L_r(G, C\ell_{0,n})(r > n)$,

$$\lim_{k\to\infty} \frac{1}{|B_{r_k}|} \int_{B_{r_k}} v(y) dB_{r_k}(y) =: \tilde{v}(x)$$

and $v = \tilde{v}$ in $L_r(G, C\ell_{0,n})$. Substituting $v = Du$ it follows by Proposition 3.19. that

$$\int_{B_{r_k}} (Du)(y) dy = \int_{S_{r_k}} \alpha(y) u(y) dS_{r_{k,y}}$$

and also

$$\frac{1}{|B_{r_k}|} \int_{B_{r_k}} (Du)(y) dy = 0 \qquad (k = 0, 1, 2, ...)$$

and, therefore, $(Du)(x) = 0$ almost everywhere on G. #

Theorem 3.36. *Let $u \in (\ker D)(\mathbb{R}^n \setminus \overline{G}) \cap C(\mathbb{R}^n \setminus \overline{G}, C\ell_{0,n})$. Suppose $\lim_{|x|\to\infty} u(x) = u(\infty)$. Then*

$$(F_\Gamma u)(x) = \begin{cases} u(\infty) - u(x) & , \quad x \in \mathbb{R}^n \setminus \overline{G} \\ u(\infty) & , \quad x \in G \end{cases} .$$

Proof. Let $B_R(x)$ be such a ball that $B_R(x) \supset G$. Write briefly $s = (\omega_x, \omega_y) = (x/|x|, y/|y|)$. On account of

$$\frac{1}{|x-y|^{n-1}} = \left(\sum_{k=0}^{\infty} C_k^{\frac{n-3}{2}}(s) \left| \frac{x}{y} \right|^k \right) \frac{1}{|y|^{n-1}} \qquad (|x| < |y|)$$

and CAUCHY's integral formula for multiply connected domains we get

$$\frac{1}{\sigma_n}\int\limits_{S_R}\frac{x-y}{|x-y|^n}\alpha(y)u(y)dS_{R,y}-\frac{1}{\sigma_n}\int\limits_{\Gamma}\frac{x-y}{|x-y|^n}\alpha(y)u(y)d\Gamma_y$$

$$=\begin{cases}0 & , \quad x\in\mathbb{R}^n\setminus\overline{(B_R\setminus\overline{G})} \\ u(x) & , \quad x\in B_R\setminus\overline{G}\end{cases} .$$

The first integral transforms into

$$\frac{1}{\sigma_n}\int\limits_{S_R}\frac{x-y}{|x-y|^n}\frac{y-x}{|y-x|}u(y)dS_{R_y}=\frac{1}{\sigma_n}\int\limits_{S_R}\frac{1}{|x-y|^{n-1}}u(y)dS_{R_y}=$$

$$=\sum_{k=0}^{\infty}\frac{1}{\sigma_n}\int\limits_{S^n}\frac{C_k^{\frac{n-3}{2}}(s)}{R^k}u(y)dS_y^n\,|x|^k.$$

For $R\to\infty$ the right-hand side converges to

$$\frac{C_0^{\frac{n-3}{2}}(s)}{\sigma_n}\int\limits_{S^n}u(\infty)dS_y^n=u(\infty)\quad (C_0^{\mu}(s)\equiv 1),$$

which provides the result. #

3.3 Function series expansions

Based on CAUCHY's integral formula, TAYLOR and LAURENT series expansions will be deduced. Furthermore, theorems of the WEIERSTRASS type will be considered. Above all we have to expand the CAUCHY kernel $e(x-y)$. For this reason we need a special type of polynomial the so-called GEGENBAUER polynomials. This topic has already been comprehensively investigated. The best references are [36], [63] and [48].

Definition 3.37. *Let $\mu \in \mathbb{C}$ with $Re\,\mu > -\frac{1}{2}$, $t, s \in \mathbb{R}$. Then the series expansion of the function*

$$(1-2st+t^2)^{-\mu}$$

into powers of t reads as follows:

$$(1-2st+t^2)^{-\mu} = \sum_{k=0}^{\infty} C_k^\mu(s) t^k \ .$$

The coefficients $C_k^\mu(s)$ are called GEGENBAUER *polynomials of degree k associated with μ.*

Remark 3.38. In the case of $n=3$ GEGENBAUER polynomials coincide with LEGENDRÉ polynomials.

Remark 3.39. There are possibilities explicitely to calculate GEGENBAUER polynomials. These read as follows:

$$C_k^\mu(s) = \frac{(-1)^k 2^k \Gamma(\mu+\frac{1}{2})\Gamma(2\mu+k)}{\Gamma(\mu+\frac{1}{2}+k)\Gamma(2\mu)k!}(1-s^2)^{-\mu+\frac{1}{2}}\left(\frac{d}{ds}\right)^k(1-s^2)^{\mu-\frac{1}{2}+k}, \quad (3.7)$$

where $\Gamma(\nu) = \int_0^\infty e^{-t}t^{\nu-1}dt \quad (Re\,\nu > 0)$. Formula (3.7) is a generalization of RODRIGUES' formula which is used to generate LEGENDRÉ polynomials. A further representation is given by the finite sum

$$C_k^\mu(s) = \sum_{i=0}^{[k/2]} \frac{(-1)^i(\mu)_{(k-i)}}{i!(k-2i)!}(2s)^{k-2i}$$

where $(\mu)_{k-i} := \Gamma(\mu+k-i)/\Gamma(\mu)$ is the so-called POCHHAMMER *symbol*. One has for instance $C_0^\mu(s) = 1$ and $C_1^\mu(s) = 2\mu s$. The following recurrence formula is sometimes very useful:

$$C_{k+1}^\mu(s) = \frac{2(k+\mu)}{k+1} s\, C_k^\mu(s) - \left(\frac{k+1+2\mu-2}{k+1}\right) C_{k-1}^\mu(s).$$

Note that GEGENBAUER polynomials form in $L_2(I,g)$ an orthogonal system where $I=(-1,1)$ and $g=(1-s^2)^{\mu-1/2}$. The norms are (cf. [166], p. 291)

$$||C_k^\mu||_{2,g}^2 = \frac{\pi\Gamma(2\mu+k)}{2^{2\mu-1}k!(k+\mu)[\Gamma(\mu)]^2}.$$

Moreover, GEGENBAUER polynomials satisfy the ordinary differential equation

$$(1-s^2)\left(\frac{d^2}{ds}\right)^2 C_k^\mu(s) - (2\mu+1)s\left(\frac{d}{ds}\right) C_k^\mu(s) + k(k+2\mu)C_k^\mu(s) = 0.$$

There is also a very close connection to JACOBI polynomials. The best reference on GEGENBAUER polynomials is, with the exception of the classic [166], the thesis of J. CNOPS [48].

Consider now the CAUCHY kernel

$$e(x-y) = \frac{1}{\sigma_n}\frac{\overline{x}-\overline{y}}{|x-y|^n} = \frac{1}{(n-2)\sigma_n}\overline{D}_x\frac{1}{|x-y|^{n-2}}.$$

Next we inspect the function $|x-y|^{2-n}$. We have

$$\frac{1}{|x-y|^{n-2}} = \frac{1}{|y|^{n-2}}\frac{1}{(1-2ts+t^2)^{\frac{n-2}{2}}}$$

if putting for $x, y \in \mathbb{R}^n$ $x = |x|\omega_x$, $y = |y|\omega_y$, $t = |x|/|y|$ and $s = (\omega_x, \omega_y)$. For $|x| < |y|$ (y is fixed!) we get the series expansion

$$\frac{1}{|x-y|^{n-2}} = \sum_{k=0}^{\infty}\frac{|x|^k}{|y|^{n+k-2}}C_k^{\frac{n-2}{2}}(s). \tag{3.8}$$

Differentiating this sum by $\overline{D}_y$, $\overline{D}_x$, respectively, we get the following formulae:

Lemma 3.40. *Let $k \geq 0$ then*

$$\overline{D}_x(C_{k+1}^{\frac{n-2}{2}}(s)|x|^{k+1}) = |x|^k(n-2)\ A_{n,k}(x,y)$$

with

$$(n-2)A_{n,k}(x,y) := [(k+1)C_{k+1}^{\frac{n-2}{2}}(s) + (2-n)C_k^{\frac{n}{2}}(s)\omega_y \wedge \omega_x]\overline{\omega}_x.$$

and

$$\overline{D}_y\left(C_k^{\frac{n-2}{2}}(s)\frac{1}{|y|^{n+k-2}}\right) = \frac{1}{|y|^{(n-1+k)}}(2-n)B_{n,k}(x,y)$$

with

$$(2-n)B_{n,k}(x,y) := [(-(n-2+k)C_k^{\frac{n-2}{2}}(s) + (2-n)C_{k-1}^{\frac{n}{2}}(s)\omega_x \wedge \omega_y]\overline{\omega}_y.$$

Proof. We follow here the proof given in [63]. Write the DIRAC operator in polar coordinates. This gives

$$D_x = \omega_x\left(\partial_{|x|} + \frac{1}{|x|}\Gamma_{\omega_x}\right)$$

with $\Gamma_{\omega_x} = -x \wedge \partial_x := \sum_{i<j} e_i e_j (x_j\partial_i - x_i\partial_j)$ (cf. Sections 2.3, 2.5, 3.7).

Now we have

$$D_x(C_{k+1}^{\frac{n-2}{2}}(s)|x|^{k+1}) = |x|^k\omega_x\left[(k+1)C_{k+1}^{\frac{n-2}{2}}(s) + \Gamma_{\omega_x}(C_{k+1}^{\frac{n-2}{2}}(s))\right]. \tag{3.9}$$

It is easy to see that

$$\frac{d}{ds}C_{k+1}^{\mu}(s) = 2\mu C_k^{\mu+1}(s).$$

Thus we obtain

$$\Gamma_{\omega_x}\left[C_{k+1}^{\frac{n-2}{2}}(s)\right] = (n-2)C_k^{\frac{n}{2}}(s)(\Gamma_{\omega_x}s).$$

Because of

$$\Gamma_{\omega_x}s = \Gamma_{w_x}(\omega_x,\omega_y) = \frac{1}{|x||y|}\Gamma_{\omega_x}(x,y)$$

we now have

$$\Gamma_{\omega_x}s = \sum_{i<j} e_ie_j(x_j\partial_i - x_i\partial_j)(x,y) = \frac{1}{|x||y|}\, y\wedge x = \omega_y\wedge\omega_x.$$

Together with (3.8) we proved our statement. #

Corollary 3.41. *We have the identity* $A_{n,k}(x,y) = B_{n,k}(x,y)$.

Theorem 3.42. (McLaurin expansion.) *Let u be* Clifford *regular in the ball $B_r(0)$ $(r>1)$ and $|y|=1, |x|\le 1-\varepsilon, (\varepsilon>0)$. Then*

$$u(x) = \sum_{k=0}^{\infty}(a_ku)(\omega_x)|x|^k \tag{3.10}$$

with

$$(a_ku)(\omega_x) = \frac{1}{\sigma_n}\int_{S^n} B_{n,k}(x,y)\omega_y u(y)dS_y^n.$$

Series (3.10) converges normally in $B_{r'}(0)\,(r'<r)$. Note that $|y|$ is being fixed.

Proof. Let $M_\delta = \{x\in\mathbb{R}^n : |x| < (1-\delta)|y|\}$. Then the series (3.8) converges normally if $0<\delta<1$ and its sum is $\mathbb{R}$-analytic in the domain M_0. Hence, for any $k\in N$ the expression

$$\frac{|x|^k}{|y|^{n+k-2}}C_k^{\frac{n-2}{2}}(s)$$

is a homogeneous harmonic polynomial of degree k relative to x. Cauchy's integral formula delivers now in $B_r(0)$

$$u(x) = \frac{1}{\sigma_n}\int_{S_\varepsilon}\frac{\overline{x}-\overline{y}}{|x-y|^n}\omega_y\, u(y)dS_{\varepsilon,y}$$

and with the help of Lemma 3.40. we get

$$u(x) = \sum_{k=1}^{\infty} \left(\frac{1}{\sigma_n} \int_{S^n} B_{n,k}(x,y)\omega_y u(y) dS_y^n \right) |x|^k . \qquad \#$$

Proposition 3.43. *Let* $x = \sum_{i=1}^n x_i e_i$, $D_x = \sum_{i=1}^n \partial_i e_i$ *then*

(i) $(x, D_x)\frac{x}{|x|} = 0$,

(ii) $(x, D_x)|x|^k = k|x|^{k-1}$.

Proof. Item (i) follows from

$$\begin{aligned} (x, D_x)\frac{x}{|x|} &= \sum_{j=1}^n x_j \partial_j \sum_{i=1}^n \frac{x_i e_i}{|x|} = \sum_{j=1}^n \sum_{i=1}^n \frac{x_j \partial_j x_i e_i |x| - \partial_j |x| x_j x_i e_i}{|x|^2} \\ &= \sum_{i=1}^n \sum_{j=1}^n \frac{(\delta_{ij} x_j e_j |x|^2 - x_j^2 x_i e_i)}{|x|^3} = 0. \end{aligned}$$

Item (ii) is now immediately clear.

Corollary 3.44. *The coefficients* $(a_k u)(\omega_x)$ *can be calculated by the formula*

$$(a_k u)(\omega_x) = \Big(\frac{(\omega_x, D_x)^k}{k!} u \Big)(0) \quad (k = 0, 1, ...).$$

Proof. Applying (x, D_x) to $(a_k u)(\omega_x)$ we get for Proposition 3.43.

$$(x, D_x)(a_k u)(\omega_x) = \frac{1}{\sigma_n} \int_{S^n} (x, D_x) B_{n,k}(x,y) \omega_y \, u(y) dS_y^n = 0$$

and thus

$$\big[(x, D_x)u\big](x) = \sum_{k=1}^{\infty} k (a_k u)(\omega_x) |x|^{k-1}$$

For $x \to 0$ we obtain $[(x, D_x)u](0) = (a_1 u)(\omega_x)$. Similarly we get the result for $k = 2, 3, \dots$. #

Theorem 3.45. (TAYLOR expansion). *Let u be* CLIFFORD *regular in the ball* $B_r(a), |y-a| = 1, |x-a| < 1 - \varepsilon, \varepsilon > 0$. *Then*

$$u(x) = \sum_{k=0}^{\infty} (a_k u)(a, \omega_{x-a}) |x-a|^k \tag{3.11}$$

with

$$(a_k u)(a, \omega_{x-a}) := \frac{1}{\sigma_n} \int_{S^n(a)} B_{n,k}^{[a]}(x,y) \omega_{y-a} u(y) dS^n(a). \tag{3.12}$$

Series (3.11) converges normally in $B_{r'}(a)$ $(r' < r)$, *where* $B_{n,k}^{[a]}(x,y)$ *is similarly defined as* $B_{n,k}(k,y)$ *and* $B_{n,k}(x,y) = B_{n,k}^{[0]}(x,y)$.

Proof. In Theorem 3.42. we have only to substitute: $x := x - a$, $y := y - a$,

$\omega_x := \omega_{x-a} = \dfrac{x-a}{|x-a|}$ and $\omega_y := \omega_{y-a} = \dfrac{y-a}{|y-a|}$. #

Corollary 3.46. (CAUCHY's inequality.) *Let* $|u(x)| \leq M$. *Then*

$$|(a_k u)(a, \omega_{x-a})| \leq C[\mathrm{dist}(a, S_\varepsilon)]^k \quad (C > 0;\ k = 0, 1, 2, ...)$$

with $\varepsilon = |y - a|$.

Proof. Since $\int_{S^n(a)} |B^{[a]}_{n,k}(x, y)| dS^n_y(a) \leq$ const. we obtain the statement from Theorem 3.42. #

Theorem 3.47. (1. WEIERSTRASS' theorem). *Let* $(u_k)_{k=1}^\infty$ *be a sequence of* CLIFFORD *regular functions in a domain* $G \subset \mathbb{R}^n$. *Furthermore, let on any compact set* $K \subset G$, $(u_k)_{k=1}^\infty$ *be a* CAUCHY *sequence, i.e., that for any* $k, k' > N(\varepsilon)$ $||u_k - u_{k'}||_{C(K)} < \varepsilon$ *where* $\varepsilon > 0$ *is arbitrarily given. If* $(u_k)_{k=1}^\infty$ *converges uniformly to a function* $u \in C(K, Cl_{0,n})$, *then on compact sets* K:

(i) $u \in (\ker D)(K)$,

(ii) $(\partial^l u_k)_{k=1}^\infty$ *converges uniformly on the interior of all compact subsets* $K \subset G$ *to* $\partial^l u$ *for any* $l = (l_1, ..., l_n)$, where

$$\partial^l := \frac{\partial^{|l|}}{\partial_1^{l_1} \partial_2^{l_2} ... \partial_n^{l_n}} \quad (|l| = l_1 + ... + l_n).$$

Proof. Let K be a compact subset of G. Now CAUCHY's integral formula yields the estimate

$$\begin{aligned} |(\partial^l u_k - \partial^l u_{k'})(x)| &\leq \frac{C}{\sigma_n} \max_{y \in \partial K} \frac{1}{|x-y|^{n-1+|l|}} \max_{y \in \partial K} |(u_k - u_{k'})(y)| \\ &\leq \frac{C\varepsilon}{d(x, \partial K)^{|l|+n-1}} \end{aligned}$$

if $k, k' > N$ and $\varepsilon > 0$ are arbitrarily given. As all components $u_{A,k}$ of $u_k = \sum_A u_{A,k} e_A$ are harmonic we conclude by using HARNACK's theorem that $(\partial^l u_{A,k})_{k=1}^\infty$ converges uniformly to $\partial^l u_A$ ($\forall A$) on the interior of all compact subsets $K \subset G$. In particular, we have Du_k as uniformly convergent to Du on compact sets K. As for $Du_k = 0$ $(k = 1, 2, ...)$ in G we find $u \in (\ker D)(G)$. We have only to note that G can be exhausted by compact sets, that means $G = \bigcup_{K \subset G} K$. #

Theorem 3.48. (2. WEIERSTRASS theorem.) *Let* $(u_k)_{k=1}^\infty$ *be a sequence of functions with* $u_k \in (\ker D)(G) \cap C(\overline{G})$ $(k = 1, 2, ...)$. *If the series* $\sum_{k=1}^\infty u_k(x)$ *converges uniformly on* Γ *then this series converges uniformly on* $\overline{G}$.

Proof. The proof follows immediately. #

Lemma 3.49. *Let $a \in \mathbb{R}^n$, $x \in \mathbb{R}^n \setminus B_1(a)$, $|y-a| = 1$. Then we have for any fixed $|y-a|$ the expansion*

$$u(x) = \sum_{k=0}^{\infty} \frac{1}{\sigma_n} \frac{1}{|x-a|^{k+n-1}} \int_{S^n} A_{n,k}^{[a]}(x,y)\omega_{y-a} u(y) dS_y^n$$

or

$$u(x) = \sum_{k=0}^{\infty} (a_k u)(\omega_{x-a}) \frac{1}{|x-a|^{k+n-1}}$$

with

$$(a_k u)(\omega_{x-a}) = \frac{1}{\sigma_n} \int_{S^n} A_{n,k}^{[a]*}(x,y)\omega_{y-a} u(y) dS_y^n, \quad \textit{where} \quad A_{n,k}^{[a]*} := A_{n,-k-n+1}^{[a]}.$$

Remark 3.50. In [63] (p. 186) it is shown that $(a_k u)(\omega_{x-a})|x-a|^k$, $k \in \mathbb{N}$ are inner spherical CLIFFORD regular functions of order k and $(b_k u)(\omega_{x-a})|x-a|^{1-k-n}$ are outer spherical CLIFFORD regular functions of order k.

Theorem 3.51. (LAURENT expansion). *Let $a \in \mathbb{R}^n$, $0 < r < 1 < R < \infty$ and $|y-a| = 1$. Furthermore, let u be CLIFFORD regular in the open annulus $B_{rR} = \overset{\circ}{B}_R(a) \setminus \overline{B}_r(a)$. Then u permits the representation*

$$u(x) = \sum_{k=-\infty}^{\infty} (a_k u)(\omega_{x-a})|x-a|^k$$

where

$$(a_k u)(\omega_x - a) := \begin{cases} \frac{1}{\sigma_n} \int_{S^n(a)} B_{n,k}^{[a]}(x,y)\omega_{y-a} u(y) dS_y^n, & k = 0,1,2,\ldots \\ 0, & k = -1,\ldots,-n+2 \\ \frac{1}{\sigma_n} \int_{S^n(a)} A_{n,k}^{[a]*}(x,y)\omega_{x-a} u(y) dS_y^n, & k = -n+1,-n,\ldots \end{cases}$$

This series converges normally on each annulus including B_{rR}.

Proof. Choose for simplification $a = 0$. Applying CAUCHY's integral formula for the annulus $B_{r'R'}$ with $0 < r < r' < R' < R$ we get

$$u(x) = \int_{S_{r'}} e(x-y)\alpha(y)u(y)dS_{r',y} - \int_{S_{R'}} e(x-y)\alpha(y)u(y)dS_{R',y} = (u_1+u_2)(x),$$

where $S_{r'} = \partial B_{r'}(0)$ and $S_{R'} = \partial B_{R'}(0)$. Taking into account Theorem 3.42. we obtain in *int* $B_R(0)$ the expansion

$$u_1(x) = \sum_{k=0}^{\infty} (a_k u)(\omega_x)|x|^k.$$

Analogously we get in $\mathbb{R}^n \setminus \overline{B}_r(0)$ following Lemma 3.49. the expansion

$$u_2(x) = \sum_{k=0}^{\infty}(a_{-k-n+1}u)(\omega_x)|x|^{k-n+1}.$$

If $\alpha(y)$ is outward directed in both integrals we obtain the result. #

Definition 3.52. *The series*

$$\sum_{k=0}^{\infty}(a_k u)(\omega_{x-a})|x-a|^k$$

is called main part of the LAURENT *expansion, while*

$$\sum_{k=-n+1}^{\infty}(a_k u)(\omega_{x-a})|x-a|^k$$

is called singular part of the LAURENT *expansion.*

Remark 3.53. Further generalizations of TAYLOR and LAURENT expansions connected with residues theory can be found in [64].

3.4 Singular integrals of Cauchy's type

The singular integral of CAUCHY type S_Γ given by

$$(S_\Gamma u)(z) = \frac{1}{\pi i} \int\limits_\Gamma \frac{u(t)}{t-z} dt$$

plays an essential role in applications. This integral was first defined on LIAPUNOV surfaces for HÖLDER continuous functions (cf. [209]). Moreover it satisfies the remarkable algebraic property

$$(S_\Gamma^2 u) = u.$$

Up to the early fifties it was completely unknown how to construct a multidimensional analogue of S_Γ. The Russian mathematician A.W. BIZADSE found in 1953 such an analogue for quaternions and studied its properties. (cf. [26], [27]). Let Γ be a compact LIAPUNOV surface and let u be a quaternionic-valued HÖLDER continuous function

$$(S_\Gamma u)(x) = \frac{1}{2\pi} \int\limits_\Gamma \frac{\overline{x} - \overline{y}}{|x-y|^3} \alpha(y) u(y) d\Gamma_y \qquad (x \in \Gamma),$$

where $\alpha(y) = (0, \alpha_1(y), \alpha_2(y), \alpha_3(y))$ and $(\alpha_1, \alpha_2, \alpha_3)$ is the unit vector of the outer normal. L.G. MAGNARADSE gave 1957 an elementary proof of the property $S_\Gamma^2 = I$. On the basis of fundamental results of G. GIRAUD the Grusinian mathematician T.G. GEGELIA proved the L_p-boundedness of the operator S_Γ. The introduction of the operator S_Γ opened the way for the study of singular integral equations on LIAPUNOV manifolds of the type

$$Au + B\, S_\Gamma u = f.$$

When A, B are real constants and $A + B \neq 0,\ A - B \neq 0$ then the solution suggested by S. PRÖSSDORF is given by the formula

$$u = \frac{(A+B)^{-1} + (A-B)^{-1}}{2} f + \frac{(A+B)^{-1} - (A-B)^{-1}}{2} S_\Gamma f.$$

For variable coefficients further results are contained in [296], [202]. Singular integrals of CAUCHY's type over compact LIAPUNOV surfaces in $\mathbb{R}^n$ one can also find in the paper [137] of V. IFTIMIE in spaces of HÖLDER continuous functions. The L_p-boundedness of the singular CAUCHY integral operator on a LIPSCHITZ curve in the complex plane $\mathbb{C}$ was first proved in 1977 by A.P. CALDERON [40] (for curves with small LIPSCHITZ constants) and in 1982 by R. COIFMAN, A. MCINTOSH, and Y. MEYER [47] in the general case. There was proved the L_2-boundedness of the double-layer potential on LIPSCHITZ surfaces in $\mathbb{R}^n$ using CALDERON's rotation method. With CLIFFORD algebra technique R. COIFMAN suggested a more natural proof. Note that for small LIPSCHITZ constants M. MURRAY [209] proved also such a theorem. We also will mention here the paper by A. RATHSFELD [240] who proved the invertibility of the double layer potential operator on a polyhedron in the space of continuous functions. One of the main problems in the transition from LIAPUNOV surfaces to LIPSCHITZ surfaces is that the double layer potential loses the property to be a weakly singular operator.

Let us start with the definitions of LIAPUNOV bounded domains and LIPSCHITZ domains. In this section we will denote the scalar product for vectors with "$\cdot$".

Definition 3.54. *A compact surface* Γ *is called* LIAPUNOV *surface, if the following conditions are satisfied:*

(i) *At each point* $x \in \Gamma$ *there is the tangential space.*

(ii) *There exists a number* r, *that for any point* $x \in \Gamma$ *the set* $\Gamma \cap B_r(x)$ *(*LIAPUNOV *ball) is connected and parallel lines to the outer normal* $\alpha(x)$ *intersect at not more than one point.*

(iii) *The normal* $\alpha(x)$ *is* HÖLDER *continuous on* Γ, *i.e., there are constants* $C > 0$ *and* $0 < \alpha \leq 1$ *such that for* $x, y \in \Gamma$

$$|\alpha(x) - \alpha(y)| \leq C|x - y|^{\alpha}.$$

Corollary 3.55. (i) *Each surface of the class* $C^2(\Gamma)$ *is also a* LIAPUNOV *surface.* (ii) *Each* LIAPUNOV *surface is a surface of the type* $C^1(\Gamma)$. (iii) *Within a* LIAPUNOV *ball* $B_\varepsilon(x_0)$ *the inequality*

$$2|x_0 - y| \geq |x - y| \geq \frac{1}{2}|x_0 - y| \qquad (x, y \in B_\varepsilon(x))$$

holds.

Proof. For the proof of (iii) we refer to [272]. #

Proposition 3.56. *Let* Γ *be the* LIAPUNOV *boundary of the domain* G. *Then*

$$\frac{1}{\sigma_n} \int_\Gamma \frac{\overline{x} - \overline{y}}{|x - y|^n} \alpha(y) d\Gamma_y = \begin{cases} e_0 & , \quad x \in G \\ \frac{1}{2} e_0 & , \quad x \in \Gamma \\ 0 & , \quad x \in \mathbb{R}^n \setminus \overline{G}, \end{cases}$$

where $\alpha(y)$ *is again the outward pointing normal at* y.

Proof. We have the decomposition

$$\frac{1}{\sigma_n}\int\limits_{\Gamma} \frac{\overline{x}-\overline{y}}{|x-y|^n}\alpha(y)d\Gamma_y = \frac{1}{\sigma_n}\int\limits_{\Gamma} \frac{x-y}{|x-y|^n}\cdot\alpha(y)d\Gamma_y$$

$$+\frac{1}{\sigma_n}\int\limits_{\Gamma} \frac{\sum_{i>j}(x_i-y_i)\alpha_j-(x_j-y_j)\alpha_i e_i e_j}{|x-y|^n}d\Gamma_y.$$

The first integral is just the double layer potential with the density one. It is well known that

$$\frac{1}{\sigma_n}\int\limits_{\Gamma} \frac{x-y}{|x-y|^n}\cdot\alpha(y)d\Gamma_y = \begin{cases} e_0 & , \quad x\in G \\ \frac{1}{2}e_0 & , \quad x\in\Gamma \\ 0 & , \quad x\in \mathbb{R}^n\setminus\overline{G} \end{cases} .$$

It remains only to show that the second integral vanishes. Let be $G_\varepsilon = \overline{G}\setminus\overline{B}_\varepsilon(x)$ and $\Gamma_\varepsilon = \partial G_\varepsilon$, then

$$\int\limits_{\Gamma} \frac{(x_i-y_i)\alpha_j-(x_j-y_j)\alpha_i}{|x-y|^n}e_ie_jd\Gamma_y$$

$$=\frac{1}{2-n}\lim_{\varepsilon\to 0}\left[\partial_j\int\limits_{\Gamma_\varepsilon}\frac{\alpha_i(y)}{|x-y|^{n-2}}d\Gamma_y - \partial_i\int\limits_{\Gamma_\varepsilon}\frac{\alpha_j(y)}{|x-y|^{n-2}}d\Gamma_y\right]$$

$$=\frac{1}{2-n}\lim_{\varepsilon\to 0}\left[\partial_{ji}\int\limits_{G_\varepsilon}\frac{1}{|x-y|^{n-2}}dy - \partial_{ij}\int\limits_{G\varepsilon}\frac{1}{|x-y|^{n-2}}dy\right]=0. \qquad \#$$

Definition 3.57. *A domain $G\subset\mathbb{R}^n$ is called strongly* LIPSCHITZ *if the boundary $\partial G=\Gamma$ can be covered by a finite number of open sets V_i $(i=1,...,k)$ such that each set $G\cap V_i$ $(i=1,...,k)$ can be represented by the inequality $x_n > g(x_1,...,x_{n-1})$, where g is* LIPSCHITZ *continuous. The surface Γ is called* LIPSCHITZ *surface if each point $x\in\Gamma$ can be represented by $x=g(\underline{x})e_0+\underline{x}\in\mathbb{R}^n$, where $\underline{x}\in\mathbb{R}^{n-1}$. Furthermore, g has to satisfy the condition*

$$||Dg||_\infty \le \tan\alpha < \infty,$$

where

$$0\le\alpha<\frac{\pi}{2} \quad \textit{and} \quad ||Dg||_\infty = \operatorname{vrai}\max_{x\in\mathbf{R}^{n-1}}\Big(\sum_{j=1}^{n-1}|\partial_j g(x)|^2\Big)^{1/2}.$$

Remark 3.58. All domains which are bounded by a C^1 surface are strongly LIPSCHITZean.

Remark 3.59. The unit vector of the outer normal at y is at almost all points on Γ given by

$$\alpha(y) = \frac{e_0-(Dg)(y)}{\sqrt{1+|Dg(y)|^2}}.$$

Definition 3.60. *Let Γ be* LIPSCHITZ *surface of a domain $G \subset \mathbb{R}^n$. Then it is possible to cover the surface Γ by a finite number of open sets $V_1, ..., V_k$. Moreover, there exist homomorphisms $h_i^{-1} : V_i \to \Omega_i \in \mathbb{R}^{n-1}$. A $Cl_{0,n}$-valued function u defined on Γ is said to belong to the* BANACH *space $B(\Gamma)$ if and only if*

$$(h_i \circ u)(x') = u(h_i(x')) \in B(\Omega_i) \quad (i = 1, ..., k).$$

A norm can be defined by

$$||u||_{B(\Gamma)} = \sum_{i=1}^{k} ||(h_i \circ u)(x)||_{B(\Omega_i)}.$$

Theorem 3.61. (A. MCINTOSH.) *Let $G \subset \mathbb{R}^{n+1}$ be a strongly* LIPSCHITZ *domain with boundary $\partial G = \Gamma$. Let $p \in (1, \infty)$ $u \in L_p(\Gamma)$. Then the singular* CAUCHY *integral*

$$(S_\Gamma u)(x) = \frac{2}{\sigma_n} \lim_{\varepsilon \to 0} \int_{\Gamma_\varepsilon} \frac{\overline{x} - \overline{y}}{|x - y|^n} \alpha(y) u(y) d\Gamma_{\varepsilon,y}$$

with $\Gamma_\varepsilon = \{y \in \Gamma : |x - y| > \varepsilon\}$ is bounded on $L_p(\Gamma)$ and it holds that

$$||S_\Gamma u||_p \le C(p, \Gamma)||u||_p.$$

Moreover, its scalar part, the singular double-layer potential operator W_Γ, is defined for all $u \in L_p(\Gamma)$ and almost all $x \in \Gamma$ by

$$(W_\Gamma u)(x) := \frac{2}{\sigma_n} \lim_{\varepsilon \to 0} \int_{\Gamma_\varepsilon} \frac{x - y}{|x - y|^n} \cdot \alpha(y) u(y) d\Gamma_{\varepsilon,y}$$

and bounded on $L_p(\Gamma)$ with the same bound.

Proof. [196] Per definition Γ can be covered with the open sets $V_1, ..., V_k$. (cf. Definition 3.57). Let $\{\mathcal{X}_j : j = 1, ..., k\}$ a decomposition of the unit. In more detail we claim the following properties: (i) $\mathcal{X}_j$ is a real-valued continuous function $(j = 1, ..., k)$ defined on Γ, (ii) $0 \le \mathcal{X}_j \le 1$, (iii) supp $(\mathcal{X}_j) \subset V_j$, (iv) $\sum_{j=1}^{k} \mathcal{X}_j = 1$.

Furthermore, we consider a second collection of continuous real-valued functions Θ_j defined on Γ with

(i) supp $(\Theta_j) \subset V_j$,

(ii) $0 \le \Theta_j \le 1$,

(iii) dist (supp $(1 - \Theta_j)$, supp $(\mathcal{X}_j)) \ge \delta \quad (\delta > 0)$.

Then

$$(S_\Gamma u)(x) = \sum_{j=1}^{k} \left(\Theta_j(x) S_\Gamma(\mathcal{X}_j u)(x) + \int_{\mathrm{supp}(\mathcal{X}_j)} (1-\Theta_j)(x) \frac{x-y}{|x-y|^n} \alpha(y) \mathcal{X}_j(y) u(y) d\Gamma_y\right).$$

The second sum contains no singularities. It is no problem to get L_p estimates. Each term of the first sum satisfies

$$||\Theta_j S_\Gamma(\mathcal{X}_j u)||_p \leq C||\mathcal{X}_j u||_p \leq C||u||_p$$

because of $\Gamma \cap V_j$ can be represented as a part of the LIPSCHITZ graph g_j. #

Remark 3.62. In HÖLDER continuous spaces the proof of the existence in the case of LIAPUNOV surfaces follows immediately from Proposition 3.56 The boundedness of S_Γ in the case of $n = 3$ and again for the LIAPUNOV surface for $u \in C^{0,\beta}(\Gamma)$ was proved in [84]. In the scale of $W_p^s(\Gamma, Cl_{0,n})(s < 0, 1 < p < \infty)$ the proof can made similarly to that in ([202], p. 421). For the proof of the boundedness on a LIPSCHITZ graph in the complex plane we refer to [47]. Also the papers in [258] and [237] are dedicated to such a topic.

3.5 Formulae of Plemelj–Sokhotzki type

The application of BOREL–POMPEIU's formula to the solution of boundary value problems is connected with an evaluation of the boundary values of the CAUCHY's type integral. In the plane case first results were obtained by the Russian mathematician J.W. SOKHOTZKI in 1873 [278]. J. PLEMELJ independently from him proved the same in 1908 [232]. Ten years later in 1918 I.I. PRIWALOV obtained the so-called formulae of PLEMELJ–SOKHOTZKI for piecewise smooth LIAPUNOV boundaries. For a quaternion model A.W. BITSADZE in 1955 [26] proved such formulas. I. IFTIMIE established in 1968 [137] PLEMELJ-SOKHOTZKI formulae for HÖLDER continuous functions defined over compact LIAPUNOV surfaces in EUCLIDean space. These results has been recently extended to L_p-spaces over boundaries of LIPSCHITZ domains in $\mathbb{R}^n$ (cf. [195]).

For simplicity we will discuss the case of a compact piecewise LIAPUNOV surface for HÖLDER continuous functions. Before we formulate the main theorem we will prove the following proposition:

Proposition 3.63. *Let $a, b \in \mathbb{R}^n$. Then*

$$\big|a|b|^n - b|a|^n\big| \leq |b||a-b| \left[2|b|^{n-1} + |b|^{n-2}|a| + \ldots + |a|^{n-1}\right]$$

Proof.

$$\begin{aligned}\big|a|b|^n - b|a|^n\big| &\leq |b|\big||b|^n - |a|^n\big| + |a-b||b|^n \\ &\leq |a-b||b| \left[2|b|^{n-1} + |a||b|^{n-2} + \ldots + |a|^{n-1}\right]. \qquad \#\end{aligned}$$

Theorem 3.64. (PLEMELJ-SOKHOTZKI formulae.)
Let $u \in C^{0,\beta}(\Gamma, Cl_{0,n})(0 < \beta \leq 1)$. Then we have in each regular point $x_0 \in \Gamma$

$$n.t. - \lim_{\substack{x \to x_0 \\ x \in G^{\pm},\, x_0 \in \Gamma}} (F_\Gamma u)(x) = \frac{1}{2}(\pm u(x_0) + (S_\Gamma u)(x_0)),$$

where $G^+ := G$ and $G^- := \mathbb{R}^n \setminus \overline{G^+}$. The notation "n.t.-lim" means "non-tangential limit".

Proof. Let x_0 be a fixed point on Γ. Proposition 3.56 yields the representation

$$(F_\Gamma u)(x) = \begin{cases} a(x) + u(x_0) &, \quad x \in G \\ a(x) &, \quad x \in \mathbb{R}^n \setminus \overline{G} \end{cases},$$

where

$$a(x) = \frac{1}{\sigma_n} \int_\Gamma \frac{y-x}{|y-x|^n} \alpha(y)[u(y) - u(x_0)] d\Gamma_y. \tag{3.13}$$

For $x \neq x_0$ we have

$$|a(x) - a(x_0)| \leq \frac{1}{\sigma_n} \int\limits_{\Gamma} \frac{|(x_0 - y)|x - y|^n - (x - y)|x_0 - y|^n}{|x_0 - y|^n |x - y|^n} |u(y) - u(x_0)| d\Gamma_y .$$

Around x_0 we cut out the ball $B_\varepsilon(x_0)$. Set $\Gamma_\varepsilon = \Gamma \backslash B_\varepsilon(x_0)$ and $\Gamma'_\varepsilon = B_\varepsilon(x_0) \cap \Gamma$. Obviously, for $\varepsilon \to 0$, $\Gamma_\varepsilon \to \Gamma$ and $\Gamma'_\varepsilon \to \{x_0\}$. We get

$$|a(x) - a(x_0)| \leq \frac{1}{\sigma_n} \Big(\int\limits_{\Gamma'_\varepsilon} ... d\Gamma_y + \int\limits_{\Gamma_\varepsilon} ... d\Gamma_y \Big).$$

On account of $x \to x_0$ we assume $x \in B_{\varepsilon/2}(x_0)$. In this way the second integral has no singularities. Using Proposition 3.63 with $a = x - y$ and $b = x_0 - y$ $(a - b = x - x_0)$ we find the estimate

$$\frac{1}{\sigma_n} \int\limits_{\Gamma_\varepsilon} ... d\Gamma_y \leq C ||u||_{C^{0,\beta}(\Gamma)} |x - x_0| \qquad (0 < \beta \leq 1).$$

The first integral can be estimated as follows:

$$|\frac{1}{\sigma_n} \int\limits_{\Gamma'_\varepsilon} ... d\Gamma_y| \leq \frac{1}{\sigma_n} \int\limits_{\Gamma'_\varepsilon} |x_0 - y|^{1-n+\beta} \times$$

$$\times \frac{2|x_0 - y|^{n-1} + |x_0 - y|^{n-2}|x - y| + ... + |x - y|^{n-1}}{|x - y|^n} d\Gamma'_y |x_0 - x|^{1-\delta} ||u||_{C^{0,\beta}(\Gamma)}$$

with $1 > \delta > 0$ and $\beta + \delta > 1$. Using Corollary 3.55 (iii) we have

$$\frac{1}{2}|x_0 - y| \leq |x - y| \leq 2|x_0 - y|$$

in each LIAPUNOV ball ($B_\varepsilon(x_0)$ is just such a ball). Finally,

$$|\frac{1}{\sigma_n} \int\limits_{\Gamma'_\varepsilon} ... d\Gamma'_y| \leq \frac{2^n n}{\sigma_n} \int\limits_{\Gamma'_\varepsilon} \frac{1}{|x_0 - y|^{n-(\beta+\delta)}} d\Gamma'_y |x_0 - x|^{1-\delta} ||u||_{C^{0,\beta}(\Gamma)}$$

and such

$$|a(x) - a(x_0)| \leq \text{ const. } ||u||_{C^{0,\beta}(\Gamma)} |x_0 - x|^{1-\delta} \to 0 \quad (x \to x_0) .$$

The definition of S_Γ and Proposition 3.56 leads to the result. #

Remark 3.65. In [18] one can find the corresponding proof for unbounded surfaces.

Corollary 3.66. *Let x_0 be lying on a wedge or corner. Then the* PLEMELJ–SOKHOZKI *formulae reads as follows:*

$$n.t. - \lim_{\substack{x \to x_0 \\ x \in G^{\pm}}} (F_\Gamma u)(x) = \frac{1}{2}[\pm u(x_0) + (S^\delta_\Gamma u)(x_0)]$$

where

$$(S^\delta_\Gamma u)(x_0) := \frac{\sigma_n - 2\delta(x)}{\sigma_n} u(x_0) + (S_\Gamma u)(x_0)$$

and $0 < \delta(x) < \sigma_n$. $\delta(x)$ denotes the inner angle at $x \in \Gamma$ taken as part of the unit ball centred at x.

Proof. We have only to generalize Proposition 3.56. Let $B_\varepsilon(x)$ be a ball centred at x with radius ε and $x \in \Gamma$ an arbitrary point. Then we decomposite

$$\frac{1}{\sigma_n}\int_\Gamma \frac{y-x}{|x-y|^n}\cdot \alpha(y) d\Gamma_y = \int_{\Gamma_\varepsilon \cup (S_\varepsilon \cap G)} ...d\Gamma_y + \int_{\Gamma \cap B_\varepsilon} ...d\Gamma_y + \int_{S_\varepsilon \cap G} dS_{\varepsilon y}\,.$$

Here Γ_ε denotes the rest of the boundary without the ball $B_\varepsilon(x)$, i.e. $\Gamma_\varepsilon = \Gamma \setminus B_\varepsilon(x)$ and $S_\varepsilon = \partial B_\varepsilon(x)$. On account that x is outside of the domain bounded by $\Gamma_\varepsilon \cup (S_\varepsilon \cap G)$, by using Proposition 3.56 the first integral has to vanish. For sufficiently small $\varepsilon > 0$ the outer normal α and $(y-x)/|y-x|$ are almost orthogonal such that

$$\left|\alpha \cdot \frac{y-x}{|y-x|}\right| = O(\varepsilon).$$

Hence, the second integral tends to zero for $\varepsilon \to 0$. For the third term we obtain

$$\frac{1}{\sigma_n}\int_{S_\varepsilon \cap G} \frac{y-x}{|x-y|^n}\cdot \alpha(y) d\Gamma_{\varepsilon,y} = \frac{1}{\sigma_n}\int_{S_\varepsilon \cap G} \frac{1}{|y-x|^{n-1}} d\Gamma_{\varepsilon,y} = \frac{1}{\sigma_n \varepsilon^{n-1}}\int_{S_\varepsilon \cap G} d\Gamma_{\varepsilon,y}.$$

All together this means that

$$\frac{1}{\sigma_n}\int_\Gamma \frac{y-x}{|x-y|^n}\cdot \alpha(y) d\Gamma_y = \frac{\delta(x)}{\sigma_n} \quad (x \in \Gamma)\,.$$

Substituting this in (3.13) we verify our statement. #

Proposition 3.67. *Let Γ be a piecewise* LIAPUNOV *surface, $u \in C^{0,\beta}(\Gamma, Cl_{0,n})$ $(0 < \beta \leq 1)$. Then*

(i) *The equality $(S^\delta_\Gamma u)(x) = u(x)$ $(x \in \Gamma)$ is valid iff $u(x)$ are just the boundary values of a* CLIFFORD *regular function in G.*

(ii) *The equality* $(S_\Gamma^\delta u)(x) = -u(x)$ $(x \in \Gamma)$ *is valid iff* $u(x)$ *are just the boundary values of a* CLIFFORD *regular function in* $\mathbb{R}^n \setminus \overline{G}$ *which vanishes at infinity.*

Proof. (i) Let v be the $Cl_{0,n}$-regular extension of u into the domain G. CAUCHY's integral theorem delivers $F_\Gamma u = v$. From Corollary 3.66 we get

$$u(x) = v(x) = n.t. - \lim_{\substack{x' \to x \\ x' \in G}} (F_\Gamma u)(x') = \frac{1}{2}[u(x) + (S_\Gamma^\delta u)(x)]$$

and so $u(x) = (S_\Gamma^\delta u)(x)$. Conversely, assume $(S_\Gamma^\delta u)(x) = u(x)$ $(x \in \Gamma)$, then $(F_\Gamma u)(x')$ is the desired $Cl_{0,n}$-regular extension; (ii) can be shown similarly by using Theorem 3.36. #

Corollary 3.68. *Let* Γ *be a piecewise* LIAPUNOV *surface,* $u \in C^{0,\beta}(\Gamma, Cl_{0,n})$ $(0 < \beta < 1)$. *We have the identity*

$$(S_\Gamma^\delta)^2(u)(x) = u(x) \quad (x \in \Gamma).$$

Proof. Putting $n.t. - \lim_{\substack{x' \to x \\ x' \in G^\pm}} (F_\Gamma u)(x') =: F^\pm(x)$.

Then it follows that $(S_\Gamma^\delta u)(x) = F^+(x) + F^-(x)$, whence

$$(S_\Gamma^\delta)^2(u)(x) = S_\Gamma^\delta F^+(x) + S_\Gamma^\delta F^-(x) = F^+(x) - F^-(x) = u(x). \qquad \#$$

To simplify our explanations we will introduce some new operators.

Definition 3.69. *Let* Γ *be a piecewise* LIAPUNOV *surface. Then we set*

$$P_\Gamma := \frac{1}{2}(I + S_\Gamma^\delta) \quad \text{and} \quad Q_\Gamma := \frac{1}{2}(I - S_\Gamma^\delta),$$

where P_Γ, Q_Γ *are called* HARDY *projections.*

Corollary 3.70. *The* HARDY *projections fulfil the properties*

(i) $P_\Gamma^2 = P_\Gamma$, $Q_\Gamma^2 = Q_\Gamma$, $P_\Gamma Q_\Gamma = Q_\Gamma P_\Gamma = 0$

(ii) P_Γ *is just the projection onto all functions which can be* $Cl_{0,n}$*-regular extended into the domain* G.

(iii) Q_Γ *is the projection onto all functions which can be* $Cl_{0,n}$*-regular extended into the domain* $\mathbb{R}^n \setminus \overline{G}$ *and vanish at infinity.*

Remark 3.71. The operators $S_\Gamma, P_\Gamma, Q_\Gamma$ have continuous extensions onto the space $L_p(G, Cl_{0,n})$. The denotation will be the same.

Remark 3.72. The results of Section 3.5 can be transferred to polynomially weighted spaces of the type

$\omega^s_{2,\mu}(\Gamma, Cl_{0,n}) =$

$$= \{u \in \mathcal{D}(\Gamma) : \|u\|^2_{\omega^s_{2,\mu}(\Gamma)} = \int_\Gamma \sum_{|\alpha| \leq s} p_{\mu-2(s-|\alpha|)}(x)|\partial^\alpha u(x)|^2 dx < \infty\},$$

where Γ is not necessarily bounded and

$p_\mu(x) = (1 + |x|^2)^{\mu/2}$.

For detailed information see in [20].

3.6 Decomposition of the $Cl_{0,n}$-valued Hilbert space

In this section we deal with the structure of the $Cl_{0,n}$-valued HILBERT space according to the DIRAC operator. We use in our considerations ideas which are known as the so-called HODGE decomposition. This means that $\ker D$ will be separated. We are able to describe the complement of $\ker D \cap L_2(G, C\ell_{0,n})$ in $L_2(G, C\ell_{0,n})$ exactly. Corresponding orthoprojections can be constructed in a more or less explicit form.

At first we will verify the subspace structure of $L_p(G, C\ell_{0,n})$.

Proposition 3.73. *Let $G \subset \mathbb{R}^n$ be a bounded domain with a piecewise smooth boundary. Then the right-linear set*

$$\mathcal{L} := L_p(G, C\ell_{0,n}) \cap \ker D \qquad (1 < p < \infty)$$

is a subspace in $L_p(G, C\ell_{0,n})$.

Proof. Let $(\phi_k)_{k=1}^{\infty}$ be a CAUCHY-sequence in $\mathcal{L}$. Then there exists a $Cl_{0,n}$-valued function $\phi \in L_p(G, C\ell_{0,n})$ with $\phi_k \to \phi$ in $L_p(G, C\ell_{0,n})$. From the mean value theorem we obtain

$$|(\phi_k - \phi_l)(x)| \leq \frac{n}{\sigma_n r^n} \int\limits_{B_r(x)} |\phi_k - \phi_l|(y) dy,$$

where $B_r(x)$ is the ball centred at x with radius r. Hence,

$$\sup_{x \in G_r} |(\phi_k - \phi_l)(x)| \leq \left(\frac{n}{\sigma_n r^n} \right)^{\frac{1}{p}} ||\phi_k - \phi_l||_p$$

with $G_r = \{x \in G : \text{ dist } (x, \Gamma) > r\}$. In this way we have the uniform convergence on compact subsets. For the validity of the WEIERSTRASS theorem we obtain
$\phi \in \ker D \cap L_p(G, C\ell_{0,n})$. #

In the following sections we will investigate elliptic boundary value problems. Preparing this investigation we restrict ourselves for the moment to the case $p = 2$.

Definition 3.74. *Let $u, v \in L_2(G, C\ell_{0,n})$. We define an inner product by*

$$(u, v)_{L_2(G)} = \int\limits_G \overline{u(x)} v(x) \, dx. \tag{3.14}$$

We remark that we will often write $(u, v)_2$ instead of $(u, v)_{L_2(G)}$. The norm induced by this inner product coincides with the usual L_2-norm for vector-valued functions. It should be mentioned that in general this inner product is $C\ell_{0,n}$-valued (and not a real or complex number) with the properties:

(i) $(u,u)_2 \geq 0 \quad \forall\, u \in L_2(G), \quad (u,u)_2 = 0 \leftrightarrow u = 0,$

(ii) $(u,v)_2 = \overline{(v,u)_2} \quad \forall\, u,v \in L_2(G),$

(iii) $(u+v,w)_2 = (u,w)_2 + (v,w)_2,$

(iv) $\lambda(u,v)_2 = (\overline{\lambda}u,v)_2, \ (u,v)_2\lambda = (u,v\lambda)_2 \quad \forall\, \lambda \in C\ell_{0,n}.$

That means all the known properties of scalar products are available. We have only to pay some attention to the properties connected with the $C\ell_{0,n}$-multiplication.
In the following we look for the orthogonal complement of the monogenic functions in L_2 with respect to the inner product (3.14). This orthogonal decomposition is the central idea of the theory described below.

Theorem 3.75. *The $C\ell_{0,n}$-valued* HILBERT *space $L_2(G, C\ell_{0,n})$ permits the orthogonal decomposition*

$$L_2(G, C\ell_{0,n}) = L_2(G, C\ell_{0,n}) \cap \ker D \oplus D \overset{\circ}{W}{}_2^1 (G, C\ell_{0,n})$$

with respect to the inner product (3.14).

Proof. Let $X_1 = L_2(G, C\ell_{0,n}) \cap \ker D$ and $X_2 = L_2(G, C\ell_{0,n}) \ominus X_1$. For $u \in X_2$ we have $v = T_G u \in W_2^1(G, C\ell_{0,n})$. Hence, $u = Dv$ follows. It remains to show that $v \in \overset{\circ}{W}{}_2^1 (G, C\ell_{0,n})$. To prove this we take $\phi \in X_1$. Then,

$$\int_G \overline{Dv}\phi dy = 0.$$

The singularity functions

$$\phi_l(x) = \frac{1}{\sigma_n} \frac{x - x^{(l)}}{|x - x^{(l)}|^n}$$

are both left and right CLIFFORD regular in G if the points $x^{(l)}$ belong to $\mathbb{R}^n \setminus \overline{G}$. Due to the assumptions

$$\int_G \overline{Dv}\phi_l dy = 0$$

holds. Hence (cf. Proposition 3.22)

$$\begin{aligned} 0 &= \int_G \overline{Dv}(y)\phi_l(y)dy = \overline{\int_G \overline{\phi}_l(Dv)(y)dy} = \\ &= \overline{\int_G (\phi_l D)(y)v(y)dy} - \overline{\int_\Gamma \phi_l(y)\alpha(y)v(y)dy} = -\overline{(F_\Gamma v)(x^{(l)})}. \end{aligned}$$

If now $\{x^{(l)}\}$ is supposed to be a dense subset in $\mathbb{R}^n \setminus \overline{G}$, then

$$(F_\Gamma v)(x) = 0 \quad \text{in} \quad \mathbb{R}^n \setminus \overline{G}$$

and, consequently, v belongs to the HARDY space $\operatorname{im} P_\Gamma \cap W_2^{\frac{1}{2}}(\Gamma, C\ell_{0,n})$. Therefore, a function h with $\operatorname{tr}_\Gamma h = v$ exists. Here, tr_Γ denotes the restriction operator (trace operator) onto the boundary Γ. Taking now

$$w := v - h \in \overset{\circ}{W}{}_2^1 (G, C\ell_{0,n})$$

then $u = Dv = Dw$ and the proof is complete. #
Of course, this orthogonal decomposition defines two orthoprojections onto the occurring subspaces. More precisely, we have the orthoprojections

$$\mathbf{P} : L_2(G) \mapsto L_2(G) \cap \ker D(G) \quad \text{and} \quad \mathbf{Q} : L_2(G) \mapsto D(\overset{\circ}{W}{}_2^1).$$

The same idea can be generalized into two directions. On the one hand, we can investigate similar decompositions for other differential operators. On the other hand, it is possible by introducing special weight functions to generalize the inner product. At first we look to decompositions with respect to other differential operators.
Analogously to Theorem 3.75. we can prove orthogonal decomposition with respect to the CAUCHY–FUETER operator and its adjoint operator, respectively.

Theorem 3.76. *Let D be the* CAUCHY–FUETER *operator and $\overline{D}$ its adjoint operator. Then, we have the orthogonal decompositions*

$$L_2(G) = L_2(G) \cap \ker D \oplus \overline{D}(\overset{\circ}{W}{}_2^1)(G)$$

and

$$L_2(G) = L_2(G) \cap \ker \overline{D} \oplus D(\overset{\circ}{W}{}_2^1)(G).$$

with respect to the inner product (3.14).

The corresponding orthoprojection will be denoted as follows:

$$\mathbf{P}^D : L_2(G) \mapsto L_2(G) \cap \ker D, \quad \mathbf{Q}^D = I - \mathbf{P}^D : L_2(G) \mapsto \overline{D}(\overset{\circ}{W}{}_2^1)(G)$$

$$\mathbf{P}^{\overline{D}} : L_2(G) \mapsto L_2(G) \cap \ker \overline{D}, \quad \mathbf{Q}^{\overline{D}} = I - \mathbf{P}^{\overline{D}} : L_2(G) \mapsto D(\overset{\circ}{W}{}_2^1)(G)$$

An analogous result can be proved in the case of the Laplacian.

Theorem 3.77. *The right* HILBERT *modul $L_2(G, Cl_{0,n})$ allows the orthogonal decomposition*

$$L_2(G, Cl_{0,n}) = \ker \Delta(G) \cap L_2(G, Cl_{0,n}) \oplus \Delta(\overset{\circ}{W}{}_2^2(G)) \tag{3.15}$$

with respect to the inner product (3.14).

Proof. We refer to [105] for the proof of (3.15). #

Remark 3.78. The decomposition (3.15) of the space $L_2(G, Cl_{0,n})$ defines the complementary orthoprojections

$$\mathbf{P}^{\Delta} : L_2(G, Cl_{0,n}) \mapsto \ker \Delta \cap L_2(G, Cl_{0,n}) \quad \text{and} \quad \mathbf{Q}^{\Delta} : L_2(G, Cl_{0,n}) \mapsto \Delta(\overset{\circ}{W}{}_2^2).$$

Having in mind boundary value problems with variable coefficients, we will also look for decompositions with respect to $D + a$, where D stands for the DIRAC operator and a is a $C\ell_{0,n}$-valued function. Under certain assumptions we can expect analogous results. But, because there are more details to consider in preparing this theorem we refer to Section 4.3.

The idea to change the inner product by the help of weight functions has already been considered in [110]. We will repeat here only the result for the quaternionic case without proof.

Theorem 3.79. *Let B be a multiplication operator defined by*

$$(Bu)(x) = \sum_{i=0}^{3} b_i(x) u_i(x) e_i$$

where the $b_i \in C^\infty(\overline{G})$, $i = 0, 1, 2, 3$ are positive (scalar) functions. Then we have the orthogonal decomposition

$$L_2(G) = B \ker D \cap L_2(G) \oplus_B D(\overset{\circ}{W}{}_2^1)(G)$$

with respect to the inner product

$$[u, v]_B = \int_G \overline{B^{-1}u(x)} B^{-1} v(x)\, dx.$$

Clearly, $B^{-1}u = \sum_{i=0}^{3} \frac{1}{b_i} u_i e_i$.

3.7 Remarks on residues in Clifford analysis

First suggestions to residues we find already in the classical paper of A.L. CAUCHY "Memoire sur les intégrales definies" (Œuvres 1, 1. Serie (1814), 319–506). The notation "residuum" is first mentioned by A.L. CAUCHY in 1826 (cf. [173]). The residue is nothing less than the coefficient a_{-1} of the LAURENT expansion

$$\sum_{k=-\infty}^{\infty} a_k(z-a)$$

of a holomorphic function f within the annulus $B_1(a) \setminus \{a\}$, which can be expressed by

$$a_{-1} = \frac{1}{2\pi i} \int_{S^n} f(\zeta) d\zeta.$$

We will shortly explain the meaning of the residue in the concept of CLIFFORD analysis.

We have already seen in Section 3.3 that also in quaternionic and CLIFFORD analysis LAURENT expansions near to isolated singularities are possible. We will consider the notion of a residue for functions which are CLIFFORD regular up to isolated singularities. Unfortunately, not all singularities of a CLIFFORD regular function are isolated. We will only explain here the notion of a residue in the simplest case. In more general situations we will refer to the literature.

Definition 3.80. *A point $a \in \mathbb{R}^n$ is called* CLIFFORD *regular of the $Cl_{0,n}$-valued function u iff there is a open ball $B_r(a)$ in which u is* CLIFFORD *regular. A point $a \in \mathbb{R}^n$ is called (left)* CLIFFORD *singular point of u iff for any $r > 0$, $u \notin (\ker D)(B_r(a))$. A singular point a is called (left) isolated* CLIFFORD *singular iff $u \in (\ker D)(B_r(a) \setminus \{a\})$ for any $r > 0$.*

Definition 3.81. *An isolated* CLIFFORD *singular point is called (left) pole of order m iff the singular part of* LAURENT*'s expansion (cf. Section 3.3) breaks off from $k \geq m$. In the opposite when the singular part does not break off we are speaking of a so-called essential* CLIFFORD *singular point.*

Remark 3.82. In [36] the notion of a CLIFFORD meromorphic function has been introduced. A generalization of MITTAG-LEFFLER's theorem on the structure of such a CLIFFORD meromorphic function is proved.

Definition 3.83. *Let u be a* CLIFFORD *regular function with a* CLIFFORD *singular point at the point a. Then*

$$(\text{res } u)(a, \omega_{x-a}) := \frac{\omega_{x-a}}{\sigma_n} \int_{S^n(a)} \alpha(y) u(y) dS_{1,y},$$

where $\alpha(y)$ is the outward pointing normal unit vector at y. In the case of a sphere around a we have $\alpha(y) = \omega_{y-a}$ which is called residue of u at the CLIFFORD *singular point a in direction $\omega(x-a)$.*

Proposition 3.84. *Let u be a* CLIFFORD *regular function with a* CLIFFORD *singular point at a. Then*

$$\text{res}\,(a, \omega_{x-a}) = (a_{-n+1}u)(x),$$

where $(a_{-n+1}u)(x)$ is the first coefficient at the singular series of the LAURENT *expansion of u at a.*

Proof. Lemma 3.40 yields for $k = -n+1$

$$\begin{aligned} A^{[a]*}_{n,-n+1} &= A^{[a]}_{n,0} = \frac{1}{n-2}(C_1^{\frac{n-2}{2}}(s) + (2-n)C_0^{\frac{n}{2}}(s)\omega_{y-a} \wedge \omega_{x-a})\overline{\omega}_{x-a} \\ &= (s - \omega_{y-a} \wedge \omega_{x-a})\overline{\omega}_{x-a} = \Big[(\omega_{y-a}, \omega_{x-a}) - \omega_{y-a} \wedge \omega_{x-a}\Big]\overline{\omega}_{x-a} \\ &= \overline{\omega}_{x-a}\overline{\omega}_{y-a}\overline{\omega}_{x-a}. \end{aligned}$$

Then we get

$$(a_{-n+1}u)(x) = \frac{1}{\sigma_n} \int\limits_{S^n(a)} \omega_{x-a}\omega_{y-a}u(y)dS_{1,y} = (\text{res}\ u)(a, \omega_{x-a}). \qquad \#$$

Remark 3.85. The coefficient $(a_{-n+1}u)(x)$ does not depend on $|x|$.

Remark 3.86. When a be a pole of order $(m+1)$ of a function u then the formula

$$(\text{res}\ u)(a) = \lim_{x \to a} \frac{1}{m!} \frac{d^m}{d|x|^m}[|x-a|^{m+n-2}(x-a)u(x)]$$

is valid and is proved in [36] (p. 101).

Immediately from CAUCHY's integral theorem we have

Theorem 3.87. *Let u be a (left)* CLIFFORD *regular function in $G \subset \mathbb{R}^n$ except for the finite number of isolated (left)* CLIFFORD *singular points $a_1, ..., a_p$ such that $a_i \notin \partial G = \Gamma$ $(i = 1, ..., p)$. Then*

$$\frac{1}{\sigma_n} \int\limits_{\Gamma} \alpha(y)u(y)d\Gamma_y = \sum_{i=1}^{p} (\text{res}\ u)(a_i, \omega_{x-a_i})$$

Remark 3.88. In [340] one can find a generalization of this type of residue to $(m-1)$-forms of the type $\omega = ud\sigma v$ with

$$d\sigma = \sum_{j=1}^{n} (-1)^{j+1} e_j d\hat{x}_j$$

where u and v are, right and left CLIFFORD regular on a domain in $\mathbb{R}^n$, respectively. Furthermore, there are proved good connections between residue calculus, mapping degree and the theorem of implicit functions.

Remark 3.89. An alternative approach by using the simplicial calculus can be found in [275]. A geometric calculus on an oriented k-surface embedded in $\mathbb{R}^n$ is developed. It is used in an oriented k-surface as the limit set of a sequence of k-chains. Furthermore, it outlines the relationship between vector derivative and residue theorem. This calculus may be interesting in finite difference calculations.

Remark 3.90. As already mentioned by R. FUETER [80] the set of singularities of a CLIFFORD regular function can have an arbitrary dimension. For this reason it is quite natural to look for a suitable notion of residues for such higher dimensional singularities and prove the corresponding residues theorems. A comprehensive and powerful theory has been developed. Results in this field are based on papers of J. LERAY [167] as well as F. NORGUET [217] from 1959. The understanding of this theory requires knowledge in homology and cohomology theory. Since we cannot explain here elements of this theories we refer to the literature. The best references for this type of investigations are the paper of R. DELANGHE/F. SOMMEN/V. SOUCEK [64] and there given bibliographic data. The authors of this paper dealt with two principal types of residues, namely residues as numbers and residues as functionals. To the first type belong the so-called LERAY–NORGUET *residue* and the *residue of a $(n-1)$-form ω at a compact oriented submanifold* $\sum \subset G \subset \mathbb{R}^n$, where G is a domain and $\dim \sum \leq n-2$. There is a close connection between both residues. The notion of a residue as functional was introduced by F. SOMMEN in [282]. He called this functional *grand residue* of ω at $\sum$.

This new type of residue can be used for computing integrals of the form $f\omega$ over surfaces of dimension $n-1$ for any CLIFFORD regular function f and $(n-1)$-form ω. As a special case it is just the "*grand residue*" of the CAUCHY kernel at $\sum$ (compact submanifold of dimension less or equal to $(n-2)$) the DIRAC functional. Further interesting papers on residues are found in [253] and [215].

Remark 3.91. In this book we will not comment on the notion of the function theoretical index in some sense. We note only here that A. SUDBURY introduced in [313] a so-called *wrapping number*, a multidimensional analogue to the winding number of a curve around a point in the plane. K. HABETHA gave in [117] the definition of the KRONECKER-*Index* for a n-dimensional closed and bounded manifold $M \subset \mathbb{R}^{n+1}$. L.S. PONTRJAGIN introduced in 1952 the notion of the *topological winding number* which can be calculated by the help of the so-called *indicatrix of cycles* which was introduced by F. SOMMEN [282]. In [39] is given the definition of the index of a point z with respect to a n-dimensional cycle Γ. In order to derive a general CAUCHY integral formula in [160] is defined for regular mapping in the Euclidean HURWITZ pairs theory, also an index formula. This paper generalizes earlier results in [15].

3.8 Möbius transformations in $\mathbb{R}^n$

In this section we will study the analogue of MÖBIUS transformations in higher dimensions. In 1853 A.F. MÖBIUS started the investigation of an equivalent class of geometrical transformations the so-called "Kreisverwandtschaften". He invented a method for the description of all EUCLIDean motions in the complex plane. For complex numbers a, b, c, d the matrix

$$\begin{pmatrix} a & b \\ c & d \end{pmatrix}$$

with $ad - bc \neq 0$ generates a mapping $L : \mathbb{C} \to \mathbb{C}$ such that

$$L(z) := \begin{pmatrix} a & b \\ c & d \end{pmatrix} (z) = \frac{az + b}{cz + d} \quad (z \in \mathbb{C}).$$

This class of mappings is called MÖBIUS transformations, which form a group. Based on their fixed points one get a corresponding classification. K.TH. VAHLEN [318] discovered in 1902 that similar transformations in $\mathbb{R}^n$ can also be expressed by 2×2–matrices but over a half century these ideas were forgotten until H. MAASS [183] in 1949 rediscovered VAHLEN's approach and proved most of the results more emphatically. Again, over thirty years later, L.V. AHLFORS dedicated some papers (cf. [4], [5]) to this topic and made this theory so popular that now higher dimensional MÖBIUS transformations are widely used in CLIFFORD analysis (cf. [74], [178], [180], [255], [258], [260]).

In order to characterize MÖBIUS transformations it is useful to introduce some groups which are included in the CLIFFORD algebra $Cl_{0,n}$.

Spoin group [Spoin(n)]: Let $\omega_i \in S^n$, $(1 \leq i \leq m)$ $(m \in \mathbb{N})$. Then the spoin group is just the group of all elements ω, which can be represented by

$$\omega = \omega_1 \cdot \ldots \cdot \omega_m.$$

Obviously, Spoin$(1) = S^1$. Moreover, $S^n \subset$ Spoin(n). Spoin(n) can be seen as the natural generalization of the circle group.

Spin group [Spin$(n+1)$]: Let $x \in Cl_{0,n}$. Furthermore, let $x_i \in \mathbb{R}^{n+1}$, $0 \leq i \leq 2m$, $(m \in \mathbb{N})$ with $x_i^2 = -1$. Spin$(n+1)$ is just the group of all elements x which can be represented by

$$x = x_1 \cdot \ldots \cdot x_{2m}.$$

It can be observed that Spin$(n+1)$=Spoin(n). By the algebra isomorphism $\phi : Cl_{0,n} \ni e_i \to e_i e_{n+1} \in Cl_{0,n}^+$, where $Cl_{0,n}^+$ is a subalgebra spanned by $e_0 = 1$ and each element of the form $e_{i_1} \cdot \ldots \cdot e_{i_{2m}}$ $(1 \leq 2m \leq n)$.

Lemma 3.92. *Let $y \in \mathbb{R}^n$ and $y^2 = -1$. Then the mapping*

$$\rho : \mathbb{R}^n \ni x \to yxy \in \mathbb{R}^n$$

is a reflection in $\mathbb{R}^n$ in the direction y. Furthermore, there is a matrix $U_y \in O(n)$ (orthogonal group) that

$$U_y(x) = yx\tilde{y},$$

where $\tilde{y}$ is the reverse involution of y (cf. Section 1.4).

Pin group [$\mathrm{Pin}(n+1)$]: Let $x_i \in S^n$, $1 \leq i \leq m \quad (m \in \mathbb{N})$. The pin group is just the group of all elements which have the representation

$$x = x_1 \cdot \ldots \cdot x_m.$$

Clearly, $\mathrm{Spin}(n+1) \subset \mathrm{Pin}(n+1)$.

Lemma 3.93. *Let $y \in$ Pin (n) and $\tilde{y}$ again the reverse involution of y. Then*

$$\rho_y : \mathbb{R}^n \ni x \to yx\tilde{y} \in \mathbb{R}^n$$

is an orthogonal transformation, i.e. $\rho_y \in O(n)$. Pin(n) *and* Spin(n) *belong to the class of so-called* Lie *groups and are of the dimension $n(n-1)/2$.*

Lipschitz group [Γ_n]: The products of all non-zero vectors in $\mathbb{R}^n$ form the Lipschitz group. Note that $\tilde{x}x$ is real and, therefore, $|x|^2 = \tilde{x}x$. Furthermore, we have $|xy| = |x||y|$. Sometimes the Lipschitz group is also called the Clifford group.

Lemma 3.94. *Let $\hat{x}$ be the main involution of x (cf. Section 1.4) then for $y \in \mathbb{R}^n$ also $\hat{x}yx$ is a vector. Moreover, the transformation $\tau : y \to \hat{x}^{-1}yx$ is a Euclidean isometry, i.e. there exists a matrix $U_x(y) \in O(n)$ such that $\hat{x}^{-1}yx = U_x(y)$.*

Möbius group: The group which is generated by translations, reflections, and the inversion $x \to -x/|x|^2$ is called the Möbius *group.*

Definition 3.95. *A matrix*

$$m = \begin{pmatrix} a & b \\ c & d \end{pmatrix}$$

which fulfils the conditions

(i) $a, b, c, d \in \Gamma_n \cup \{0\}$,

(ii) $a\tilde{d} - b\tilde{c} = \pm 1$,

(iii) $a^{-1}b, ca^{-1} \in \mathbb{R}^n \quad (a \neq 0)$ *or* $ac^{-1}, c^{-1}d \in \mathbb{R}^n \quad (c \neq 0)$

is called VAHLEN *matrix in* $\mathbb{R}^n$. *With the usual matrix multiplication the* VAHLEN *matrices form a group the so-called* VAHLEN *group.*

Remark 3.96. It can be proved that the VAHLEN group is connected.

Remark 3.97. The CLIFFORD number $a\tilde{d} - b\tilde{c}$ is sometimes named *pseudo-determinant.* Note that in a VAHLEN matrix both the diagonal elements a, d are either of even or of odd order.

Theorem 3.98. [178] *The* VAHLEN *matrices of* $\mathbb{R}^n$ *with even diagonal and pseudo-determinant 1 form a two-fold covering group of the sense-preserving* MÖBIUS *group of* $\mathbb{R}^n$. MÖBIUS *transformations which change the sense of orientation are related to the* VAHLEN *matrices with odd diagonal and pseudo-determinant -1.*

The following list gives an overview on all basic sense-preserving MÖBIUS transformations and their corresponding VAHLEN matrices.

transformation	action	explanation	matrix
rotation	$x \to axa^{-1}$	$a \in \text{Spin}(n)$	$\begin{pmatrix} a & 0 \\ 0 & a \end{pmatrix}$
dilatation	$x \to x\frac{1}{d^2}$	$d \in \mathbb{R}$	$\begin{pmatrix} 1/d & 0 \\ 0 & d \end{pmatrix}$
translation	$x \to x + b$	$b \in \mathbb{R}^n$	$\begin{pmatrix} 1 & b \\ 0 & 1 \end{pmatrix}$
inversion	$x \to \frac{-\bar{x}}{\lvert x \rvert^2}$	-	$\begin{pmatrix} 0 & -1 \\ 1 & 0 \end{pmatrix}$
transversion	$x \to x(cx+1)^{-1}$	$c \in \mathbb{R}^n$	$\begin{pmatrix} 1 & 0 \\ c & 1 \end{pmatrix}$

Corollary 3.99. *The transversion can also be written in the form*

$$(x^{-1} + c)^{-1} = x(cx+1)^{-1} = \frac{x + x^2 c}{1 + 2x \cdot c + x^2 c^2}.$$

In this way a transversion is a combination of two inversions and one translation

Proof. This can be checked by a straightforward calculation. #

It may immediately proved:

Lemma 3.100. *Each* VAHLEN *matrix permits the factorization*

$$\begin{pmatrix} a & b \\ c & d \end{pmatrix} = \begin{pmatrix} 1 & ac^{-1} \\ 0 & 1 \end{pmatrix} \begin{pmatrix} \tilde{c}^{-1} & 0 \\ 0 & c \end{pmatrix} \begin{pmatrix} 0 & -1 \\ 1 & 1 \end{pmatrix} \begin{pmatrix} 1 & c^{-1}d \\ 0 & 1 \end{pmatrix},$$

where the second factor is a combination of a rotation and a dilatation.

Remark 3.101. The factorization properties of VAHLEN matrices are detailed studied in [185] and J. RYAN's book [259] which is in preparation.

Proposition 3.102. *The inverse mapping of the* MÖBIUS *transformation is given by*

$$m^{-1} : \mathbb{R}^n \ni y \to x \in \mathbb{R}^n \quad \textit{with} \quad m^{-1} = (-\tilde{d}y + \tilde{b})(\tilde{c}y - \tilde{a})^{-1}.$$

Proof. Acting from the right by $cx + d$ then we obtain $y(cx + d) = ax + b$. As x, y are vectors in $\mathbb{R}^n$ the reversion provides $\tilde{x} = x$ and $\tilde{y} = y$. As such we get

$$(c\tilde{x} + \tilde{d})y = x\tilde{a} + \tilde{b}$$

and

$$x\tilde{c}y - x\tilde{a} = x(\tilde{c}y - \tilde{a}) = \tilde{b} - \tilde{d}y \quad x = (\tilde{b} - \tilde{d}y)(\tilde{c}y - \tilde{a})^{-1}. \qquad \#$$

Let us now have a closer look at connections between the DIRAC operator, CLIFFORD regular functions and MÖBIUS transformations. We write in abbreviated form: $m(x) := (ax + b)(cx + d)^{-1}$.

Lemma 3.103. [253] *Let u be* CLIFFORD *regular with respect to the variable $m(x)$. Then the function*

$$v(x) = \frac{\widetilde{(cx + d)}}{|cx + d|^n} u(m(x))$$

is also CLIFFORD *regular.*

Definition 3.104. [253] *A function $J_k(cx + d)$ is called a conformal weight of k-th order if the function*

$$v(x) = J_k(cx + d)u(m(x))$$

is annihilated by the k-th iterate D^k of the DIRAC *operator.*

Example 3.105. (J. RYAN) Let $k = 1$ then

$$J_1(cx+d) = \frac{\widetilde{cx+d}}{|cx+d|^n}.$$

Lemma 3.106. [31] *For each k there is a conformal weight function J_k such that*

$$J_k(cx+d)D^k u(m(x)) = D^k J_k(cx+d)u(m(x)).$$

Remark 3.107. J. PEETRE and T. QIAN gave in [224] a geometric proof of this result of B. BOJARSKI.

Remark 3.108. J. RYAN succeeded in applying MÖBIUS transformations to transfer the treatment of boundary value problems over bounded domains (cf. [110]) to classes of unbounded manifolds.

Remark 3.109. MÖBIUS transformations can also be considered in more general CLIFFORD algebras $Cl_{p,q}$. For this reason we refer for instance to [180], [253] and [8]. A discussion on fixed points and corresponding classifications in elliptic, loxodromic, and parabolic cases the reader can find in [5]. We also mention here the very extensive investigations on this topic by G. ZÖLL (cf. [340]).

3.9 Fourier transform and related transforms

This part is dedicated to the consideration of integral transforms which are acting in higher dimensions. Doubtless the FOURIER transform will be the most famous among all the others. Already the mathematicians of the 19*th* century need such a kind of transform in order to solve integral equations which appear in physics. Probably J. FOURIER's inversion of the integral

$$f(\xi) = \frac{1}{\sqrt{2\pi}} \int_{-\infty}^{\infty} e^{i\xi x} \hat{f}(x)dx$$

in 1811 was the starting-point of a rapid development of theories of integral transforms. Enforced investigations into integral equations formed these theories into powerful instruments of the analytical calculus in the EUCLIDean spaces, the so-called transform analysis. We regret having only a few pages on this topic. Nevertheless, we hope by giving important definitions at least to convey some ideas in this field. The relevant literature is quoted.

Let us start with considerations of the FOURIER transform. We will use the abbreviation $||\cdot||_p := ||\cdot||_{L_p}$. The equivalence of two norms $||\cdot||$ and $||\cdot||'$ means that there exist positive constants c, C with

$$c||\cdot||' \leq ||\cdot|| \leq C||\cdot||'.$$

Definition 3.110. *Let $u \in L_p(\mathbb{R}^n)$ be a function defined with values in $Cl_{0,n} \otimes \mathbb{C}$. Then the* FOURIER *transform is defined by*

$$\hat{u}(\xi) = (\mathcal{F}u)(\xi) := \int_{\mathbf{R}^n} e^{-i(\xi,x)} u(x)dx \quad (\xi \in \mathbb{R}^n),$$

where "$(\cdot,\cdot)$" denotes the scalar product in $\mathbb{R}^n$.

Example 3.111. Let p_{n+1} be the function

$$p_{n+1}(x) = \frac{2}{\sigma_{n+1}} \frac{x_{n+1}}{\left(|x|^2 + x_{n+1}^2\right)^{\frac{n+1}{2}}}$$

with

$$\frac{2}{\sigma_{n+1}} = \frac{\Gamma(\frac{n+1}{2})}{\pi^{\frac{n+1}{2}}}.$$

Then one gets immediately

$$\hat{p}_{n+1}(\xi) = c_n e^{-x_{n+1}|\xi|}.$$

The function p_{n+1} is called POISSON *kernel* in $\mathbb{R}^n$.

Lemma 3.112. *$\mathcal{F}$ provides an isometry of the space $L_2(\mathbb{R}^n, Cl_{0,n} \otimes \mathbb{C})$ on itself.*

Proof. It is well known that for $u \in L_2(\mathbb{R}^n) \cap L_1(\mathbb{R}^n)$ the function $\hat{u} \in L_2(\mathbb{R}^n)$ and there exist constants c and C with $c\|u\|_2 \leq \|\hat{u}\|_2 \leq C\|u\|_2$. Hence u can be extended to an isomorphism from $L_2(\mathbb{R}^n)$ to itself. Using the representation

$$u(x) = \sum_{\alpha} u_\alpha(x) e_\alpha$$

with $u_\alpha \in L_2(\mathbb{R}^n, \mathbb{R}^1)$ one gets by PARSEVAL's equation for these scalar functions that $\|u\|_2 = \|\hat{u}\|_2$. #

A further important transform is the so-called RIESZ transform.

Definition 3.113. *Let $u \in L_p(\mathbb{R}^n)$. Then the relation*

$$\widehat{R_k u}(\xi) := i\frac{\xi_k}{|\xi|}\hat{u}(\xi) \quad (k = 1, ..., n) \quad (\xi \in \mathbb{R}^n \setminus \{0\})$$

defines the k-th RIESZ *transform R_k.*

Lemma 3.114. (cf. [88]) *R_k is a linear operator in $L_p(\mathbb{R}^n)$ and permits the representation*

$$(R_k u)(x) = \frac{2}{\sigma_{n+1}} \int_{R^n} \frac{x_k - y_k}{|x-y|^{n+1}} u(y) dy \quad (k = 1, ..., n).$$

Moreover, the operator R_k is bounded in $L_p(\mathbb{R}^n)$ $(1 < p < \infty)$.

Example 3.115. The following identity is valid:

$$(R_k p_{n+1})(x) = \frac{2}{\sigma_{n+1}} \frac{x_k}{[|x|^2 + x_{n+1}^2]^{\frac{n+1}{2}}} \quad (k = 1, ..., n).$$

Note that the images of the POISSON kernel under the transforms R_k are called *conjugated* POISSON *kernels*, which are often denoted by $q_{n+1}^{(k)}(x)$.

Definition 3.116. *Let*

$$u * v := \int_{R^n} u(x-y) v(y) dy$$

be the convolution integral. Then

$$(P_{n+1} u)(x) := (p_{n+1} * u)(x)$$

is called POISSON *transform.*

Corollary 3.117. POISSSON *transform defines a harmonic extension onto* $\mathbb{R}^{n+1}_+$ *by setting*

$$u(z) := (P_{n+1}u)(x) \quad \textit{with} \quad z := (x, x_{n+1})$$

and a function $u \in L_p(\mathbb{R}^n)$ *is (left)* CLIFFORD *regular if and only if*

$$\Big(\sum_{k=1}^{n} i\xi_k e_k - |\xi|\Big)\hat{u}(\xi) = 0.$$

Definition 3.118. *Let* $u \in L_p(\mathbb{R}^n)$. *Then the linear operator*

$$H = \sum_{k=1}^{n} R_k e_k$$

is called HILBERT *transform. Furthermore, the linear operator*

$$(Cu)(z) := \frac{1}{\omega_{n+1}} \int\limits_{\mathbf{R}^n} \frac{y-z}{|y-z|^{n+1}} e_{n+1} u(y) dy \quad (x \notin \mathbb{R}^n)$$

is called n-dimensional CAUCHY *transform.*

These operators were also used to define HARDY spaces. Z. WU [331] proposed $H^p(\mathbb{R}^n)$ to be the space defined by

$$\{u \in L_p(\mathbb{R}^n) : Hu \in L_p(\mathbb{R}^n)\}.$$

A survey about several approaches to HARDY spaces can be found in [203].

Corollary 3.119. CAUCHY*'s transform permits the representation*

$$(Cu)(z) = \frac{1}{2}[p_{n+1} * (I + He_{n+1})u](x) \quad (x \notin \mathbb{R}^n).$$

For $x \to x' \in \mathbb{R}^n$ *we have the formula*

$$\lim_{\substack{x \to x' \\ x \in \Gamma_\alpha(x')}} (Cu)(x) = \frac{1}{2}[(I + He_{n+1})u](x'),$$

where $\Gamma_\alpha(x') = \{z \in \mathbb{R}^{n+1}_+ : |x - x'| < \alpha x_{n+1}\}$.

It is well known in the complex case that FOURIER transform permits a holomorphic extension into a half space. The question is now to find a kernel that the image of the corresponding multidimensional hypercomplex FOURIER transform permits a CLIFFORD regular extension onto the half space $\mathbb{R}^{n+1}_+$. For this reason let $\mathcal{S}(\mathbb{R}^n)$ be the ordinary SCHWARTZ space and $\mathcal{S}(\mathbb{R}^n, \mathcal{A})$ the corresponding SCHWARTZ space which contains test functions with values in the algebra $\mathcal{A}$.

Definition 3.120. *Let*

$$e(\xi, x) := e^{i(\underline{\xi}, x)}[e_0 \cosh\, \xi_0|x| - i\omega(x) \sinh\, \xi_0|x|],$$

where

$$\xi = (\xi_0, \xi_1, ..., \xi_n) \in \mathbb{R}^{n+1}, \quad \underline{\xi} = (\xi_1, ..., \xi_n),\ x = (x_1, ..., x_n) \in \mathbb{R}^n.$$

Further, let $u \in \mathcal{S}(\mathbb{R}^n, Cl_{0,n} \otimes \mathbb{C})$ be a suitable test function, which is holomorphic extendable to $\mathbb{C}^n$. Then the integral transform

$$(\mathbf{F}u)(\xi) := \int_{\mathbb{R}^n} e(\xi, -x)u(x)dx \tag{3.16}$$

is called hypercomplex FOURIER *transform.*

Remark 3.121. F. SOMMEN in [283] was the first to introduce this hypercomplex version of a FOURIER transform. Independent of these ideas A. MCINTOSH found another in some sense more suitable hypercomplex FOURIER transform (cf. [171]).

Now, we extend the kernel function $e(\xi, x)$ to $\mathbb{R}^{n+1} \times \mathbb{C}^n$, that means we have $x = (x_1, ..., x_n) \in \mathbb{C}^n$, $\omega(x) = x/|x|_*$, $|x|_*^2 = |x'|^2 - |x''|^2 - 2i(x' \cdot x'')$ with $x' = (x'_1, ..., x'_n)$, $x'' = (x''_1, ..., x''_n) \in \mathbb{R}^n$ and $x_k = x'_k + ix''_k \quad (k = 1, ..., n)$. On the basis of PAINLEV's theorem (cf. [36]) one gets:

Theorem 3.122. [129] *The kernel function $e(\xi, x)$ is the unique function which fulfils the following properties:*

(i) $e^*(\xi, x) : \mathbb{R}^{n+1} \times \mathbb{C}^n \to Cl_{0,n} \otimes \mathbb{C}$,

(ii) $e^*(\underline{\xi}, x) = e(\underline{\xi}, x) = e^{i(\underline{\xi}, x)}$ *for* $(\xi, x) \in \mathbb{R}^{n+1} \times \mathbb{R}^n$,

(iii) $e^*(\xi, x)$ *is (left)* CLIFFORD *regular in* $\xi \in \mathbb{R}^{n+1}$ *for an arbitrary* $x \in \mathbb{C}^n$,

(iv) *Each complex-valued component of the function $e^*(\xi, x)$ is holomorphic relative to x for any fixed $\xi \in \mathbb{R}^n$.*

Incorporating the ordinary FOURIER transform we find the notion of the hypercomplex inverse FOURIER transform. It holds that:

Definition 3.123. *A measurable function* $m : \mathbb{R}^n \to Cl_{0,n} \otimes \mathbb{C}$ *is called hypercomplex inverse* FOURIER *transform of a bounded measurable function* $b : \mathbb{R}^n \to Cl_{0,n} \otimes \mathbb{C}$, *if the following relation is valid for any* $u \in \mathcal{S}(\mathbb{R}^n, Cl_{0,n} \otimes \mathbb{C})$

$$\frac{1}{(2\pi)^n} \int_{\mathbf{R}^n} b(\xi)\hat{u}(-\xi)d\xi = \int_{\mathbf{R}^n} m(x)u(x)dx \quad (\xi = \underline{\xi}!),$$

where $\hat{u}$ *is the ordinary* FOURIER *transform. This hypercomplex inverse* FOURIER *transform is denoted by* $m(x) =: (\mathbf{F}^{-1}b)(\xi)$.

Corollary 3.124. *For* $b \in L_\infty(\mathbb{R}^n)$ *and* $||b||_\infty \leq const.\, e^{-\alpha|x|}$ *the inverse hypercomplex* FOURIER *transform has a (right)* CLIFFORD *regular extension into the domain* $\{\xi = \xi_0 e_0 + \underline{\xi} : |\xi_0| < \alpha\}$. *This extension is defined by*

$$(\mathbf{F}^{-1}b)(\xi) = \frac{1}{(2\pi)^n} \int_{\mathbf{R}^n} b(x) e^{i(\underline{\xi},x)} [\cosh\,(\xi_0|x|) - i\omega(x)\sinh(\xi_0|x|)]dx$$

and satisfies the the following estimate:

$$||\mathbf{F}^{-1}b||_2 \leq \frac{c}{|\xi_0|^n} ||b||_\infty.$$

Remark 3.125. In an important paper by A. McIntosh [196] the reader can find a detailed description of classes of functions which play a role in the study of the hypercomplex Fourier transform, especially in order to obtain mapping properties. It seems to be necessary in order to discuss advantageously properties of the Fourier transform to integrate over suitable cones in $\mathbb{C}^n$ (cf. [171]).

At present J. Ryan [262] defined a Fourier transform on the unit sphere. In order to explain this construction we have to introduce the Möbius transform (more detailed in Section 3.1).

Definition 3.126. *Let* $a, b, c, d \in Cl_{0,n}$ *with* $a\tilde{d} - b\tilde{c} \neq 0$. *Then the mapping*

$$M(x) = (ax + b)(cx + d)^{-1}$$

is called MÖBIUS *transform. In the special case if the coefficients are* $a = e_1, b = 1, c = 1$ *and* $d = e_1$ *this transform is called* CAYLEY *transform.*

Lemma 3.127. *Let* u *be a (left)* CLIFFORD *regular function, then* $J(\Phi, x)u(x)$ *is also (left)* CLIFFORD *regular, where*

$$J(\Phi, x) := \frac{\widetilde{cx + d}}{||cx + d||^n}.$$

Definition 3.128. *Let be ψ the inverse of the* CAYLEY *transform. Then spherical* FOURIER *transform is defined by*

$$(\mathcal{F}_{S^n}u(\xi) = \int\limits_{S^n} E_{S^n}(\xi,x)u(x)dS_x,$$

where

$$E_{S^n}(\xi,x) := J(\psi,x)e^{-i(\psi(\xi),\psi(x))}\overline{J}(\psi,\xi).$$

Theorem 3.129. *([262]) The transform*

$$\mathcal{F}_{S^n} : L_2(S^n, Cl_{0,n}\otimes\mathbb{C}) \to L_2(S^n, Cl_{0,n}\otimes\mathbb{C})$$

is an isometry.

Remark 3.130. A detailed discussion of this spherical FOURIER transform is given in [262].

Remark 3.131. There are a large number of other transforms, which play an important role. In this way RADON transform has many applications, especially in computer tomography and geophysics. The problem is to reconstruct a function defined on a domain in $\mathbb{R}^n$ by integrals over $(n-1)$-dimensional hyperplanes. First papers in this field are written by J. RADON. There are also some very interesting monographs (cf. [212], [325], [130]). Connections with CLIFFORD analysis arise in papers by F. SOMMEN [285], [284], [287], [290] and in the thesis of J. CNOPS ([48]). Also nonlinear transforms as the Inverse Scattering Transform (IST) were developed and investigated (cf. [129]). More recently curved RADON transforms were studied (cf. [290]).

Chapter 4

Boundary value problems

4.1 First-order boundary value problems

The application of results of the classical function theory for a qualitative description of various plane physical processes which is given in [209] and [165] leads to the investigation of generalized CAUCHY–RIEMANN systems. In the papers [23], [24], [233] essential results were transferred to systems of the type

$$\begin{aligned} \partial_x u &= \sigma(x,y)\partial_y v \\ \partial_y u &= \sigma(x,y)\partial_x v \end{aligned}$$

for a suitable class of functions σ. New types of applications could be considered by I.N. VEKUA's generalized holomorphic functions (cf. Section 4.3). Consequent to the initiating papers [208] and [78], multidimensional generalizations of CAUCHY–RIEMANN systems and their boundary value problems found a growing interest. Among others, contributions by R. DELANGHE [61], B. BOJARSKI [30], E. SAAK [263], and R. HILE [131] should be mentioned. In the paper [307] from 1968 the American mathematicians E. STEIN and G. WEISS described a general construction (in some sense) of CAUCHY–RIEMANN equations by the help of an algebraical classification. Each irreducible representation of the n-dimensional group $SO(n)$ may be related to a first-order system with constant coefficients. All of these systems form the class of multidimensional CAUCHY–RIEMANN systems. The reader can find a continuation of these ideas in [88]. There are a large number of papers, above all by Russian mathematicians (cf. [12], [65], [279], [323]) but also by B. BOOSS [33] who asked for the existence of Fredholmean boundary value problems. In the dissertation paper of I. STERN [308] these kinds of questions were almost completely investigated. The remaining cases could be solved by reference to [151].
In this section we will give only some important results – most of them without proof. This is because our aim is to solve special classes of boundary value problems more or less explicitly. If we are successful in this process, results about the Fredholm property are already included. Of course, it is not possible to obtain explicit representation for general systems as now discussed.

4.1.1 On Fredholmean boundary value problems

We consider the following first-order system

$$A_0\partial_0 u + A_1\partial_1 u + \ldots + A_n\partial_n u = 0, \tag{4.1}$$

where $A_1, \ldots, A_n$ are constant, complex $m \times m$ matrices. For simplification we write $A_0 = E$ for the identity matrix. We follow initially explanations in [308].

Definition 4.1. *If each solution u of system (4.1) has harmonic components u_k $(k = 1, ..., m)$ then the system (4.1) is called generalized* CAUCHY–RIEMANN *system.*

Proposition 4.2. *Each generalized* CAUCHY–RIEMANN *system is elliptic, i.e.*

$$\det \sum_{k=0}^{n} \lambda_k A_k \neq 0 \quad \text{for any } \lambda = (\lambda_0, ..., \lambda_n) \in \mathbb{R}^{n+1} \setminus \{0\}.$$

Proof. Let $\lambda \neq 0$ and assume that there exists a vector $u \in \mathbb{C}^m$ with $\sum_{k=0}^{n} \lambda_k A_k u = 0$, then

$$v := e^{\sum_{x=0}^{n} \lambda_k x_k} u \tag{4.2}$$

is also a solution of (4.1). But v is not harmonic. #

Proposition 4.3. *System (4.1) is a generalized* CAUCHY–RIEMANN *system if and only if the coefficients $A_1, ..., A_n$ fulfil the relations*

$$A_i^2 = -E \qquad (i = 1, ..., n) \tag{4.3}$$

$$A_i A_j + A_j A_i = 0 \quad (i, j = 1, ..., n; i \neq j) \tag{4.4}$$

and the solution u belongs to $W_2^1(\mathbb{R}^{n+1})^m$.

Proof. We consider the function

$$u := 2x_i x_j b - (x_i^2 A_i^{-1} A_j b + x_j^2 A_j^{-1} A_i b) \quad (i \neq j)$$

where $b \in \mathbb{C}^m$. Clearly, u is a solution of (4.1). Furthermore, we have

$$\Delta u = -2(A_i^{-1} A_j + A_j^{-1} A_i) b.$$

Suppose now $\Delta u = 0$. Then it follows that $A_i^{-1} A_j + A_j^{-1} A_i = 0 \quad (i \neq j)$. For $i = 0$ we obtain

$$A_j^{-1} = -A_j \Leftrightarrow A_j^2 = -E \quad (j = 1, ..., n).$$

and henceforth, also

$$A_i A_j + A_j A_i = 0 \quad (i \neq j).$$

On the other hand, we get under the assumption of the relations (4.3) and (4.4)

$$(E\partial_0 - \sum_{i=1}^{n} A_i \partial_i)(E\partial_0 + \sum_{j=1}^{n} A_j \partial_j) = \Delta$$

and this is our statement for any $u \in C^2(\mathbb{R}^{n+1}, \mathbb{R}^m)$. By the help of GAUSS' theorem and WEYL's lemma (cf. [309]) one gets the result for $u \in W_1^2(\mathbb{R}^{n+1})^m$.
#

Theorem 4.4. *Let C be an orthogonal matrix in $\mathbb{R}^{n+1}$. Then for each solution $u(x)$ of (4.1) also $Mu(Cx)$ solves (4.1) with a non-singular matrix M in $\mathbb{C}^m$.*

Remark 4.5. In the case of quaternions one can find this result in [313]. For quadratic coefficient matrices the rotation invariance is proved in [307].

Now we will formulate the following boundary value problem:

Let $G \subset \mathbb{R}^{n+1}$ be a bounded domain with a piecewise smooth LIAPUNOV *boundary Γ. The number m is assumed to be even. For any system $\{A_i : i = 1, ..., n\}$ of nonsingular $m \times m$-matrices which fulfil the properties (4.3) and (4.4) we consider the problem*

$$E\partial_0 u + A_1\partial_1 u + ... + A_n\partial_n u = f \quad \text{in } G \subset \mathbb{R}^{n+1} \tag{4.5}$$

$$(B_1 B_2)u = g \qquad \text{on } \Gamma \tag{4.6}$$

where f, g are $\mathbb{C}^{m/2}$ valued functions and B_1, B_2 are complex $m/2 \times m/2$-matrix functions.

Remark 4.6. The boundary condition (4.6) permits the following representation

$$\operatorname{Re} B\, u = h \qquad \text{on } \Gamma \tag{4.7}$$

where $h = (\operatorname{Re} g, \operatorname{Im} g)^T$ and

$$B = \begin{pmatrix} B_1 & B_2 \\ -iB_1 & -iB_2 \end{pmatrix}.$$

On the other hand, if a boundary condition of type (4.7) is given then we immediately have

$$u = \begin{pmatrix} \operatorname{Re} u \\ \operatorname{Im} u \end{pmatrix}, B_1 = \operatorname{Re} B \quad \text{and} \quad B_2 = \operatorname{Im} B \text{ and } g = h.$$

Additionally, suppose that $A_i^* = -A_i$, B has orthonormal lines and its entries belong to $C^k(\Gamma)$.

Theorem 4.7. *Let $f \in W_2^{k-1}(G)$, $g \in W_2^{k-1/2}(\Gamma)$ $(k \in \mathbb{N})$. Under the above conditions the boundary value problem (4.5) and (4.6) is Fredholmean if and only if for all $x \in \Gamma$*

$$det\Big((B_1 B_2)(\sum_{i=0}^{n} \alpha_i A_i^*)(\sum_{i=0}^{n} \mu_j A_j)(B_1^* B_2^*)^T\Big) \neq 0, \tag{4.8}$$

where $\alpha = (\alpha_0, ..., \alpha_n)$ is the unit vector of the inner normal at $x \in \Gamma$ and $\mu = (\mu_0, ..., \mu_n) \in S^{(n+1)}(x)$.

Remark 4.8. Condition (4.8) is equivalent to the well-known SHAPIRO – LOPATINSKIJ condition (cf. [1]).

Example 4.9. [309] We consider the boundary value problem

$$\begin{matrix} \operatorname{div} \underline{u} = 0 \\ \operatorname{rot} \underline{u} = 0 \end{matrix} \quad \text{in } G \tag{4.9}$$

$$b_1 u_1 + b_2 u_2 + b_3 u_3 = g \quad \text{on } \Gamma = \partial G \tag{4.10}$$

where b_i $(i = 1, 2, 3)$ are real functions and $|\underline{b}| = 1$ with $\underline{b} = (b_1, b_2, b_3)$.

Using corresponding matrix notations (cf. Section 1.1) we get the equivalent problem

$$E\,\partial_1 u + \begin{pmatrix} 0 & 1 & 0 & 0 \\ -1 & 0 & 0 & 0 \\ 0 & 0 & 0 & 1 \\ 0 & 0 & -1 & 0 \end{pmatrix} \partial_2 u + \begin{pmatrix} 0 & 0 & 1 & 0 \\ 0 & 0 & 0 & -1 \\ -1 & 0 & 0 & 0 \\ 0 & 1 & 0 & 0 \end{pmatrix} \partial_3 u = 0 \quad \text{in } G$$

and

$$\begin{pmatrix} b_1 & b_2 & b_3 & 0 \\ 0 & 0 & 0 & 1 \end{pmatrix} u = \begin{pmatrix} g \\ 0 \end{pmatrix}, \qquad |\underline{b}| = 1 \qquad \text{on } \Gamma,$$

where $u = (\underline{u}, u_0)$ and $\underline{b} = (b_1, b_2, b_3)$. In this special case the condition (4.8) reads as follows:

$$\det\left[\begin{pmatrix} b_1 & b_2 & b_3 & 0 \\ 0 & 0 & 0 & 1 \end{pmatrix}\begin{pmatrix} \alpha_1 & -\alpha_2 & -\alpha_3 & 0 \\ \alpha_2 & \alpha_1 & 0 & \alpha_3 \\ \alpha_3 & 0 & \alpha_1 & -\alpha_2 \\ 0 & -\alpha_3 & \alpha_2 & \alpha_1 \end{pmatrix}\begin{pmatrix} \mu_1 & \mu_2 & \mu_3 & 0 \\ -\mu_2 & \mu_1 & 0 & -\mu_3 \\ -\mu_3 & 0 & \mu_1 & \mu_2 \\ 0 & \mu_3 & -\mu_2 & \mu_1 \end{pmatrix} \times\right.$$

$$\left.\times \begin{pmatrix} b_1 & 0 \\ b_2 & 0 \\ b_3 & 0 \\ 0 & 1 \end{pmatrix} - iE\right]$$

$$= \det\left[\begin{pmatrix} (\underline{b}, \underline{\alpha})(\underline{b}, \underline{\mu}) + (\underline{b} \times \underline{\alpha}, \underline{b} \times \mu) & (\underline{\mu}, \underline{b} \times \underline{\alpha}) \\ (\underline{\alpha}, \underline{b} \times \underline{\mu}) & (\underline{\alpha}, \underline{\mu}) \end{pmatrix} - iE\right]$$

$$= \det\begin{pmatrix} -i & -(\underline{b}, \underline{\mu} \times \underline{\alpha}) \\ (\underline{b}, \underline{\mu} \times \underline{\alpha}) & -i \end{pmatrix} = (\underline{b}, \underline{\mu}, \underline{\alpha})^2 - 1 \neq 0.$$

Here is $\underline{\alpha} = (\alpha_1, \alpha_2, \alpha_3)$ and $\underline{\mu} = (\mu_1, \mu_2, \mu_3)$. This condition means that $\underline{b}$ is not allowed to be tangential. We obtained that the problem (4.9) and(4.10) is Fredholmean if and only if the vector $\underline{b}$ is not tangential on Γ.

Remark 4.10. Methods developed in [308] provide approvable conditions for investigating the validity of SHAPIRO – LOPATINSKIJ conditions.

4.1.2 On a special Div–Rot–Grad system.

Now, we generalize our considerations from Example 4.9. to a wider class of boundary value problems.
Let $G \subset \mathbb{R}^3$ be a bounded domain with a piecewise LIAPUNOV surface Γ. We will investigate the following problem:

$$a \operatorname{div} \underline{u} + b \operatorname{rot} \underline{u} + c \operatorname{grad} u_0 + \underline{d} u_0 + e\underline{u} = f \quad \text{in } G. \tag{4.11}$$

We look for such functions as u which fulfil on the boundary Γ an extension condition. The coefficients $a, b(b(x) \neq 0, x \in G), c, e$ are real functions and $\underline{d}$ a vector-valued function. Let $J : u \to \overline{u}$ be the operator of conjugation. Obviously we have $J^2 = I$. Then the operators $\mathcal{P} := 1/2(J+I)$ and $\mathcal{Q} := 1/2(I-J)$ are projections onto $\mathbb{R} \subset \mathbb{H}$ and $\operatorname{Vec} \mathbb{H} \subset \mathbb{H}$, respectively. Consequently, the following representations are valid:

$$\mathcal{P}Du = -\operatorname{div} \underline{u}\ , \quad \mathcal{Q}D\mathcal{P}u = \operatorname{grad} u_0, \quad \mathcal{Q}D\mathcal{Q}u = \operatorname{rot} \underline{u}. \tag{4.12}$$

For any real function c we have $\mathcal{Q}c\mathcal{P} = 0, \mathcal{P}c\mathcal{P} = c\mathcal{P}, \mathcal{Q}c\mathcal{Q} = c\mathcal{Q}$. Setting now:

$$Mu := \frac{c}{b} u_0 + \underline{u}.$$

Theorem 4.11. *Let $a(x), b(x), c(x) \neq 0$ in G, $e = 0$ and $\underline{d} = b \operatorname{grad}(c/b)$. There exists a unique solution u^* of equation (4.11) such that Mu^* permits a $\mathbb{H}$-regular extension into the exterior domain $\mathbb{R}^3 \setminus \overline{G}$ which vanishes at infinity. This solution is given by*

$$u^* = M^{-1}T(-a^{-1}\mathcal{P} + b^{-1}\mathcal{Q})f.$$

Proof. Using the notations (4.12) equation (4.11) obtains the form

$$-a\mathcal{P}DMu + b\mathcal{Q}DMu = f.$$

Using $DTu = u$ in G we get

$$(-a\mathcal{P}DM + b\mathcal{Q}DM)M^{-1}T(-a^{-1}\mathcal{P} + b^{-1}\mathcal{Q})f = \mathcal{P}f + \mathcal{Q}f = f,$$

such that u^* is a solution of (4.11). It remains to describe the kernel of the operator

$$A = ((-a)\mathcal{P} + b\mathcal{Q})DM.$$

Because the operator in the brackets has only a trivial kernel we have:

$$\ker A = \{\phi : M^{-1}\phi \text{ is } \mathbb{H} - \text{regular}\}.$$

CAUCHY's formula delivers the assumption. #

Remark 4.12. The general case has to be dealt with by the methods of Section 4.2.

4.1.3 On boundary value problems for the Dirac equation

The simplest boundary value problem is of the form

$$Du = f \quad \text{in } G \tag{4.13}$$

$$u = g \quad \text{on } \Gamma = \partial G. \tag{4.14}$$

We already know that this problem cannot be solved without restrictions. One obtains immediately by BOREL–POMPEIU's formula

$u = \phi + Tf,\ \ \phi \in \ker D.$

Clearly, $\phi = F_\Gamma g$. PLEMELJ–SOKHOTZKIJ's formulae deliver now the necessary condition

$$Q_\Gamma g = \mathrm{tr}_\Gamma Tf \tag{4.15}$$

Proposition 4.13. *Let $f \in L_2(G, Cl_{0,n}), g \in W_2^{1/2}(\Gamma, Cl_{0,n})$. Then the problem (4.13,4.14) is solvable in $W_2^1(G, Cl_{0,n})$ if and only if*

$Q_\Gamma g = \mathrm{tr}_\Gamma Tf$.

If a solution u exists, then it can be represented by

$$u = F_\Gamma g + Tf \in W_2^1(G, Cl_{0,n}). \tag{4.16}$$

Proposition 4.14. *For any given right-hand side $f \in L_2(G, Cl_{0,n})$ there always exists a function $g \in W_2^{1/2}(\Gamma, Cl_{0,n})$, such that condition (4.15) is satisfied.*

Proof. $\mathrm{tr}_\Gamma Tf$ permits a $Cl_{0,n}$-regular extension into the domain $\mathbb{R}^n \setminus \overline{G}$. Moreover,

$$\lim_{|x|\to\infty} (Tf)(x) = 0.$$

In this way $\mathrm{tr}_\Gamma Tf \in \mathrm{im} Q_\Gamma$ and then exists a function g with $Q_\Gamma g = \mathrm{tr}_\Gamma Tf$. #

Example 4.15. Let $g \in \mathrm{im} P_\Gamma$,i.e., g is $Cl_{0,n}$-extendable into the domain G and $f \in \mathrm{im}\mathbf{Q}^\mathbf{D}$ (cf. Section 3.6). Then the condition (4.15) is fulfilled and problem (4.13,4.14) has the unique solution (4.16).

Remark 4.16. The equations $\mathrm{rot}\, u = j \ \ \mathrm{div}\, u = \lambda$ are sometimes called *prae-*MAXWELL *equations*, which play a role in the magnetohydrodynamics and in the computation of electric generators (cf. [190]). For the use of collocation techniques we refer to Section 5.8 and also to [124]. An interesting approach to the MOISIL–TEODORESCU system by complex techniques one can find in [73]. For further results we refer to the papers [16], [310], [220].

4.2 Dirichlet problems and orthoprojections

The main topic of this section is to deduce representation formulas for the solutions of certain second-order boundary value problems of partial differential equations under DIRICHLET conditions. The advantage of our represention is just the separation of the influences on the right-hand side and the boundary data to the solution. Furthermore, by the help of a unified operator calculus answers can be obtained to the questions of existence, uniqueness, and regularity as well as also representation formulas. Besides, we prove explicit representations for the orthoprojections introduced in Section 3.5.

Let G be a bounded domain whose boundary is a compact piecewise smooth LIAPUNOV surface. Furthermore, we define for in G defined $C\ell_{0,n}$-valued functions the space

$$W_2^{k,\text{loc}}(G, C\ell_{0,n}) := \{f : f \in W_2^k(G_1)\ \forall G_1 : \overline{G}_1 \subset G\}.$$

Lemma 4.17. *Let $f \in W_2^k(G, Cl_{0,n})$. Then the boundary value problem*

$$\begin{aligned} D\,Du &= f && \text{in } G \\ u &= 0 && \text{on } \Gamma \end{aligned}$$

has a solution

$$u = T_G \mathbf{Q} T_G f \in W_2^{k+2,\text{loc}}(G, Cl_{0,n}).$$

Proof. Using the mapping properties of T_G we have $T_G f \in W_2^{k+1}(G)$. Because $\mathbf{Q} = \mathbf{I} - \mathbf{P}$ and $\operatorname{im}\mathbf{P} = \ker D \subset C^\infty(G)$ we obtain $\mathbf{Q}T_G f \in W_2^{k+1,\text{loc}}(G)$. Looking again to the mapping properties of T_G we now obtain $u \in W_2^{k+2,\text{loc}}(G)$ and also $\operatorname{tr}_\Gamma T_G \mathbf{Q} T_G f = 0$. Using now the properties $DT_G = \mathbf{I}$ and $D\mathbf{Q} = D$ we see by inserting u into the differential equation that indeed $u = T_G \mathbf{Q} T_G f$ solves the boundary value problem. #

Lemma 4.18. *Let $g \in W_2^{k+3/2}(G, Cl_{0,n})$ $(k \geq 0)$. The boundary value problem*

$$\begin{aligned} D\,Du &= 0 && \text{in } G \\ u &= g && \text{on } \Gamma \end{aligned}$$

permits a solution $u \in W_2^{k+2,\text{loc}}(G, Cl_{0,n})$ by

$$u = F_\Gamma g + T_G \mathbf{P} Dh,$$

where h is an arbitrary $W_2^{k+2}(G, Cl_{0,n})$-extension of g.

Proof. The function g allows a $W_2^{k+2}(G, Cl_{0,n})$-extension h (h is not necessarily unique!) with $\mathrm{tr}_\Gamma h = g$. Put now $u = v + h$ we obtain

$$DDv = -DDh \quad \text{in } G, \quad v = 0 \quad \text{on } \Gamma.$$

Applying now Lemma 4.17. we have the representation

$$v = -T_G \mathbf{Q} T_G D\, Dh.$$

BOREL–POMPEIU's formula implies

$$v = -T_G \mathbf{Q} Dh + T_G \mathbf{Q} F_\Gamma Dh = -T_G \mathbf{Q} Dh = -h + F_\Gamma g + T_G \mathbf{P} Dh$$

and so

$$u = v + h = F_\Gamma g + T_G \mathbf{P} Dh. \qquad \#$$

Theorem 4.19. *Let $f \in W_2^k(G, Cl_{0,n}), g \in W_2^{k+3/2}(\Gamma, Cl_{0,n})$. The boundary value problem*

$$D\,Du = f \qquad \text{in } G \tag{4.17}$$

$$u = g \qquad \text{on } \Gamma \tag{4.18}$$

has the unique solution

$$u = F_\Gamma g + T_G \mathbf{P} Dh + T_G \mathbf{Q} T_G f \in W_2^{k+2,\mathrm{loc}}(G, Cl_{0,n}) \tag{4.19}$$

where h denotes a $W_2^{k+2}(G, Cl_{0,n})$-extension of g.

Proof. We have only to add the solutions described in Lemma 4.17. and Lemma 4.18. For the uniqueness we have to show that the problem

$$\begin{aligned} D\,Du &= 0 \qquad \text{in } G \\ u &= 0 \qquad \text{on } \Gamma \end{aligned}$$

has the only solution $u = 0$. From the boundary condition we obtain $Du \in \operatorname{im} \mathbf{Q}$. On the other hand, $DDu = 0$ implies that $Du \in \ker D = \operatorname{im} \mathbf{P}$. Consequently,

$$Du \in \operatorname{im} \mathbf{P} \cap \operatorname{im} \mathbf{Q} = \{0\} \ ,$$

i.e., $Du = 0$ and $\mathrm{tr}_\Gamma u = 0$. By the help of BOREL–POMPEIU's formula we immediately obtain $u = T_G Du = 0$. $\qquad \#$

Lemma 4.20. *Let $1 < p < \infty$; $k > 0 \, (k \in \mathbb{N})$. Then the* CAUCHY – BITSADZE *operator F_Γ is a continuous map from $W_p^{k-1/p}(G, Cl_{0,n})$ to $W_p^k(G, Cl_{0,n}) \cap (\ker D)(G, Cl_{0,n})$.*

Proof. Let $u \in W_p^{k-1/p}(G, Cl_{0,n})$. Then there exists a $Cl_{0,n}$-valued extension $v \in W_p^k(G, Cl_{0,n})$ with $\mathrm{tr}_\Gamma v = u$. Using BOREL-POMPEIU's formula (cf. Remark 3.26.) we get

$$F_\Gamma u = v - T_G D v.$$

From Theorem 3.14. and the following remark we see that $v - T_G D v$ belongs to $W_p^k(G, Cl_{0,n})$ and $I - T_G D$ is continuous within $W_p^k(G, Cl_{0,n})$. #

Remark 4.21. BOREL–POMPEIU's formula also holds for $u \in W_p^1(G, Cl_{0,n})$. The action of DIRAC's operator has to be understood in the sense of SOBOLEV. Furthermore, by using FUBINI's theorem we have for any test function $\varphi \in C_0^1(G, Cl_{0,n})$

$$\int_G \overline{(T_G u)}(x) D\varphi(x) dx = \int_G \left(\int_G \overline{e(y-x) u(y)} dy \right) D\varphi(x) dx =$$

$$= \int_G \overline{u}(y) \left(\int_G \overline{e(y-x)} D\varphi(x) dx \right) dy = -\int_G \overline{u}(y)(T_G D\varphi)(y) dy$$

$$= -\int_G \overline{u}(y)\varphi(y) dy,$$

and so for any $u \in L_p(G, Cl_{0,n})$ the equation $DT_G u = u$ holds.

Theorem 4.22. *Let $k > 0, k \in N$. Then the operator*

$$\mathrm{tr}_\Gamma T_G F_\Gamma : W_p^{k-\frac{1}{p}}(\Gamma, Cl_{0,n}) \cap \mathrm{im}\, P_\Gamma \to W_p^{k+1-\frac{1}{p}}(\Gamma, Cl_{0,n}) \cap \mathrm{im}\, Q_\Gamma$$

is an isomorphism.

Proof. First, we have by the help of Lemma 4.20., Theorem 3.14. and the well-known theorems on traces that

$$(\mathrm{tr}_\Gamma T_G F_\Gamma)(W_p^{k-\frac{1}{p}}(\Gamma, Cl_{0,n})) \subset W_p^{k+1-\frac{1}{p}}(\Gamma, Cl_{0,n}).$$

Now, we show that $\ker(\mathrm{tr}_\Gamma T_G F_\Gamma) = \{0\}$. For this reason let v be a function belonging to $W_p^{k-\frac{1}{p}}(\Gamma, Cl_{0,n}) \cap \mathrm{im} P_\Gamma$ and $\mathrm{tr}_\Gamma T_G F_\Gamma v = 0$. On account of $DDT_G F_\Gamma v = 0$ and Theorem 4.19. we obtain $T_G F_\Gamma v = 0$ in G. From this we obtain $F_\Gamma v = 0$ and for $v \in \mathrm{im} P_\Gamma$ also $v = 0$.
Now, let $w \in \mathrm{im}\, Q_\Gamma$, i.e., $w = Q_\Gamma v$. From Theorem 4.19. we know that the boundary value problem

$$DDu = 0 \quad \text{in } G, u = w \quad \text{on } \Gamma$$

has the solution

$$u = F_\Gamma w + T_G \mathbf{P} D h$$

with a $W_2^{k+2}(G)$-extension of the boundary values w. Using now that, of course, $F_\Gamma w = 0$ for $w \in \operatorname{im} Q_\Gamma$ and $T_G \mathbf{P} Dh = T_G F_\Gamma \mathbf{P} Dh$, we have $w \in \operatorname{im} \operatorname{tr}_\Gamma T_G F_\Gamma$. This proves the theorem. #

Corollary 4.23. *The orthoprojections* $\mathbf{P}$ *and* $\mathbf{Q}$ *have the representations*

$$\mathbf{P} = F_\Gamma(\operatorname{tr}_\Gamma T_G F_\Gamma)^{-1}\operatorname{tr}_\Gamma T_G\,, \quad \mathbf{Q} = I - \mathbf{P}\,.$$

For $u \in W_2^k(G, Cl_{0,n})$, $k > 0$ *we get* $\mathbf{P}u, \mathbf{Q}u \in W_2^k(G, Cl_{0,n})$.

Proof. We have to prove that the operators expressed in the corollary coincide with the orthoprojections from Section 3.6. Clearly, $\mathbf{P}^2 = \mathbf{P}$ and $\mathbf{Q}^2 = \mathbf{Q}$. It is sufficient to show that $\operatorname{im} \mathbf{P} = \ker D(G) \cap L_2(G)$ and $\ker \mathbf{P} = D(\overset{\circ}{W_2^1}(G))$. Then we can conclude that $\mathbf{P}$ is the orthoprojection onto $\ker D \cap L_2(G)$. Obviously, $\operatorname{im} \mathbf{P} \subset \ker D(G) \cap L_2(G)$. On the other hand, taking a function f from $\ker D \cap W_2^1(G)$ we get

$$\mathbf{P}f = F_\Gamma(\operatorname{tr}_\Gamma T_G F_\Gamma)^{-1}\operatorname{tr}_\Gamma T_G f = F_\Gamma(\operatorname{tr}_\Gamma T_G F_\Gamma)^{-1}\operatorname{tr}_\Gamma T_G F_\Gamma f = F_\Gamma f = f.$$

From the density of $W_2^1(G) \cap \ker D$ in $L_2(G) \cap \ker D$ and the continuity of $F_\Gamma(\operatorname{tr}_\Gamma T_G F_\Gamma)^{-1}\operatorname{tr}_\Gamma T_G$ in L_2 we obtain the first of the properties we look for.
If we now have a function $g \in D(\overset{\circ}{W_2^1}(G))$ then we conclude as follows:

$$\begin{aligned} g = Dv,\ v \in \overset{\circ}{W_2^1}(G) \quad &\Longrightarrow \quad \mathbf{P}g = \mathbf{P}Dv = F_\Gamma(\operatorname{tr}_\Gamma T_G F_\Gamma)^{-1}\operatorname{tr}_\Gamma T_G Dv \\ &\Longrightarrow \quad F_\Gamma(\operatorname{tr}_\Gamma T_G F_\Gamma)^{-1}\operatorname{tr}_\Gamma v = 0 \end{aligned}$$

Here, we used $v = T_G Dv \quad \forall\ v \in \overset{\circ}{W_2^1}(G)$.
Finally, from $\mathbf{P}f = 0$ it immediately follows that $\operatorname{tr} T_G f = 0$ and we already know that this is equivalent to $f \in D(\overset{\circ}{W_2^1}(G))$.
The statement about the mapping properties within SOBOLEV spaces directly follows from the mapping properties of F_Γ, T_G, tr, and $(\operatorname{tr}_\Gamma T_G F_\Gamma)$. #

These representations and its properties now allow the improvement of the local regularity result from Theorem 4.19. to a global result.

Corollary 4.24. *The solution (4.19) of the boundary value problem (4.17, 4.18) belongs to* $W_2^{k+2}(G, Cl_{0,n})$.

The ideas demonstrated above to connect the investigation of boundary value problems with some suitable orthoprojections is not restricted to the case considered here. If we use another inner product, e.g., the weighted inner product $[\cdot,\cdot]_B$ from Theorem 3.79., the whole program of this section can be repeated.

Remark 4.25. The following special cases of equations should be mentioned:

We consider the boundary value problem

$$\begin{aligned} DBDH &= f \quad \text{in} \quad G \\ H &= g \quad \text{on} \quad \Gamma, \end{aligned}$$

where the multiplication operator B is defined as in Theorem 3.79.

(i) We assume that $b_i(x) = \gamma(x) \quad (i = 0, 1, 2, 3)$. If $H = \underline{H}$ is the magnetic field, $\gamma(x) = \varepsilon/(\mu\kappa(x))$, where ε is the dielectric constant, κ the electric conductivity and μ the permeability then this boundary value problem corresponds with the first boundary value problem for the time-harmonic MAXWELL equations

$$\begin{aligned} \operatorname{div} \varepsilon\underline{E}(x) &= \rho(x) & \operatorname{div} \mu\underline{H}(x) &= 0 \\ \operatorname{rot} \underline{E}(x) &= 0 & \operatorname{rot} \underline{H}(x) &= \kappa(x)\underline{E}(x). \end{aligned}$$

where we mention that $f := -\rho(x)e_0$ and $\underline{E}(x)$ is the electric field.

(ii) Let $b_i(x) = \underline{1}$ $(i = 1, 2, 3)$ and $b_0 = b_0(x) > 0$. Then our boundary value problem is just a generalized LAMÉ system. The classical LAMÉ equation arises if $b_0(x) = 2(m-1)/(m-2)$ and m denotes the POISSON number. It reads as follows

$$\begin{aligned} \Delta\underline{u} + \tfrac{m}{m-2} \operatorname{grad} \operatorname{div}\underline{u} &= -\underline{f} \quad \text{in} \quad f \\ \underline{u} &= \underline{g} \quad \text{on} \quad \Gamma. \end{aligned}$$

In this way it seems to be possible to treat some important boundary value problems of mathematical physics by the approach described above. Finally it should be mentioned that the LAPLACE equation (we used it above only as a helpful tool) is included in a very natural way. The corresponding DIRICHLET problem can be seen, e.g., as a linear torsion problem (cf. [120]).

4.3 Vekua type problems

In 1956 the famous classical papers [319] and [320] of I.N. Vekua appeared. Basic properties of holomorphic functions were transferred to partial differential equations of the type

$$\begin{pmatrix} \partial_x & -\partial_y \\ \partial_y & \partial_x \end{pmatrix} \begin{pmatrix} u \\ v \end{pmatrix} = \begin{pmatrix} a & b \\ c & d \end{pmatrix} \begin{pmatrix} u \\ v \end{pmatrix} - \begin{pmatrix} f \\ g \end{pmatrix}.$$

A large number of interesting physical applications lead to this type of equation. We emphasize the equations of gas dynamics or of elastic shells. In 1944 a paper of Von Mises [324] appeared concerning a multidimensional generalization of Vekua's system. Solutions could be obtained by series expansions. Furthermore, we should mention the papers [71], [218], [219], [90], and [91]. As we could see in Section 2.4 Vekua systems occur in the treatment of axial-symmetric domains (cf. [286], [181]). In the thesis of Xu Zhenyan a function theory of the disturbed Dirac operator $D - \lambda$ ($\lambda \in \mathbb{C}$) is developed. More recently we have to mention the papers [159] and [21]. In this section we will deal with boundary value problems of Dirac operators which are disturbed by $P\mathrm{Vec}\, C\ell_{0,n}$-valued constant potentials and in special cases also with $C\ell_{0,n}$-regular potentials. Applications to Yukawa and Helmholtz problems will be outlined. A large number of further applications are contained in [158].

Let $a \in \mathbb{R}^n \oplus \mathbb{R}$. $\mathbb{R}^n \oplus \mathbb{R}$ should always be understood as the subspace of paravectors in $C\ell_{0,n}$. The paravector a permits a representation as a sum of the scalar part $a_0 := \mathrm{Sc}\ a$ and the vector part $\underline{a} = \mathrm{Vec}\, a$ with $\underline{a} = \sum_{i=1}^n a_i e_i$. The main involution in $C\ell_{0,n}$ is defined by $\hat{a} = a_0 - \underline{a}$. As an immediate consequence from the generalized Leibniz formula

$$D(uv) = \hat{u}Dv + (Du)v + 2\mathrm{Sc}(uD)v$$

we get with $x = x_1 e_1 + \ldots + x_n e_n$ that a fundamental solution of the operator $D + a$ is given by

$$e_a(x) = e^{-\langle \underline{a}, x\rangle} e_{a_0}(x),$$

where

$$e_{a_0}(x) = \frac{a_0}{(2\pi)^{n/2}} \frac{(a_0|x|)^{n/2-1}}{|x|^{n-2}} \left(\frac{a_0\, x}{|x|} K_{n/2}(a_0|x|) + K_{n/2-1}(a_0|x|) \right)$$

and $K_\nu(x)$ denotes the so-called Macdonald *function*. There is a close relation to the Bessel function $J_\nu(x)$. It reads as follows:

$$\int_0^\infty \frac{t^{n/2}}{(1+t^2)^{\alpha/2}} J_{\frac{n-2}{2}}(t|\xi|)dt = \left(\frac{|\xi|}{2}\right)^{\alpha/2-1} \frac{1}{\Gamma(\alpha/2)} K_{\frac{n-\alpha}{2}}(|\xi|)$$

(cf. [21]). In this context should be also mentioned the paper [251]. A straightforward calculation shows that we have

$$(D+\overline{a})(D+a) = -\Delta + 2\overline{a}D - 2\partial_{\underline{a}} + |a|^2.$$

We introduce the following abbreviations: $D_a := D + a$, $(T_a u)(x) := -\int_G e_a(x-y)u(y)dy$ and $(F_a u)(x) := \int_\Gamma e_a(x-y)\alpha(y)u(y)d\Gamma_y$. Here again $\alpha(y) = \underline{\alpha}(y) = \sum_{i=1}^n \alpha_i(y)e_i$ denotes the the outward pointing normal unit vector to Γ at the point y. As above demonstrated we immediately obtain that $D_a T_a = I$.

Let $u \in C^1(G, C\ell_{0,n}) \cap C(\overline{G}, C\ell_{0,n})$. Then it is easy to generalize from Section 3.2 formulae of the BOREL–POMPEIU type :

$$(F_a u)(x) + (T_a D_a u)(x) = \begin{cases} (D_a T_a u)(x) = u(x) & , \quad x \in G \\ 0 & , \quad x \in \mathbb{R}^n \setminus \overline{G}. \end{cases}$$

This is also valid for $u \in W_2^1(G, C\ell_{0,n})$. Using this formula and $D_a T_a = I$ one immediately obtains that $F_a T_a u(x) = 0$. Without difficulty PLEMELJ–SOKHOTZKI's formulae follow also for the operator F_a. Therefore, we have for $u \in C^{0,\alpha}(\Gamma, C\ell_{0,n})$

(i) $\lim\limits_{\substack{x\to x_0\\ x\in G}} (F_a u)(x) = (P_a u)(x_0) := (\frac{1}{2}(I + S_a)u)(x_0)$,

(ii) $\lim\limits_{\substack{x\to x_0\\ x\notin \overline{G}}} (F_a u)(x) = -(Q_a u)(x_0) := -(\frac{1}{2}(I - S_a)u)(x_0)$,

where $(S_a u)(x_0) = 2\int_\Gamma e_a(x_0 - y)\alpha(y)u(y)d\Gamma_y$.

Clearly, P_a is the HARDY projection onto the space of all $C\ell_{0,n}$-valued functions which can be $C\ell_{0,n}$ – a-regular extended into the domain G, i.e. that the extension u has to fulfil the equation $(D+a)u = 0$ in G, and Q_a is the HARDY projection onto the space of all $C\ell_{0,n}$-valued functions which can be $C\ell_{0,n}$-a-regular extended into the exterior domain $\mathbb{R}^n \setminus \overline{G}$ and vanish at infinity, respectively. We mention that for $1 < p < \infty$, and $k > 0$ holds

(i) $T_a : W_p^k(G, C\ell_{0,n}) \longmapsto W_p^{k+1}(G, C\ell_{0,n})$,

(ii) $F_a : W_p^{k-1/p}(\Gamma, C\ell_{0,n}) \longmapsto W_p^k(G, C\ell_{0,n}) \cap \ker D_a$.

We will now pay attention to a suitable decomposition formula with the operators D_a and $D_{\overline{a}}$ similar to Section 3.6.
Let G be a symmetric domain with respect to the origin.

Theorem 4.26. *The* HILBERT *space* $L_2(G, C\ell_{0,n})$ *permits the following orthogonal decomposition:*

$$L_2(G, C\ell_{0,n}) = \ker D_a \cap L_2(G, C\ell_{0,n}) \oplus D_{\overline{a}} \overset{\circ}{W}{}_2^1 (G, C\ell_{0,n}).$$

Proof. The proof is given in [301] and uses the same main ideas as the proof of Theorem 3.75. #

Owing to the validity of this theorem, we instantly have the existence of orthoprojections $\mathbf{P}_a$ and $\mathbf{Q}_a$ with

$$\mathbf{P}_a : L_2(G, C\ell_{0,n}) \longmapsto \ker\ D_a \cap L_2(G, C\ell_{0,n}),$$

$$\mathbf{Q}_a = I - \mathbf{P}_a : L_2(G, C\ell_{0,n}) \longmapsto D_{\overline{a}} \overset{\circ}{W}{}_2^1 (G, C\ell_{0,n}) \cap L_2(G, C\ell_{0,n}).$$

We note for $u \in W_2^1(G, C\ell_{0,n})$ the property $D_a \mathbf{Q}_a u = D_a u$. Furthermore, it is an exercise to show that the boundary value problem

$$\begin{aligned} D_{\overline{a}} u &= f \text{ in } G, \\ u &= 0 \text{ on } \Gamma, \end{aligned} \tag{4.20}$$

has a solution if and only if $\mathbf{Q}_a f = f$, which can be found by $u = T_{\overline{a}} f$. Our methods from Section 4.2 lead to the result

Theorem 4.27. *Let $f \in W_2^k(G, C\ell_{0,n})$, $g \in W_2^{k+3/2}(\Gamma, C\ell_{0,n})$. Then* DIRICHLET*'s problem*

$$(-\Delta + 2\overline{a}D + \partial_{\underline{a}} + |a|^2)u = f \quad \text{in } G, \quad u = g \quad \text{on } \Gamma \tag{4.21}$$

has the unique solution

$$u = F_{\overline{a}} g + T_{\overline{a}} \mathbf{P}_a D_{\overline{a}} h + T_{\overline{a}} \mathbf{Q}_a T_a f \in W_2^{k+1,\mathrm{loc}}(G, C\ell_{0,n})$$

where h denotes a $W_2^k(G, C\ell_{0,n})$-extension of g.

Proof. We restrict our considerations to show the uniqueness. Assuming that u is a solution of the homogeneous problem we immediately get from the boundary condition that $D_a u \in \operatorname{im} \mathbf{Q}_a$. From $D_{\overline{a}} D_a u = 0$ we conclude that $D_a u \in \operatorname{im} \mathbf{P}_a$. Therefore, $D_a u = 0$ and from the BOREL–POMPEIU formula we obtain $T_a D_a u = u \equiv 0$ #.

Corollary 4.28. *There are no* CLIFFORD*-a-regular eigenfunctions u of the problem*

$$(-\Delta + 2\overline{a}D + 2\partial_{\underline{a}}, \mathrm{tr}).$$

Proof. Let u be an eigenfunction of that problem which belongs to $\ker D_a$ for some $a \in R^n$. Then $D_a u = 0$ in G and $T_a D_a u = u = 0$. This contradicts our assumption. #

Now we intend to clarify the structure of the orthogonal projections $\mathbf{P}_a$ and $\mathbf{Q}_a$. For this reason we have to investigate the operator $\mathrm{tr}_\Gamma T_{\overline{a}} F_a$.

Proposition 4.29. *We have*

$$\mathrm{tr}_\Gamma T_{\overline{a}} u = 0 \quad \textit{iff} \quad u \in \ \operatorname{im} \mathbf{Q}_a.$$

Proof. Let $u \in \text{im } \mathbf{Q}_a$. Then a function ω exists such that $u = D_{\overline{a}}w$, $w \in \overset{\circ}{W}{}_2^1 (G, C\ell_{0,n})$. This leads to $\text{tr}_\Gamma u = \text{tr}_\Gamma T_{\overline{a}} D_{\overline{a}} w = \text{tr}_\Gamma w = 0$.
Now we assume that $\text{tr}_\Gamma T_{\overline{a}} u = 0$. Then u can be decomposed into the sum $u = u_1 + u_2$, with $u_1 \in \ker D_a$ and $u_2 \in \text{im } \mathbf{Q}_a$. We get

$$0 = \text{tr}_\Gamma T_{\overline{a}} u_1 + \text{tr}_\Gamma T_{\overline{a}} u_2 \ .$$

The first term belongs to $\ker(-\Delta + 2\overline{a}D + 2\partial_{\underline{a}} + |a|^2)$. For the second term we obtain the representation

$$u_2 = D_{\overline{a}}w, w \in \overset{\circ}{W}{}_2^1 (G, C\ell_{0,n}).$$

Hence, $\text{tr}_\Gamma T_{\overline{a}} D_{\overline{a}} w = \text{tr}_\Gamma w = 0$. From Theorem 4.27. we conclude that the only solution of the boundary value problem

$$(-\Delta + 2\overline{a}D + 2\partial_{\underline{a}} + |a|^2) T_{\overline{a}} u_1 = 0 \quad \text{in } G, \quad T_{\overline{a}} u_1 = 0 \quad \text{on } \Gamma. \tag{4.22}$$

is $T_{\overline{a}} u_1 \equiv 0$, whence $u_1 \equiv 0$. Therefore, we get $u = u_2 \in \text{im } \mathbf{Q}_a$. #

Proposition 4.30. $\ker \text{tr}_\Gamma T_{\overline{a}} F_a \cap (W_2^{\frac{1}{2}}(\Gamma, C\ell_{0,n}) \cap \text{ im } P_a) = \{0\}$

Proof. Let $\text{tr}_\Gamma T_{\overline{a}} F_a u = 0$. Then we have to show that for $u \in W_2^{\frac{1}{2}}(\Gamma, C\ell_{0,n}) \cap \text{im } P_a$ it follows that $u \equiv 0$. Setting now $v = F_a u$, then we immediately obtain $\text{tr}_\Gamma T_{\overline{a}} v = 0$. Following Proposition 4.29. we get $v \in \text{im } \mathbf{Q}_a$. Because, on the other hand $v = F_a u \in \text{im} \mathbf{P}_a$ we have that $F_a u = u = 0$ ($u \in \text{im } P_a!$). #

Proposition 4.31. *The operator* $\text{tr}_\Gamma T_{\overline{a}} F_a$ *defined on im* $P_a \cap W_2^{\frac{1}{2}}(\Gamma, C\ell_{0,n})$ *is a surjective mapping onto the space* $\text{im } Q_{\overline{a}} \cap W_2^{\frac{3}{2}}(\Gamma, C\ell_{0,n})$.

Proof. Let $w \in \text{im} Q_{\overline{a}}$, i.e., $w = Q_{\overline{a}} v$. From Theorem 4.27. we know that the boundary value problem

$$D_a D_{\overline{a}} u = 0 \quad \text{in } G, u = w \quad \text{on } \Gamma$$

has the solution

$$u = F_{\overline{a}} w + T_{\overline{a}} \mathbf{P}_a D_{\overline{a}} h$$

with an $W_2^{k+2}(G)$-extension of the boundary values w. Using now that $F_{\overline{a}} w = 0$ for $w \in \text{im} Q_{\overline{a}}$ and $T_{\overline{a}} \mathbf{P}_a D_{\overline{a}} h = T_{\overline{a}} F_a \mathbf{P}_a D_{\overline{a}} h$ we have $w \in \text{im} \, \text{tr}_\Gamma T_{\overline{a}} F_a$. The theorem is proved. #
Now we have

Theorem 4.32. *The operator*

$$\text{tr}_\Gamma T_{\overline{a}} F_a : \text{im} P_a \cap W_2^{\frac{1}{2}}(\Gamma, C\ell_{0,n}) \to \text{im} Q_{\overline{a}} \cap W_2^{\frac{3}{2}}(\Gamma, C\ell_{0,n})$$

is an isomorphism.

The integral representation formulas obtained above are not completely different to those known from the classical potential theory. For connecting both theories we introduce the following operators: Let $u \in W_2^{\frac{1}{2}}(\Gamma, C\ell_{0,n})$. Then we will denote by V_a the operator

$$(V_a u)(x) := -\int_\Gamma \frac{1}{a_0}\left(\frac{a_0}{2\pi}\right)^{n/2} \frac{K_{n/2-1}(a_0|x-y|)}{|x-y|^{n/2-1}} u(y)dy,$$

which is a potential operator of single-layer type, and K_a the volume potential

$$(K_a u)(x) := -\int_G \frac{1}{a_0}\left(\frac{a_0}{2\pi}\right)^{n/2} \frac{K_{n/2-1}(a_0|x-y|)}{|x-y|^{n/2-1}} u(y)dy.$$

Immediately, we get for $u \in W_p^1(G, C\ell_{0,n}), p > 1$, the identity $V_a u = T_a u + K_a D_a u$, $F_a V_a u = 0$ for $u \in \ker D_a$ and also $\mathrm{im}\, D_a V_a \subset \ker D_a$.

Theorem 4.33. *The operator*

$$\mathrm{tr}_\Gamma V_a : \mathrm{im}\, P_a \cap W_2^{\frac{1}{2}}(\Gamma, C\ell_{0,n}) \to \mathrm{im}\, Q_{\overline{a}} \cap W_2^{\frac{3}{2}}(\Gamma, C\ell_{0,n})$$

is an isomorphism.

Remark 4.34. It is worth noticing that both Theorem 4.32. and Theorem 4.33. remain valid also in the pair of spaces

$$\left[\mathrm{im}\, P_a \cap W_2^{k-1/2}(\Gamma, C\ell_{0,n}), \mathrm{im}\, Q_{\overline{a}} \cap W_2^{k+1/2}(\Gamma, C\ell_{0,n})\right], \ (k \geq 1).$$

Now we intend to consider the operators

$$\mathbf{P}'_a := F_a(\mathrm{tr}_\Gamma T_{\overline{a}} F_a)^{-1} \mathrm{tr}_\Gamma T_{\overline{a}} \qquad \mathbf{Q}'_a := I - \mathbf{P}'_a.$$

A straightforward calculation as in Section 4.2 immediately leads to descriptions of the kernel and the image of the operator $\mathbf{P}'_a$. Hence, we get the formulae

$$\ker \mathbf{P}'_a = D_{\overline{a}} \overset{\circ}{W}{}_2^1 (G, C\ell_{0,n}), \quad \mathrm{im}\, \mathbf{P}'_a = \ker D_a \cap L_2(G, C\ell_{0,n}).$$

Theorem 4.35. *The operators $\mathbf{P}'_a$ and $\mathbf{Q}'_a$ are orthoprojections onto the subspaces $\ker D_a$ and $D_{\overline{a}} \overset{\circ}{W}{}_2^1 (G, C\ell_{0,n})$, respectively. Furthermore, we have for $u \in W_2^k(G, C\ell_{0,n})$ also $\mathbf{P}'_a u, \mathbf{Q}'_a u \in W_2^k(G, C\ell_{0,n})$. Finally there is*

$$\mathbf{P}'_a = \mathbf{P}_a \quad \text{and} \quad \mathbf{Q}'_a = \mathbf{Q}_a.$$

Proof. It can be easily seen that $\mathbf{P}'^2_a = \mathbf{P}'_a$, $\mathbf{Q}'^2_a = \mathbf{Q}'_a$ and $\mathbf{P}'_a \mathbf{Q}'_a = \mathbf{Q}'_a \mathbf{P}'_a = 0$. Immediately, $T_{\overline{a}} u \in W_2^{k+1}(G, C\ell_{0,n})$ and $\mathrm{tr}_\Gamma T_{\overline{a}} u \in W_2^{k+1/2}(\Gamma, C\ell_{0,n})$ follows. Owing to Theorem 4.33. we obtain

$$(\mathrm{tr}_\Gamma T_{\overline{a}} F_a)^{-1} \mathrm{tr}_\Gamma T_{\overline{a}} u \in W_2^{k-1/2}(\Gamma, C\ell_{0,n})$$

and, finally,

$F_a(\mathrm{tr}_\Gamma T_{\overline{a}} F_a)^{-1}\mathrm{tr}_\Gamma T_{\overline{a}} u \in W_2^k(G, C\ell_{0,n})$.

Without problems we obtain

$$\begin{aligned}\mathbf{Q}'_a u &= (I - \mathbf{P}'_a)u = u - F(\mathrm{tr}_\Gamma T_{\overline{a}} F_a)^{-1}\mathrm{tr}_\Gamma T_{\overline{a}} u \\ &= D_{\overline{a}}(T_{\overline{a}} u - T_{\overline{a}} F_a(\mathrm{tr}_\Gamma T_{\overline{a}} F_a)^{-1}\mathrm{tr}_\Gamma T_{\overline{a}} u) = D_{\overline{a}} w\end{aligned}$$

with

$w = T_{\overline{a}} u - T_{\overline{a}} F_a(\mathrm{tr}_\Gamma T_{\overline{a}} F_a)^{-1}\mathrm{tr}_\Gamma T_{\overline{a}} u$.

Obviously, $\mathrm{tr}_\Gamma w = 0$. Therefore, $\mathbf{Q}'_a u \in D_{\overline{a}} \overset{\circ}{W}{}_2^1 (G, C\ell_{0,n})$.
For the uniqueness of the orthoprojections we also obtain $\mathbf{P}'_a = \mathbf{P}_a$ and $\mathbf{Q}'_a = \mathbf{Q}_a$. #

Corollary 4.36. *Each solution u of the problem (4.21) belongs to the space $W_2^{k+2}(G, C\ell_{0,n})$.*

Remark 4.37. The result of this Corollary can be generalized to $W_p^k(G)$-spaces: Let $1 < p < \infty, k \geq 0$. For $f \in W_p^k(G, C\ell_{0,n}), g \in W_p^{k+2-1/p}(\Gamma, C\ell_{0,n})$ we understand that the unique solution (4.21) belongs to $W_p^{k+2}(G, C\ell_{0,n})$.

Let us now have a brief look for possible generalizations to classes of boundary value problems with partial differential equations which have variable coefficients. We will only discuss here one example:

Example 4.38.
Let

$$b(x) = \underline{b}(x) = \underline{a}(x) + \sum_{i=1}^{n} x_i \partial_i a_i(x_i) e_i,$$

where

$$\underline{a}(x) = \sum_{i=1}^{n} a_i(x_i) e_i.$$

Then the fundamental solution of the operator $D + \underline{b}$ is given by

$e_{b(x)}(x) := e^{-\langle a(x), x\rangle} e_{a_0}(x)$.

Indeed, we get

$$\begin{aligned}(D + \underline{b})e_{b(x)}(x) &= \left[De^{-\langle \underline{a}(x), x\rangle}\right]e(x) + \underline{b}(x)e_a(x) \\ &= D\left[e^{-\langle \underline{a}(x), x\rangle}\right]e(x) - e^{-\langle \underline{a}(x), x\rangle} De(x) + \underline{b}(x)e_{b(x)}(x) \\ &= \delta(x). \qquad \#\end{aligned}$$

In order to construct $\underline{a}(x)$ from a given function $\underline{b}(x)$, we have only to solve ordinary differential equations of the type

$$ts'(t) + s(t) = r(t).$$

Summarizing, we can state that our theory also works for boundary value problems of the type

$$\begin{aligned} -\Delta\, u + 2b(x)D\, u - 2\partial_{\underline{b}} u + |b(x)|^2 u &= f \text{ in } G \\ u &= g \text{ on } \Gamma \end{aligned}$$

with $b(x) = \sum_{i=1}^{n} [a_i(x_i) + x_i \partial_i a_i(x_i)] e_i$ and $a_i(t) \in C^1(\mathbb{R})$ are arbitrarily given functions.

Let us now answer the following questions. Let different potentials a and b, where a, b are paravectors in $C\ell_{0,n}$ be given. Does a representation of the HILBERT space $L_2(G, C\ell_{0,n})$ exist in the form of a direct sum

$$L_2(G, C\ell_{0,n}) = \ker D_a \cap L_2(G, C\ell_{0,n}) \oplus D_b \overset{\circ}{W}{}_2^1 (G, C\ell_{0,n})\ ?$$

Up to now we know only that in the case of $b = \overline{a}$ such a decomposition of $L_2(G, C\ell_{0,n})$ is possible, and that in this case the subspaces are orthogonal to each other. We start our considerations with the construction of a suitable projection onto $\ker D_a$. For this reason we will have a closer look at the operator $\mathrm{tr}_\Gamma T_b F_a$.

Proposition 4.39. *The operator* $\mathrm{tr}_\Gamma T_b F_a$ *is a surjective mapping onto the subspace* $\mathrm{im}\, \mathbf{Q}_b \cap W_2^{3/2}(\Gamma, C\ell_{0,n})$, *if each boundary value problem* $D_a D_b u = 0, \mathrm{tr}_\Gamma u = g \in \mathrm{im}\, \mathbf{Q}_b \cap W_2^{3/2}(\Gamma)$ *has a solution in* $W_2^1(G, C\ell_{0,n})$.

Proof. Let w be an element from $\mathrm{im}\, \mathbf{Q}_b \cap W_2^{\frac{3}{2}}(\Gamma, C\ell_{0,n})$. It is well known that then exists a function $v \in W_2^2(G, C\ell_{0,n})$ such that

$$w = \mathbf{Q}_b g = \mathrm{tr}_\Gamma T_b D_b v\ ,\ \ \mathrm{tr}_\Gamma v = g\ .$$

The proof will be complete if a function $u \in \mathrm{im}\, P_a \cap W_2^{\frac{1}{2}}(\Gamma, C\ell_{0,n})$ can be found with

$$D_b v = F_a u - u \quad \text{and} \quad \mathrm{tr}_\Gamma v = g.$$

That means v has to be a solution of

$$D_a D_b v = 0 \quad \text{and} \quad \mathrm{tr}_\Gamma v = g.$$

If the latter boundary value problem is solvable, then the continuation v can be chosen such that $u = \mathrm{tr}_\Gamma D_b v$ with $\mathrm{tr}_\Gamma v = g$. #

Proposition 4.40. *We have* $\ker \mathrm{tr}_\Gamma T_b F_a \cap W_2^{\frac{1}{2}}(\Gamma, C\ell_{0,n}) \cap\ \mathrm{im}\ P_a = \{0\}$ *iff* $\ker D_a D_b \cap \overset{\circ}{W}{}_2^1 (G, C\ell_{0,n}) = \{0\}$

Proof. Firstly, let $\mathrm{tr}_\Gamma T_b F_a u = 0$. With $v = F_a u \in W_2^1(G, C\ell_{0,n})$ it immediately follows that $\mathrm{tr}_\Gamma T_b v = 0$. From Proposition 4.29. we know that for this is necessary and sufficient $v \in \mathrm{im}\ \mathbf{Q}_b$, which means that $v = D_b w, w \in \overset{\circ}{W}{}_2^1 (G, C\ell_{0,n})$ and so $F_a u = D_b w$. Hence, we obtain $D_a D_b w = 0$. Assuming that now $u = 0$ we get $D_b w = 0$ and also $w = 0$. Conversely, let $D_a D_b w = 0, w \in \overset{\circ}{W}{}_2^1 (G, C\ell_{0,n})$ it follows that $D_b w \in \mathrm{im}\ F_a$ and so $D_b w = F_a u$. Clearly, $T_b D_b w = w = T_b F_a u$ and $\mathrm{tr}_\Gamma T_b F_a u = 0$. For $w = 0$ it is necessary that $F_a u = 0$ and for $u \in \mathrm{im}\ P_a$ also $u = 0$. #

Now let us have a closer look at the structure of the operator $D_a D_b$. It is easy to get

$$(D + a)(D + b) = -\Delta + (a + \bar{b})D - 2\langle b, D\rangle + ab\ .$$

For all further considerations we assume that the boundary value problem

$$(-\Delta + (a + \bar{b})D - 2\langle b, D\rangle + ab)u \ = \ f \text{ in } G, \tag{4.23}$$

$$u \ = \ g \text{ on } \Gamma, \tag{4.24}$$

has a unique solution. Now we are able to consider the operators $\mathbf{P}_{ab}$ and $\mathbf{Q}_{ab} := I - \mathbf{P}_{ab}$ with

$$\mathbf{P}_{ab} := F_a(\mathrm{tr}_\Gamma T_b F_a)^{-1}\mathrm{tr}_\Gamma T_b.$$

It is clear that $\mathbf{P}_{ab}$ fulfils the projection property

$$\mathbf{P}_{ab}^2 = \mathbf{P}_{ab}.$$

In a natural way $\mathbf{Q}_{ab}^2 = \mathbf{Q}_{ab}$ and $\mathbf{P}_{ab}\mathbf{Q}_{ab} = \mathbf{Q}_{ab}\mathbf{P}_{ab} = 0$ also follows.

Theorem 4.41. *For $k \geq 1$ the projection $\mathbf{P}_{ab}$ it is a continuous operator within the spaces $W_2^k(G, C\ell_{0,n})$.*

Proof. Obviously, the operator $T_b : W_2^k(G, C\ell_{0,n}) \to W_2^{k+1}(G, C\ell_{0,n})$ is bounded. A well known embedding theorem shows that $\mathrm{tr}_\Gamma T_b$ is continuously embedded in $W_2^{k+\frac{1}{2}2}(\Gamma, C\ell_{0,n})$. Using Proposition 4.29. we get the mapping

$$\mathrm{tr}_\Gamma T_b F_a : W_2^{k-\frac{1}{2}}(\Gamma, C\ell_{0,n}) \cap \text{ im } P_a \to W_2^{k+\frac{1}{2}}(\Gamma, C\ell_{0,n}) \cap \text{ im } Q_b$$

continuous for $k \geq 1$. Now, we obtain from BOREL–POMPEIU's formula the boundedness of F_a in the pair of spaces $\left[W_2^{k-\frac{1}{2}}(\Gamma, C\ell_{0,n}), W_2^k(G, C\ell_{0,n})\right]$. #

Corollary 4.42. *We have*

$$\mathrm{im}\ \mathbf{P}_{ab} = \ker D_a \quad \text{in} \quad W_2^k(G, C\ell_{0,n})\ (k \geq 1).$$

Proof. Let $u \in \text{im}\,\mathbf{P}_{ab}$, then it is immediately clear that $u \in \text{im}\,F_a$ and, therefore, $\text{im}\,\mathbf{P}_{ab} \subset \ker D_a \cap W_2^k(G, C\ell_{0,n})$. Now let $u \in \ker D_a \cap W_2^k(G, C\ell_{0,n})$. It is clear, that then $u = F_a u$. Then there exists an element $w \in \text{im}\,Q_b \cap W_2^{k-\frac{1}{2}}(\Gamma, C\ell_{0,n})$ with $u = (\text{tr}_\Gamma T_b F_a)^{-1} w$. Hence, $w = Q_b g = \text{tr}_\Gamma T_b D_b v$. We get now

$$u = F_a(\text{tr}_\Gamma T_b F_a)^{-1} \text{tr}_\Gamma T_b D_b v$$

and so $u \in \text{im}\,\mathbf{P}_{ab}$. #

Corollary 4.43. *We have*

$$\ker \mathbf{P}_{ab} = D_b \overset{\circ}{W_2^1}(G, C\ell_{0,n}) \quad \text{in } W_2^k(G, C\ell_{0,n})\ (k \geq 1)\,.$$

Proof. Let $u \in \ker \mathbf{P}_{ab} \cap W_2^k(G, C\ell_{0,n})$, i.e. $F_a(\text{tr}_\Gamma T_b F_a)^{-1} \text{tr}_\Gamma T_b u = 0$. The proof of Theorem 4.41. and the fact that F_a coincides on im P_a with the identity leads to the necessary condition $\text{tr}_\Gamma T_b u = 0$. Proposition 4.29. yields now $u \in \text{im}\,\mathbf{Q}_{\bar{b}}$, hence $u = D_b v$ for $v \in \overset{\circ}{W_2^1}(G, C\ell_{0,n}) \cap W_2^k(G, C\ell_{0,n})$ and $\ker \mathbf{P}_{ab} \subset D_b \overset{\circ}{W_2^1}(G, C\ell_{0,n})$ in $W_2^k(G, C\ell_{0,n})$. Conversely, let $u \in D_b \overset{\circ}{W_2^1}(G, C\ell_{0,n}) \cap W_2^k(G, C\ell_{0,n})$. There exists a function

$$v \in \overset{\circ}{W_2^1}(G, C\ell_{0,n}) \cap W_2^{k+1}(G, C\ell_{0,n}),$$

with $u = D_b v$. Then we have

$$\mathbf{P}_{ab} D_b v = F_a(\text{tr}_\Gamma T_b F_a)^{-1} \text{tr}_\Gamma T_b D_b v = 0,$$

so $\text{tr}_\Gamma T_b D_b v = \text{tr}_\Gamma v = 0$. #

Theorem 4.44. *Let $f \in W_2^k(G, C\ell_{0,n}), g \in W_2^{k+\frac{3}{2}}(\Gamma, C\ell_{0,n})$. In the case of the unique solvability of the boundary value problem (4.21) the solution $u \in W_2^{k+2}(G, C\ell_{0,n})$ can be represented by the formula*

$$u = F_b g + T_b F_a (\text{tr}_\Gamma T_b F_a)^{-1} \mathbf{Q}_b g + T_b \mathbf{Q}_{ab} T_a f\,.$$

Proof. It is necessary to show that u permits the above mentioned representation. For this reason we have to prove that the represention for u satisfies our partial differential equation. Firstly, we have

$$D_b u = F_a(\text{tr}_\Gamma T_b F_a)^{-1} \mathbf{Q}_b g + \mathbf{Q}_{ab} T_a f \quad \text{and}$$
$$D_a D_b u = D_a \mathbf{Q}_{ab} T_a f = D_a T_a f - D_a \mathbf{P}_{ab} T_a f = f\,.$$

For the boundary condition we compute $\text{tr}_\Gamma T_b \mathbf{Q}_{ab} T_a f = 0$ and

$$\text{tr}_\Gamma F_b g + \text{tr}_\Gamma T_b F_a (\text{tr}_\Gamma T_b F_a)^{-1} \mathbf{Q}_b g = P_b g + \mathbf{Q}_b g = g. \qquad \#$$

Example 4.45. Consideration of complex valued quaternions enables us to generalize our results to the YUKAWA equation. For that reason we deal with the problem

$$\begin{aligned}(D+ai)(D-\overline{a}i)u &= f \quad \text{in } G\\ u &= g \quad \text{on } \Gamma,\end{aligned}$$

where i is the complex unit and $a \in \mathbb{R}^{\mathbf{n}} \oplus \mathbb{R} \subset C\ell_{0,n}$.
Using the generalized LEIBNIZ rule we get from the latter equation

$$\begin{aligned}-\Delta u + 2i\partial_{\underline{a}} u + |a|^2 u &= f \quad \text{in } G\\ u &= g \quad \text{on } \Gamma.\end{aligned}$$

The fundamental solution for the operator $D-\mu$ ($\mu \in \mathbb{C}$) (cf. [335]) reads as follows :

$$e_{-\mu}(x) = -\frac{\pi}{\sigma_n \Gamma(\frac{n}{2})}\left(\frac{\mu}{2}\right)^{n/2} |x|^{1-n/2}\left(Y_{(n/2)-1}(\mu|x|) - \frac{x}{|x|}Y_{n/2}(\mu|x|)\right) \tag{4.25}$$

with

$$Y_\nu(\mu|x|) = \frac{(\cos\nu\pi)J_\nu(\mu|x|) - J_{-\nu}(\mu|x|)}{\sin\nu\pi},$$

where $J_\nu(t)$ are BESSEL functions of the first kind. In the special case of $n=3$ we get after a straightforward calculation the representation

$$\underline{e}_\mu(x) = e(x)\big[\cos(\mu|x|) + |x|\mu\sin(\mu|x|)\big] + \frac{\mu\cos(\mu|x|)}{4\pi|x|}e_0$$

with $e(x) = \frac{-x}{\sigma_n|x|^n}$, which is the fundamental solution of the DIRAC operator D in $\mathbb{R}^n \setminus \{0\}$.
The representation of the solution reads as follows

$$u = F_{ai}g + T_{ai}(F_{-\overline{ai}}(\mathrm{tr}_\Gamma T_{ai}F_{-\overline{ai}})^{-1}Q_{ai}g + T_{ai}\mathbf{Q}_{ai(-\overline{ai})}T_{-\overline{ai}}f.$$

Assuming $f, g \in C\ell_{0,n}$ then for $\underline{a}=0$ the function u solves YUKAWA's problem

$$\begin{aligned}-\Delta u + |a|^2 u &= f \quad \text{in } G\\ u &= g \quad \text{on } \Gamma\end{aligned}$$

in G.

Remark 4.46. Further applications can be found for instance in [335], [158] and [157].

Remark 4.47. Note here that boundary value problems of the YUKAWA equation arise naturally in the treatment of parabolic equations with the method of semi-discretization. The corresponding coefficient a_0 is just the inverse mesh width of the time-discretization.

Another typical application is connected with the factorization problem of the SCHRÖDINGER equation. The problem consists in finding a certain potential included in the occurring operators of the kind $D+a$.

The aim of our next considerations is to investigate a certain class of linear boundary value problems by using methods of an adapted hypercomplex function theory. At first we briefly present corresponding function theoretic results. Hereby it seems to be useful to stress the class of the so-called (p, μ)-regular functions. Then we introduce a suitable operator calculus which enables us to solve a certain class of boundary value problems. In cases of such problems of second order we permit the occurrence of derivatives in the direction of a given irrotational field. In a unified manner questions of existence, uniqueness and representation formulae will be answered.

Each construction of a function theory means to fix a special class of so-called "analytic" functions and to study their properties. Now let $p = p(x) > 0$ be a scalar function of $C^1(G)$ and $\mu \in \mathbb{C}$. Further, let $D(p, \mu) := D - \frac{\mathrm{grad} p}{p} + \mu$. We call functions of the class $C^1(G, C\ell_{0,n}) \cap \ker D(p, \mu)$ *(p, μ)-regular*. It is easy to see that $D(p, \mu) = p(D + \mu)p^{-1}$.

Example 4.48. For

$$p(x) = \frac{1}{n-2} \exp\left(\frac{1}{|x - x^{(i)}|}\right), \; x^{(i)} \in \mathbb{R}^n \setminus \overline{G}$$

we have

$$D(p, 0) = D + \frac{x - x^{(i)}}{|x - x^{(i)}|^n}.$$

The class of vector potentials of the form $(\mathrm{grad} p)/p$ is sufficiently wide. Indeed, similarly to Section 5.1 one can show that any $(1, 0)$-regular function belonging to $L_p(G, C\ell_{0,n})$ can be approximated by right-linear combinations of functions of the system

$$\left\{ \frac{x - x^{(i)}}{|x - x^{(i)}|^n}, \; i = 1, 2, \ldots \right\}$$

if the set $\{x^{(i)}, i = 1, 2, \ldots\}$ is dense on a closed surface Γ' outside of G and parallel to $\partial G = \Gamma$. In this way, the investigation of DIRAC operators disturbed by $(1, 0)$-regular potentials can be reduced to the consideration of operators of the type

$$D + \frac{x - x^{(i)}}{|x - x^{(i)}|^n}, \; x^{(i)} \in \Gamma' \subset \mathbb{R}^n \setminus \overline{G}.$$

Now for $\mu \in \mathbb{R}$ and $x \neq y$ let $e^p_\mu(x, y) = p(x) e_\mu(x - y) p^{-1}(y)$. Then we can introduce the following operators:

(i) $(T(p, \mu)u)(x) := -\int_G e^p_\mu(x, y) u(y) dG_y$ (TEODORESCU transform),

(ii) $(F(p, \mu)u)(x) := \int_\Gamma e^p_\mu(x, y) \alpha(y) u(y) d\Gamma_y$ (CAUCHY-type operator) for $x \notin \Gamma$, where α denotes the quaternionic function $\alpha = \sum_{i=1}^{n} \alpha_i(y)e_i$ and $(\alpha_1, ..., \alpha_n)$ is the unit vector of the outer normal to Γ at the point y. Immediately, from [110] and [111] we get the following statements, which we will summarize in Lemma 4.49.

Lemma 4.49. *Let $p^{-1}u \in C^1(G, C\ell_{0,n}) \cap C(\overline{G}, C\ell_{0,n})$. Then we have the formulas*

(i) $$(F(p,\mu)u)(x) + (T(p,\mu)D(p,\mu)u)(x) = \begin{cases} u(x) & x \in G \\ 0 & x \in I\!R^n \setminus \overline{G} \end{cases},$$

(ii) $$(D(p,\mu)T(p,\mu)u)(x) = \begin{cases} u(x) & , \quad x \in G \\ 0 & , \quad x \in I\!R^n \setminus \overline{G} \end{cases},$$

(iii) $(D(p,\mu)F(p,\mu)u)(x) = 0$, *in* $G \cup I\!R^n \setminus \overline{G}$.

Remark 4.50. These formulas can be generalized to the case $p^{-1}u \in W_p^1(G, C\ell_{0,n})$ (see [110]; 2.5.4).

Lemma 4.51. (PLEMELJ–SOKHOTZKIJ-type formulas).
Let $p^{-1}u \in C^{0,\alpha}(\Gamma, C\ell_{0,n})$, $0 < \alpha < 1$. Then we have

(i) $$\lim_{\substack{x \to x_0 \in \Gamma \\ x \in G}} (F(p,\mu)u)(x) = (P_\Gamma(p,\mu)u)(x_0),$$

(ii) $$\lim_{\substack{x \to x_0 \in \Gamma \\ x \in \mathbb{R}^n \setminus \overline{G}}} (F(p,\mu)u)(x) = -(Q_\Gamma(p,\mu)u)(x_0),$$

for any $x_0 \in \Gamma$, where

$$P_\Gamma(p,\mu) = \frac{1}{2}(I + S_\Gamma(p,\mu)), Q_\Gamma(p,\mu) = \frac{1}{2}(I - S_\Gamma(p,\mu))$$

and

$$(S_\Gamma(p,\mu)u)(x) := 2\int_\Gamma e_\mu^p(x,y)\alpha(y)u(y)d\Gamma_y,\ x \in \Gamma.$$

The integral $(S_\Gamma(p,\mu)u)(x)$ exists in the sense of CAUCHY's principal value.

Corollary 4.52. *Let $p^{-1}u \in C^{o,\alpha}(\Gamma, C\ell_{0,n})$. Then the relations*

(i) $(S_\Gamma^2(p,\mu)u)(x) = u(x)$,

(ii) $(F(p,\mu)P_\Gamma(p,\mu)u)(x) = (F(p,\mu)u)(x)$,

(iii) $(P_\Gamma^2(p,\mu)u)(x) = (P_\Gamma(p,\mu)u)(x)$,

(iv) $(Q_\Gamma^2(p,\mu)u)(x) = (Q_\Gamma(p,\mu)u)(x)$,

are valid for any $(x \in \Gamma)$.

Now we will sum up some applications. For proofs we refer to [301]. Let $G \subset \mathbb{R}^n$ be a bounded domain and $\Gamma = \partial G$ be a LIAPUNOV surface. We consider the following boundary value problem of first order:

$$D(p,\mu)u = f \quad \text{in} \quad G \tag{4.26}$$

$$u = g \quad \text{on} \quad \Gamma. \tag{4.27}$$

Theorem 4.53. *Let* $g \in W_2^{k-\frac{1}{2}}(\Gamma, C\ell_{0,n}), f \in W_2^k(G, C\ell_{0,n})$, *and* $p \in C^\infty(G)$ *with* $p(x) > 0, \mu \in \mathbb{R}$. *Then the boundary value problem (4.26–4.27) has a solution if and only if the condition*

$$Q_\Gamma(p,\mu)g = \operatorname{tr}_\Gamma T(p,\mu)f.$$

is fulfilled. The only solution $u \in W_2^{k+1}(G, C\ell_{0,n})$ *admits the representation*

$$u = F(p,\mu)g + T(p,\mu)f$$

Now we deal with second-order boundary value problems:

$$-\Delta u + c\,\partial_\beta u + 2\mu Du + \mu^2 u = f \quad \text{in } G \tag{4.28}$$

$$u = g \quad \text{on } \Gamma, \tag{4.29}$$

where c is a real constant.
This result is very useful for further discussions to have the following decomposition of the quaternionic HILBERT space $L_2(G, C\ell_{0,n})$.

Lemma 4.54. *For any real-valued function* $p = p(x) > 0$, $p \in C^\infty(G)$ *and real number* μ *we get the orthogonal decomposition*

$$L_2(G, C\ell_{0,n}) = \ker D(p,\mu) \cap L_2(G, C\ell_{0,n}) \oplus D(p,\mu) \overset{\circ}{W}{}_2^1 (G, C\ell_{0,n}),$$

where "⊕" stands for the topological sum relative to the inner product (3.14).

Now we have to normalize the system (4.28–4.29). Since rot $\underline{\beta}(x) = 0$ there is a scalar function q with $\underline{\beta} = \operatorname{grad} q$. Taking for $q = \ln p^{2/c}$ then the system (4.28–4.29) leads to

$$-\Delta u + \frac{2}{p}\partial_{\operatorname{grad} p} u + 2\mu Du + \mu^2 u = f \tag{4.30}$$

$$u = g \tag{4.31}$$

Adding to both sides of equation (4.28) the expression

$$\mathcal{H}(u) := \left[\frac{\Delta p}{p} - 2\frac{|Dp|^2}{p^2} - 2\mu\frac{Dp}{p}\right] u$$

and we get the problem

$$\begin{aligned} D^2(p,\mu)u = \mathcal{H}(u) \qquad & \text{in} \qquad G, \\ u = g \qquad & \text{on} \qquad \Gamma. \end{aligned}$$

Denote the orthoprojections with respect to the inner product $(\cdot,\cdot)$ by

$$\mathbf{P}(p,\mu) : L_2(G, C\ell_{0,n}) \longrightarrow \ker D(p,\mu) \cap L_2(G, C\ell_{0,n})$$

$$\mathbf{Q}(p,\mu) = I - \mathbf{P}(p,\mu) : L_2(G, C\ell_{0,n}) \longrightarrow D(p,\mu) \overset{\circ}{W_2^1} (G, C\ell_{0,n}).$$

Similarly to Theorem 4.22. we obtain that the operator $\mathrm{tr}_\Gamma T(p,\mu)F(p,\mu)$ is an isomorphism for the pair of spaces

$$\left[W_2^{k-\frac{1}{2}}(\Gamma, C\ell_{0,n}) \cap \mathrm{im}\, P_\Gamma(p,\mu), W_2^{k+\frac{1}{2}}(\Gamma, C\ell_{0,n}) \cap \mathrm{im}\, Q_\Gamma(p,\mu)\right]$$

and the orthoprojection $\mathbf{P}(p,\mu)$ allows the algebraic representation

$$\mathbf{P}(p,\mu) = F(p,\mu)(\mathrm{tr}_\Gamma T(p,\mu)F(p,\mu))^{-1}\mathrm{tr}_\Gamma T(p,\mu).$$

Now we get the following operator integral equation

$$\begin{aligned} u = {} & F(p,\mu)g + T(p,\mu)F(p,\mu)(\mathrm{tr}_\Gamma T(p,\mu)F(p,\mu))^{-1}Q_\Gamma(p,\mu) + \\ & + T(p,\mu)\mathbf{Q}(p,\mu)T(p,\mu)\mathcal{H}(u). \end{aligned}$$

Applying $D^2(p,\mu)$ from the left we get immediately

$$D^2(p,\mu)u = \mathcal{H}(u).$$

Since $\mathbf{Q}(p,\mu)T(p,\mu)\mathcal{H}(u) = D(p,\mu)w(u)$
with $w(u) \in \overset{\circ}{W_2^1} (G, C\ell_{0,n})$ it follows for each fixed u

$$u = F(p,\mu)g + T(p,\mu)F(p,\mu)(\mathrm{tr}_\Gamma T(p,\mu)F(p,\mu))^{-1}Q_\Gamma(p,\mu)g + w(u)$$

and, therefore, $\mathrm{tr}_\Gamma u = P_\Gamma(p,\mu)g + Q_\Gamma(p,\mu)g = g$.
Conversely, let u be a solution of the boundary value problem (4.30). Applying $T(p,\mu)$ from the left we get

$$D(p,\mu)u = T(p,\mu)\mathcal{H}(u) + \Phi_1, \Phi_1 \in \ker D(p,\mu)$$

and

$$u = T^2(p,\mu)\mathcal{H}(u) + T(p,\mu)\Phi_1 + \Phi_0, \Phi_0 \in \ker D(p,\mu).$$

Using Lemma 4.49. it follows that

$$u - F(p,\mu)g = T^2(p,\mu)\mathcal{H}(u) + T(p,\mu)F(p,\mu)(\mathrm{tr}_\Gamma \Phi_1).$$

From Lemma 4.54. we get

$$\begin{aligned} Q_\Gamma(p,\mu)g = \mathrm{tr}_\Gamma(u - F_\Gamma(p,\mu)g) = \mathrm{tr}_\Gamma T^2(p,\mu)\mathcal{H}(u) + \\ + \mathrm{tr}_\Gamma T(p,\mu)F(p,\mu)(\mathrm{tr}_\Gamma \Phi_1). \end{aligned}$$

Hence,

$$\begin{aligned}\mathrm{tr}_\Gamma \Phi_1 &= (\mathrm{tr}_\Gamma T(p,\mu)F(p,\mu))^{-1}Q_\Gamma(p,\mu)g - \\ &-(\mathrm{tr}_\Gamma T(p,\mu)F(p,\mu))^{-1}\mathrm{tr}_\Gamma T^2(p,\mu)\mathcal{H}(u).\end{aligned}$$

Finally, we have

$$\begin{aligned}u = F(p,\mu)g + T(p,\mu)F(p,\mu)(\mathrm{tr}_\Gamma T(p,\mu)F(p,\mu))^{-1}Q_\Gamma(p,\mu)g + \\ +T(p,\mu)\mathbf{Q}(p,\mu)T(p,\mu)\mathcal{H}(u).\end{aligned}$$

In this way we proved

Theorem 4.55. *The operator integral equation given above is equivalent to the boundary value problem (4.30).*

Write for brevity

$$r(g) := F(p,\mu)g + T(p,\mu)F(p,\mu)(\mathrm{tr}_\Gamma T(p,\mu)F(p,\mu))^{-1}Q_\Gamma(p,\mu)g.$$

Obviously, it is true that $\mathrm{tr}_\Gamma r(g) = g$. Setting now

$$B_1(r(g)) = \{u \in W_2^1(G, C\ell_{0,n}) : \|u - r(g)\|_{W_2^1(G,C\ell_{0,n})} \le 1\}.$$

Further, let

$$f \in L_2(G, C\ell_{0,n}) \text{ and } g \in W_2^{\frac{1}{2}}(\Gamma, C\ell_{0,n}),\ \beta = \underline{\beta}(x) = \sum_{i=1}^n \beta_i(x)e_i$$

with rot $\underline{\beta}(x) \equiv 0$ and ∂_β denotes the derivative in the direction of the vector function $\underline{\beta}$. Assume now that u shall satisfy the conditions $\|u - r(g)\|_1 \le 1$ and

$$\left\| \frac{\Delta p}{p} - 2\frac{|Dp|}{p^2} - 2\mu\frac{Dp}{p} \right\|_{\left[W_2^1(G,C\ell_{0,n}),L_2(G,C\ell_{0,n})\right]} \le \frac{L}{2} \tag{4.32}$$

with a constant $L > 0$. Then we get

$$\|T(p,\mu)\|_{[\mathrm{im}\ \mathbf{Q}(p,\mu)\cap L_2(G),W_2^1(G,)]}\|T(p,\mu)\|_{[L_2(G),L_2(G)} \le s/L$$

with $0 < s < 1$.

Now, we consider the following iteration procedure:

$$u_n = r(g) + T(p,\mu)\mathbf{Q}(p,\mu)T(p,\mu)\mathcal{H}(u_{n-1}) \quad (n = 1,2,...) \tag{4.33}$$

with $u_0 = r(g)$. It is easy to see that $\mathrm{tr}_\Gamma u_n = g$. We have to show that the limit $\lim u_n \in W_2^1(G, C\ell_{0,n})$ exists. Let

$$u_n = u_1 + (u_2 - u_1) + ... + (u_n - u_{n-1}). \tag{4.34}$$

Abbreviating $||T(p,\mu)||_{\left[\mathrm{im}\, Q_\Gamma(p,\mu)\cap L_2(G,C\ell_{0,n}),W_2^1(G,C\ell_{0,n})\right]} =: ||T||_1$ and $||T(p,\mu)||_{\left[L_2(G,C\ell_{0,n}),L_2(G,C\ell_{0,n})\right]} =: ||T||_2$, we have the estimates: let

$$u_{n-1} \in B_1(r(g)) \quad \text{then} \quad ||u_n - r(g)||_1 \leq ||T||_1||T||_2 L \leq s < 1.$$

Furthermore,

$$\begin{aligned}||u_n - u_{n-1}||_1 &\leq ||T||_1||T||_2 L||u_{n-1} - u_{n-2}||_1 \leq (||T||_1||T||_2 L)^{n-1}||u_1 - r(g)||_1 \\ &\leq s^{n-1}||u_1 - r(g)||_1.\end{aligned}$$

Now, the $W_2^1(G, C\ell_{0,n})$-convergence of the sequence (4.34) is easy to verifiy. The statement considered above can be summarized in the following

Theorem 4.56. *Let $f \in L_2(G, C\ell_{0,n})$, let $\mathcal{H}$ satisfy condition (4.32), $g \in W_2^{\frac{1}{2}}(\Gamma, C\ell_{0,n}), \alpha \in C^\infty(G)$. Then the boundary value problem (4.30) has in the ball $B_1(r(g))$ a unique $W_2^1(G, C\ell_{0,n})$-solution, which can be obtained by the iteration procedure (4.33).*

4.4 Higher order boundary value problems

As mentioned above the development of spatial function theories was inspired by physicists with their interests in first-order systems of partial differential equations. But beginning in 1980 only second-order elliptic boundary value problems were investigated systematically in a self-contained theory. Questions of existence, uniqueness, and regularity are included in this theory, and as an important advantage also convenient representations of the solutions can be obtained. In [110] and [158] and here in the authors study a large number of applications of quaternionic analysis. Higher order boundary value problems are not included in the cited books. Also, in the consideration of first-order boundary problems several problems remain still open. One fundamental (and one of the simplest) example is the generalized $\overline{\partial}$-problem (see [236]).

In this section we first study a system of biharmonic equations. It is known that the biharmonic equation plays an important role. On the one hand it is directly connected with mathematical models. On the other hand it is needed as a mathematical tool for the treatment of other equations. The hypercomplex notation allows not only considerations in higher dimensions but also the treatment of more complicated boundary conditions. As an example we will deal with a boundary value problem for the biharmonic equation which includes on the boundary the values, divergence, gradient, and curl of the unknown vector function. Boundary operators of this kind are known from the theory of elasticity. Furthermore, we will investigate more general boundary value problems of differential operators of order $2n$. For some types of equations it is always possible to reduce to a system of differential equations of the first order.

First it should be remarked that these studies also show by example that for the applicability of the described methods maximum principles are not necessary.

1. The biharmonic problem

The main tools in the theory described up to now (see also [110] and [158]) are the above-mentioned orthogonal decompositions (3.75), (3.76) and (analogously to the complex case) the operators D, T_G, and F_Γ with their algebraical properties. We will investigate here a system of biharmonic equations coupled by the boundary conditions using analogous ideas. Preparing this investigation we need a result on special extensions of vector functions defined on the boundary of our domain G. With $\alpha(y)$ we again denote the outward pointing normal unit vector to Γ at the point y.

Theorem 4.57. *Suppose that G is a domain with a sufficiently smooth boundary Γ. For each pair of functions $w_1 \in W_2^{\frac{3}{2}}(\Gamma, Cl_{0,n}), w_2 \in W_2^{\frac{1}{2}}(\Gamma, Cl_{0,n})$ there exists an extension $h \in W_2^2(G, Cl_{0,n})$ with $h|_\Gamma = w_1$ and $Dh|_\Gamma = w_2$.*

Remark 4.58. It should be mentioned that the boundary data $\operatorname{tr}_\Gamma u, \operatorname{tr}_\Gamma Du$ and $\operatorname{tr}_\Gamma u, \alpha \operatorname{tr}_\Gamma Du$ include in essential cases of mathematical physics. The existence of extensions from Theorem 4.57. is not trivially clear without proof because we have to look for normal and tangential components in our boundary conditions (for vector functions) and their interaction.

Proof of Theorem 4.57. For each function $u \in W_2^2(G, Cl_{0,n})$ the following GREEN formulae:

$$u(x) + K\Delta u(x) = \overline{V}_\alpha Du(x) + F_\Gamma u(x) \tag{4.35}$$

and

$$u(x) + K\Delta u(x) = V\partial_\alpha u(x) - \frac{1}{(n-1)\sigma_n} \int_\Gamma \partial_\alpha \frac{1}{|y-x|^{n-1}} u(y)\, d\Gamma_y,$$

with

$$Vu(x) = \frac{1}{(n-1)\sigma_n} \int_\Gamma \frac{1}{|x-y|^{n-1}} u(y)\, d\Gamma_y,$$

hold. We remember that V_α was defined by

$$V_\alpha u(x) = \frac{1}{(n-1)\sigma_n} \int_\Gamma \frac{\alpha(y)}{|x-y|^{n-1}} u(y)\, d\Gamma_y.$$

If we put these two equations together then we obtain

$$\overline{V}_\alpha Du(x) + F_\Gamma u(x) = V\partial_\alpha u(x) - \frac{1}{(n-1)\sigma_n} \int_\Gamma \partial_\alpha \frac{1}{|y-x|^{n-1}} u(y)\, d\Gamma_y.$$

Let us consider $x \mapsto x_0 \in \Gamma$. Applying PLEMELJ–SOKHOTZKIJ's formula, the jump formula for the double layer potential, and the continuity of the single layer potential, we get a boundary integral equation of the form

$$\begin{aligned}
&\overline{V}_\alpha Du(x_0) + \frac{1}{2}u(x_0) + \frac{1}{2}S_\Gamma u(x_0) \\
&\quad = V\partial_\alpha u(x_0) + \frac{1}{2}u(x_0) - \frac{1}{(n-1)\sigma_n} \int_\Gamma \partial_\alpha \frac{1}{|y-x_0|^{n-1}} u(y)\, d\Gamma_y.
\end{aligned}$$

With other words we have

$$V\partial_\alpha u = g,$$

if we set $u = w_1$ and $Du = w_2$ on the boundary Γ. Moreover, by using the well-known mapping properties of the boundary operators in the above boundary integral equation we have $g \in W_2^{\frac{3}{2}}(\Gamma, Cl_{0,n})$. Now, we use the invertibility of V as a mapping from $W_2^{\frac{1}{2}}(\Gamma, Cl_{0,n})$ to $W_2^{\frac{3}{2}}(\Gamma, Cl_{0,n})$ (see [53], [213], and elsewhere). Let us consider the problem $\Delta u = 0, u|_\Gamma = g, g \in W_2^s(\Gamma, Cl_{0,n}), s \geq \frac{1}{2}$. Following arguments of the potential theory, it is a simple matter to see that $u = Vv$ with $v = [\partial_\alpha u]$ ($[\partial_\alpha u]$ means the jump of $\partial_\alpha u$). The result of J. NEČAS [213] now gives us $v \in W_2^{s-1}(\Gamma, Cl_{0,n}), s \geq 1/2$, and we get the invertibility of V as a mapping between $W_2^{s-1}(\Gamma, Cl_{0,n}) \mapsto W_2^s(\Gamma, Cl_{0,n}), (s \geq \frac{1}{2})$. Applying our results above we can calculate $\partial_\alpha u|_\Gamma$. It may be observed that there exists an $W_2^2(G, Cl_{0,n})$-extension h with $h|_\Gamma = u|_\Gamma = w_1$ and $\partial_\alpha h|_\Gamma = \partial_\alpha u|_\Gamma$. Moreover, this extension h has to satisfy

$$h(x) + K\Delta h(x) = V\partial_\alpha h(x) - \frac{1}{(n-1)\sigma_n}\int_\Gamma \partial_\alpha \frac{1}{|y-x|^{n-1}} u(y)\, d\Gamma_y$$

and

$$h(x) + K\Delta h(x) = \overline{V}_\alpha Dh(x) + F_\Gamma u(x).$$

Using the same ideas as above, we find that

$$\begin{aligned}
&\overline{V}_\alpha Dh(x_0) + \frac{1}{2}u(x_0) + \frac{1}{2}S_\Gamma u(x_0) \\
&= V\partial_\alpha u(x_0) + \frac{1}{2}u(x_0) - \frac{1}{(n-1)\sigma_n}\int_\Gamma \partial_\alpha \frac{1}{|y-x_0|^{n-1}} u(y)\, d\Gamma_y.
\end{aligned}$$

Consequently, we obtain

$$\overline{V}_\alpha Dh = \overline{V}_\alpha Du.$$

From the injectivity of $\overline{V}_\alpha$ we immediately get our assertion, because we now have

$$Dh = Du = w_2 \quad \text{on } \Gamma. \qquad \#$$

Now, we use (4.35) for harmonic functions u. If we take the traces of u and Du then we get the following boundary integral equations ($x_0 \in \Gamma$):

$$\begin{aligned}
u(x_0) &= \overline{V}_\alpha Du(x_0) + \tfrac{1}{2}u(x_0) + \tfrac{1}{2}S_\Gamma u(x_0), \\
Du(x_0) &= \tfrac{1}{2}Du(x_0) + \tfrac{1}{2}\overline{S}_\Gamma Du(x_0)
\end{aligned}$$

or in other words

$$\begin{pmatrix} u \\ Du \end{pmatrix} = \begin{pmatrix} P_\Gamma & \overline{V}_\alpha \\ 0 & \overline{P}_\Gamma \end{pmatrix} \begin{pmatrix} u \\ Du \end{pmatrix}.$$

From this we can derive the projections

$$P_\Gamma^\Delta = \begin{pmatrix} P_\Gamma & \overline{V}_\alpha \\ 0 & \overline{P}_\Gamma \end{pmatrix} \quad \text{and} \quad Q_\Gamma^\Delta = \begin{pmatrix} Q_\Gamma & -\overline{V}_\alpha \\ 0 & \overline{Q}_\Gamma \end{pmatrix}.$$

Introducing the boundary operator $F_\Gamma^\Delta u := (F_\Gamma u, \overline{V}_\alpha u)$ we have a BOREL–POMPEIU formula for the Laplacian in the form

$$F_\Gamma^\Delta \mathrm{Tr}\, u = u - K\Delta u \tag{4.36}$$

with

$$\mathrm{Tr}\, u := \begin{pmatrix} \mathrm{tr}_\Gamma\, u \\ \mathrm{tr}_\Gamma\, Du \end{pmatrix}.$$

Now we are looking for solutions of the problem

$$\begin{aligned} \Delta^2 u &= 0 \quad \text{in } G \\ \mathrm{Tr}\, u &= g \quad \text{on } \Gamma \end{aligned} \tag{4.37}$$

Theorem 4.59. *If $k \in \mathbf{N} \cup \{0\}, g \in W_2^{k+\frac{3}{2}}(\Gamma, Cl_{0,n}) \times W_2^{k+\frac{1}{2}}(\Gamma, Cl_{0,n})$ then the problem (4.37) has the solution $u \in W_2^{k+2}(G, Cl_{0,n})$*

$$u = F_\Gamma^\Delta g + K\mathbf{P}^\Delta \Delta h,$$

where h is an arbitrary W_2^{k+2}-extension of g.

Proof. Because $g \in W_2^{k+\frac{3}{2}}(\Gamma, Cl_{0,n}) \times W_2^{k+\frac{1}{2}}(\Gamma, Cl_{0,n})$ there exists an $h \in W_2^{k+2}(G, Cl_{0,n})$ with $\mathrm{Tr}\, h = g$. Let $u = v + h$ then our problem (4.37) has the form

$$\Delta^2 v = \Delta^2 h \quad \text{in } G, \quad \mathrm{Tr}\, v = 0 \quad \text{on } \Gamma \tag{4.38}$$

Obviously, (4.38) has a solution $v = -K\mathbf{Q}^\Delta K\Delta^2 h \in W_2^{k+2}(G, Cl_{0,n})$. Using formula (4.36) we get

$$v = -K\mathbf{Q}^\Delta \Delta h + K\mathbf{Q}^\Delta F^\Delta \Delta h = -K\Delta h + K\mathbf{P}^\Delta \Delta h = -h + F^\Delta h + K\mathbf{P}^\Delta \Delta h.$$

Consequently, it may be observed that

$$v + h = F_\Gamma^\Delta g + K\mathbf{P}^\Delta \Delta h.$$

Hence,

$$u = F_\Gamma^\Delta g + K\mathbf{P}^\Delta \Delta h. \qquad \#$$

Proposition 4.60. *The problem*

$$\begin{aligned} \Delta^2 u &= f \quad \text{in } G \\ \mathrm{Tr}\, u &= g \quad \text{on } \Gamma \end{aligned} \tag{4.39}$$

with $f \in W_2^k(G, Cl_{0,n})$ *and* $g \in W_2^{k+\frac{3}{2}}(\Gamma, Cl_{0,n}) \times W_2^{k+\frac{1}{2}}(\Gamma, Cl_{0,n}), k \in \mathbf{N} \cup \{0\}$, *has a solution of the form*

$$u = F_\Gamma^\Delta g + K\mathbf{P}^\Delta \Delta h + K\mathbf{Q}^\Delta K f,$$

where h *denotes a* W_2^{k+2}*-extension of* g.

Proposition 4.61. *Suppose that*
$f \in W_2^k(G, Cl_{0,n})$ *and* $g \in W_2^{k+\frac{3}{2}}(\Gamma, Cl_{0,n}) \times W_2^{k+\frac{1}{2}}(\Gamma), k \in \mathbf{N} \cup \{0\}$,
then the solution of (4.39) is unique.

Proof. Let us consider the boundary value problem $\{\Delta^2 u = 0, \mathrm{Tr}\ u = 0\}$. From our BOREL–POMPEIU formula for the Laplacian and the fact that $\Delta u \in \mathrm{im}\ \mathbf{Q}^\Delta$ we obtain $u = K\Delta u = K\mathbf{Q}^\Delta \Delta u$. Using Δu is harmonic, $u = K\Delta u = K\mathbf{P}^\Delta \Delta u$ follows. Therefore, $\mathbf{P}^\Delta \Delta u = \mathbf{Q}^\Delta \Delta u$, whence $\Delta u = 0$ and $u = 0$. #

Theorem 4.62. *Assume that* $k \in \mathbf{N} \cup \{0\}$. *Then the operator*

$$\begin{aligned}&\mathrm{Tr}\ KF_\Gamma^\Delta : W_2^{k+\frac{3}{2}}(\Gamma, Cl_{0,n}) \times W_2^{k+\frac{1}{2}}(\Gamma, Cl_{0,n}) \cap \mathrm{im}\ \mathrm{P}_\Gamma^\Delta \\ &\mapsto W_2^{k+\frac{5}{2}}(\Gamma, Cl_{0,n}) \times W_2^{k+\frac{3}{2}}(\Gamma, Cl_{0,n}) \cap \mathrm{im}\ \mathrm{Q}_\Gamma^\Delta\end{aligned}$$

is an isomorphism.

Proof. The proof of this theorem uses the same ideas as the proof of Theorem 4.22. for the operator $\mathrm{tr}_\Gamma\ T_G F$. #

Consequently, analogous to this case, we get an algebraic representation of the projection $\mathbf{P}^\Delta$.

Proposition 4.63. *The orthoprojection* $\mathbf{P}^\Delta$ *has the algebraic representation*

$$\mathbf{P}^\Delta u = F_\Gamma^\Delta(\mathrm{Tr}\ KF_\Gamma^\Delta)^{-1}\mathrm{Tr}\ Ku$$

for each function $u \in W_2^k(G, Cl_{0,n}), k \in \mathbf{N}$.

4.4.1 Boundary value problems of order 2n

We consider the boundary value problem of the following partial differential equation of order $2n$ in $\mathbb{R}^3$. We use the abbreviation $F := F_\Gamma$. Let

$$\prod_{i=1}^{n} D_{a_i} D_{\overline{a}_i} u = f \text{ in } G,\ \ a_i \in \mathbb{H}\ f \in L_2(G, \mathbb{H}), g_i \in W_2^{2n-\frac{4i+1}{2}}(\Gamma, \mathbb{H}). \tag{4.40}$$

On the boundary Γ we have the conditions

$$u = g_0,\quad D_{a_1} D_{\overline{a}_1} u = g_1, ..., D_{a_{n-1}} D_{\overline{a}_{n-1}} \cdot ... \cdot D_{a_1} D_{\overline{a}_1} u = g_{n-1}. \tag{4.41}$$

Theorem 4.64. *The unique solution $u \in W_2^{2n}(G, \mathbb{H})$ of the problem (4.40–4.41) is explicitly given by the formula*

$$u = r_1(g) + T_{a_1}\mathbf{Q}_{a_1}T_{\overline{a}_1}r_2(g) + \ldots + \prod_{i=1}^{n-1} T_{a_i}\mathbf{Q}_{a_i}T_{\overline{a}_i}r_n(g_{n-1}) + \prod_{i=1}^{n} T_{a_i}Q_{a_i}T_{\overline{a}_i}f,$$

where

$$r_k(g_{k-1}) := F_{\overline{a}_k}g_{k-1} + T_{\overline{a}_k}F_{a_k}(\operatorname{tr} T_{\overline{a}_k}F_{a_k})^{-1}Q_{\overline{a}_k}g_{k-1}.$$

This theorem allows a generalization in the following manner:

Corollary 4.65. *Let $f \in L_2(G, Cl_{0,n}), g_i \in W_2^{2m-\frac{4i+1}{2}}(\Gamma, Cl_{0,n})$, and a_i, b_i as in Proposition 4.39. Then the unique solution of the boundary value problem*

$$\prod_{i=1}^{m} D_{a_i}D_{b_i}u = f,\ a_i, b_i \in \mathbb{R}^\mathbf{n} \oplus \mathbb{R} \subset Cl_{0,n} \tag{4.42}$$

with the boundary conditions

$u = g_0, D_{a_1}D_{b_1}u = g_1, \ldots, \prod_{i=1}^{m-1} D_{a_i}D_{b_i}u = g_{m-1}$

admits the representation

$$\begin{aligned}
u \;=\;& F_{b_1}g_0 + T_{b_1}F_{a_1}(\operatorname{tr}_\Gamma T_{b_1}F_{a_1})^{-1}Q_{b_1}g_0 + \\
& T_{b_1}\mathbf{Q}_{a_1b_1}T_{a_1}(F_{b_2}g_1 + T_{b_2}F_{a_2}(\operatorname{tr}_\Gamma T_{b_2}F_{a_2})^{-1}Q_{b_2}g_1) + \\
& - \\
& \prod_{j=1}^{m-1} T_{b_j}\mathbf{Q}_{a_jb_j}T_{a_j}(F_{b_m}g_{m-1} + T_{b_m}F_{a_m}(\operatorname{tr}_\Gamma T_{b_m}F_{a_m})^{-1}Q_{b_m}g_{m-1} + \\
& \prod_{j=1}^{m} T_{b_j}\mathbf{Q}_{a_jb_j}T_{a_j}f.
\end{aligned}$$

Theorem 4.66. *Let $f \in L_2(G, Cl_{0,n})$, $g_i \in W_2^{2m-\frac{4i+1}{2}}(\Gamma, Cl_{0,n})$. Then the boundary value problem $(\Delta^m, \operatorname{tr}_\Gamma, \operatorname{tr}_\Gamma \Delta, \ldots, \operatorname{tr}_\Gamma \Delta^{m-1})$ has a unique solution $u \in W_2^{2m}(G, Cl_{0,n})$, which can be represented by*

$$\begin{aligned}
u \;=\;& Fg_0 + TF(\operatorname{tr}_\Gamma TF)^{-1}Qg_0 + T\mathbf{Q}T(Fg_1 + TF(\operatorname{tr}_\Gamma TF)^{-1}Qg_1) + \\
& + \ldots + (T\mathbf{Q}T)^{m-1}(Fg_{m-1} + TF(\operatorname{tr}_\Gamma TF)^{-1}Qg_{m-1}) + (T\mathbf{Q}T)^m f. \quad (4.43)
\end{aligned}$$

Proof. Clearly, it is valid $-\Delta(T\mathbf{Q}T) = DDT\mathbf{Q}T = I$. Besides, we have $\Delta(Fg_i + TF(\operatorname{tr}_\Gamma TF)^{-1}Qg_i) = 0$ $(i = 1, \ldots, m)$. Therefore, it is also clear that for $i = 1, \ldots, m-1$

$$(-1)^m\Delta^m(T\mathbf{Q}T)^m f = f$$

$$(-1)^m\Delta^m(T\mathbf{Q}T)^i(Fg_i + TF(\operatorname{tr}_\Gamma TF)^{-1}Qg_i) = 0.$$

For the smoothness of u we refer to the mapping properties of F, T, Q_Γ, and $\mathbf{Q}$. Considering now

$$\begin{aligned} &\Delta^m u = 0 \quad \text{in} \quad G \\ &u = 0,\ \Delta u = 0,\ \Delta^2 u = 0, ...\Delta^{m-1} u = 0 \quad \text{on} \quad \Gamma \end{aligned}$$

and setting

$$u = v_0,\ \Delta u = v_1,\ \Delta^2 u = v_2\ , ...,\ \Delta^{m-1} u = v_{m-1},$$

then we obtain the system

$$\Delta v_i = 0 \text{ in } G, v_i = 0 \text{ on } \Gamma \qquad (i = 0, ..., m-1) \tag{4.44}$$

and so the uniqueness can be taken from the classical DIRICHLET's problem. #

Remark 4.67. J. RYAN considered in [257] polynomial DIRAC equations of the type

$$(D^k + \sum_{i=0}^{k-1} b_i D^i) u = 0$$

with complex numbers b_m in $\mathbb{R}^n$. He could get generalized CAUCHY formulas. In the same topic there are also results in [293].

4.5 Weak time-dependent Maxwell equations

There are a lot of connections between general field theories and quaternionic analysis. We will consider in this section the application of a quaternionic operator calculus to the representation of the solution of MAXWELL equations.

4.5.1 Deduction of Maxwell equations

Let $G \subset \mathbb{R}^3$ be a bounded domain symmetric with respect to the origin. An electric charge is distributed with the density $\rho(x)$. Further, let the vectors $\underline{E} = \underline{E}(t,x), \underline{H} = \underline{H}(t,x), \underline{D} = \underline{D}(x,t), \underline{B} = \underline{B}(x,t)$ and $\underline{J} = \underline{J}(x,t)$ describe the *electric field, magnetic field, electric displacements, magnetic field induction,* and the *current density vector* in this order. Two fundamental laws manifest the connection between all these physical terms.

$$\int_{\Gamma} \underline{E} \cdot \underline{al} d\underline{x} = -\partial_t \int_{G} \underline{B} \cdot \underline{\alpha} dG_x \quad \text{(FARADAY's induction law)}$$

and

$$\int_{\Gamma} \underline{H} \cdot d\underline{x} = \int_{G} \underline{J} \cdot \underline{al} dG_x + \int_{G} \partial_t \underline{D} \cdot \underline{\alpha} dG_x .$$

The second term was named by J.C. MAXWELL, who (1865) corrected BIOT–SAVARD–AMPERÉ's law, electric *displacement current.* From STOKES' theorem we get

$$\int_{G} \operatorname{rot} \underline{E} \cdot \underline{\alpha} dG_x = -\partial_t \int_{G} \underline{B} \cdot \underline{\alpha} dG_x \ , \quad \int_{G} \operatorname{rot} \underline{H} \cdot \underline{\alpha} dG_x = \int_{G} (\underline{J} + \partial_t \underline{D}) \cdot \underline{\alpha} dG_x .$$

After interchanging ∂_t with the integral over G we have in G the equations

$\operatorname{rot} \underline{E} + \partial_t \underline{B} = 0, \quad \operatorname{rot} \underline{H} = \underline{J} + \partial_t \underline{D}.$

As for $\operatorname{div} \operatorname{rot} v = 0$ we immediately obtain

$\partial_t \operatorname{div} \underline{B} = 0 \qquad \operatorname{div} (\underline{J} + \partial_t \underline{D}) = 0.$

Finally, the following relations holds:

$\operatorname{div} \underline{D} = \rho$ (COULOMB's law)

$\operatorname{div} \underline{B} = 0$ (no free magnetic charge)

$\operatorname{div} \underline{J} + \partial_t \rho = 0$ (continuity equation)

These equations should be supplemented by the so-called *constraint equations* reflecting the electro-magnetic properties of the material. For a linear and isotropic media the constraint equations read as follows:

$$\underline{D} = \varepsilon \underline{E} \qquad \underline{B} = \mu \underline{H} \qquad \underline{J} = \kappa \underline{E} \qquad \text{where} \qquad \kappa = \frac{1}{\rho}.$$

Here ε denotes the *permittivity*, μ the magnetic *permeability* and κ the *conductivity*. The latter relation is well known as OHM's *law*. In the general case the non-linear relations $\underline{D} = \underline{D}(\underline{E}), \underline{B} = \underline{B}(\underline{H}), \underline{J} = \underline{J}(\underline{E})$ are valid. Sometimes it is written

$$\underline{D} = \varepsilon(\underline{E}), \qquad \underline{B} = \mu(\underline{H}), \qquad \underline{J} = \kappa(\underline{E})$$

4.5.2 Deduction of a hypercomplex representation of Maxwell equations

Now we shall compute the electric field $\underline{E}$ and the magnetic field $\underline{H}$. Assume that the progress in time is known. That means we have the time harmonic case. Let the electric field and the magnetic field permit the representation as a product of a scalar function which only depends on the time and a vector function which depends on the location. Then we have $\underline{E}(t,x) = E_0(t)\underline{E}_1(x)$ and the magnetic field $\underline{H}(t,x) = H_0(t)\underline{H}_1(x)$. Assume that the permittivity ε, the permeability μ and the electric conductivity κ depend on the position x and the time t, i.e, $\varepsilon = \varepsilon(t,x), \mu = \mu(t,x)$ and $\kappa = \kappa(t,x)$.

MAXWELL's equations now reads as follows:

$$\operatorname{rot} \underline{E} = \partial_t(\mu \underline{H}) \qquad \operatorname{rot} \underline{H} = \partial_t(\varepsilon \underline{E}) + \kappa \underline{E} \quad \operatorname{div}(\varepsilon \underline{E}) = \rho \qquad \operatorname{div}(\mu \underline{H}) = 0.$$

Using now $\underline{E} = E_0(t)\underline{E}_1(x)$ and $\underline{H} = H_0(t)\underline{H}_1(x)$ and inserting these expressions in MAXWELL's equations we obtain

$$\begin{aligned}
E_0 \operatorname{rot} \underline{E}_1 &= -\partial_t(\mu H_0)\underline{H}_1 \\
H_0 \operatorname{rot} \underline{H}_1 &= (\partial_t(\varepsilon E_0) + \kappa E_0)\underline{E}_1, \\
E_0 \operatorname{grad} \varepsilon \cdot \underline{E}_1 &+ \varepsilon \operatorname{div} \underline{E}_1) = \rho \\
\operatorname{grad} \mu \cdot \underline{H}_1 &+ \mu \operatorname{div} \underline{H}_1 = 0.
\end{aligned}$$

In this way we have with $D = \sum_1^3 \partial_i e_i$

$$\operatorname{rot} \underline{E}_1 = -\frac{\partial_t(\mu H_0)}{E_0}\underline{H}_1, \qquad \operatorname{rot} \underline{H}_1 = \frac{\partial_t(\varepsilon E_0) + \kappa \underline{E}_0}{H_0}\underline{E}_1,$$

$$-\operatorname{div} \underline{E}_1 = -\frac{\rho}{\varepsilon E_0} + \left(\frac{D\,\varepsilon}{\varepsilon}\right) \cdot \underline{E}_1, \qquad -\operatorname{div} \underline{H}_1 = \left(\frac{D\,\mu}{\mu}\right) \cdot \underline{H}_1.$$

Note that we denote by " $\cdot$ " the usual $\mathbb{R}^3$-scalar product, i.e., $\underline{u} \cdot \underline{v} = \sum_1^3 u_i v_i$. Writing for brevity

$$\underline{a} = \frac{D\,\mu}{\mu}, \quad \underline{d} = \frac{D\,\varphi}{\varphi}, \qquad \rho' = -\frac{\rho}{\varepsilon E_0},$$

$$b_0 = \frac{\partial_t(\varepsilon\, E_0) + \kappa E_0}{H_0}, \qquad c_0 = -\frac{\partial_t\,(\mu H_0)}{E_0}$$

and adding two further trivial equations

$$DE_{10} = 0, \qquad DH_{10} = 0$$

we obtain by using the denotations

$$E_1 := E_{10} + \underline{E_1},\ H_1 = H_{10} + \underline{H_1},\ d := \underline{d},\ a := \underline{a}$$

the new system

$$D\,E_1 = c_0 H_1 + d \cdot E_1 + \rho' \qquad D\,H_1 = a \cdot H_1 + b_0 E_1.$$

We get the matrix equation

$$\begin{pmatrix} D & 0 \\ 0 & D \end{pmatrix} \begin{pmatrix} E_1 \\ H_1 \end{pmatrix} = \begin{pmatrix} d\cdot & c_0 \\ b_0 & a\cdot \end{pmatrix} \begin{pmatrix} E_1 \\ H_1 \end{pmatrix} + \begin{pmatrix} \rho' \\ 0 \end{pmatrix}.$$

Assume now that $d \times E_1 = 0$ and also $a \times H_1 = 0$, then one has $d \cdot E_1 = -dE_1$ and also $a \cdot H_1 = -aH_1$. From above we obtain the matrix equation

$$\begin{pmatrix} D-d & 0 \\ 0 & D-a \end{pmatrix} \begin{pmatrix} E_1 \\ H_1 \end{pmatrix} = \begin{pmatrix} 0 & c_0 \\ b_0 & 0 \end{pmatrix} \begin{pmatrix} E_1 \\ H_1 \end{pmatrix} + \begin{pmatrix} \rho' \\ 0 \end{pmatrix}.$$

4.5.3 Deduction of an integral representation

Let $a \in \mathbb{H}$. The quaternion a permits a representation as a sum of the scalar part $a_0 := \mathrm{Sc}a$ and the vector part $a = \mathrm{Vec}\ a$ with $a = \sum_{i=1}^{3} a_i e_i$. Furthermore, we remember that in Section 4.3 we discussed the mapping properties of the operators used below, T_a and F_a.
In the case of a and d do not depend on the position, i.e., $a = a(t)$ and also $d = d(t)$ then we have only to use the generalized BOREL–POMPEIU formula for a vector which belongs to the real quaternions. By multiplication from the left with the matrix operator

$$\begin{pmatrix} T_{-d} & 0 \\ 0 & T_{-a} \end{pmatrix}$$

it provides

$$\begin{pmatrix} E_1 \\ H_1 \end{pmatrix} = \begin{pmatrix} 0 & T_{-d}c_0 \\ T_{-a}b_0 & 0 \end{pmatrix} \begin{pmatrix} E_1 \\ H_1 \end{pmatrix} + \begin{pmatrix} T_{-d}\rho' \\ 0 \end{pmatrix} + \begin{pmatrix} \Phi_{-d} \\ \Phi_{-a} \end{pmatrix},$$

where the functions $\Phi_\alpha \in \ker D_\alpha$. This leads now to the integral representations

$$E_1 = T_{-d}\, c_0 H_1 + T_{-d}\rho' + \Phi_{-d} \qquad H_1 = T_{-a}\, b_0 E_1 + \Phi_{-a}.$$

Separating these equations we find

$$E_1 = T_{-d}c_0 T_{-a} b_0\, E_1 + T_{-d}c_0\Phi_{-a} + T_{-d}\rho' + \Phi_{-d}.$$

Remark 4.68. One can also use the integral operator $T := T_0$, then one get integral equations with a more simple kernel. Nevertheless, it seems that the use of the kernels $e_{-a}(x-y)$ and $e_{-d}(x-y)$ gives more information on the behaviour of the solutions than the use of $e(x-y)$.

4.5.4 Solution of a boundary value problem of Maxwell's equations

In Subsection 4.5.2 we obtained the following representation of MAXWELL equations:

$$(D-d)\,E_1 = c_0 H_1 + \rho'$$
$$(D-a)\,H_1 = b_0 E_1.$$

On the boundary is given the values of H by g. For $b_0 \neq 0$ we have

$$(D-d)b_0^{-1}(D-a)H_1 = c_0\underline{H_1} + \rho'.$$

Theorem 4.44. now delivers the following theorem:

Theorem 4.69. *The solution of the boundary value problem of* MAXWELL*'s equations formulated above permits the following representation:*

$$\begin{aligned} H_1 &= F_{-a}g + T_{-a}(F_{-a}(\mathrm{tr}_\Gamma T_{-d}b_0F_{-a})^{-1}Q_{-a}g + T_{-a}\mathcal{Q}_{(-d)(-a)}b_0T_{-d}\rho' \\ &\quad + T_{-a}\mathcal{Q}_{(-d)(-a)}b_0T_{-d}c_0H_1. \end{aligned}$$

Remark 4.70. If one of these functions in some sense is sufficiently small, then a unique solution can be found by BANACH's fixed-point theorem. In order to formulate these conditions more detailed we have to study operators of the type T_a. These considerations are made in a more general context. Conditions are known under which it is possible to prove a strong convergence in $W_2^1(G,\mathbb{H})$ (cf. [301]).

Remark 4.71. Sometimes it is more convenient to pose the condition $\underline{\alpha}\cdot\underline{H} = g_0$ on the boundary. This can be easily realized. We have then to replace g by $g(-\underline{\alpha})$, where $g = g_0 + \underline{g}$ and $\underline{\alpha}$ denotes, as previously the outward pointing normal unit vector to Γ.

4.5.5 On time-harmonic Maxwell equations

In this special case MAXWELL equations can be given by the formulae

$$\mathrm{div}(\varepsilon\tilde{E}) = \rho \qquad \mathrm{div}(\mu\tilde{H}) = 0$$

$$\mathrm{rot}(\varphi\tilde{E}) = \mathrm{grad}\,\varepsilon\times\tilde{E} \qquad \mathrm{rot}(\mu\tilde{H}) = \mu\kappa\tilde{E} + \mathrm{grad}\mu\times\tilde{H}.$$

Here we use the denotation $\tilde{E} := \underline{E}$, $\tilde{H} := \underline{H}$. Now put $E := \varepsilon\tilde{E}$, $H = \mu\tilde{H}$, $a = \varepsilon^{-1}\mathrm{grad}\,\mu$, $b := \mu^{-1}\mathrm{grad}\,\mu$, $f = -\rho$ and $\sigma = \varepsilon^{-1}\mu\kappa$. We get the new formulation in a quaternionic structure

$$DE = f + \mathrm{Vec}(aE) \qquad DH = \sigma E + \mathrm{Vec}(bH)$$

If φ and μ are constants we solved this system under the boundary condition $H = g$ on Γ in [110]. Our system reduces to

$$\begin{aligned} D\sigma^{-1}DH &= f \quad \text{in } G \\ H &= g \quad \text{on } \Gamma \end{aligned}$$

Then the solution can be obtained explicitly by the formula

$$H = F_\Gamma g + T_G \mathbf{P}_\sigma Dh + T_G \mathbf{Q}_\sigma \sigma T_G f.$$

Here h denotes a sufficiently smooth extension of g into G. It was supposed that $f \in W_2^k(G, \mathbb{H})$, $g \in W_2^{k+\frac{3}{2}}(G, \mathbb{H})$. Then the solution belongs to $W_2^{k+2}(G, \mathbb{H})$. The orthoprojections $\mathbf{P}_\sigma$ and $\mathbf{Q}_\sigma$ are given by

$$\mathbf{P}_\sigma = \sigma F_\Gamma (\mathrm{tr}_\Gamma T_G \sigma F_\Gamma)^{-1} \mathrm{tr}_\Gamma T_G \qquad \mathbf{Q}_\sigma = I - \mathbf{P}_\sigma$$

In the original notation with a constant κ we have

$$\tilde{H} = \frac{1}{\mu} F_\Gamma g + \frac{1}{\mu} T_G F_\Gamma (\mathrm{tr}_\Gamma T_G F_\Gamma)^{-1} Q_\Gamma g - \frac{\kappa}{\varepsilon} T_G F_\Gamma (\mathrm{tr}_\Gamma T_G F_\Gamma)^{-1} T_G \rho + \frac{\kappa}{\varepsilon} T_G^2 \rho$$

$$\tilde{E} = \frac{1}{\mu^2 \kappa} F_\Gamma (\mathrm{tr}_\Gamma T_G F_\Gamma)^{-1} Q_\Gamma g - \frac{\varphi \mu}{F}_\Gamma (\mathrm{tr}_\Gamma T_G F_\Gamma)^{-1} T_G \rho - \frac{1}{\mu} T_G \rho.$$

Remark 4.72. The more general system is solved by an iteration method under a smallness condition for a and b in [110].

4.6 A coupled problem of thermo-elasticity

With this example we will demonstrate how the hypercomplex calculus works in solving a coupled problem. We will consider a situation where not only the vector of displacements is unknown but also the temperature distribution.

Let $G \subset \mathbb{R}^3$ be a bounded domain with a smooth boundary. The problem can be described by the following equations:

$$\Delta \underline{u} + \frac{m}{m-2} \operatorname{grad} \operatorname{div} \underline{u} = 2\frac{m+1}{m-2} \alpha_t \operatorname{grad} \omega \quad \text{in} \quad G \times [0,T], \tag{4.45}$$

$$\frac{\partial u}{\partial \alpha} + \frac{1}{m-2} \underline{\alpha} \operatorname{div} \underline{u} + \frac{1}{2} \underline{\alpha} \times \operatorname{rot} \underline{u} = \frac{m+1}{m-2} V \omega \underline{\alpha} \quad \text{on} \quad \Gamma \times [0,T], \tag{4.46}$$

$$\Delta \omega = \frac{1}{\kappa} \partial_t \omega + \eta \omega_0 \operatorname{div}(\partial_t \underline{u}) \quad \text{in} \quad G \times [0,T], \tag{4.47}$$

$$\omega = \psi \text{ on } \Gamma \times [0,T] \quad \text{and} \quad \omega(\cdot, 0) = \omega_0 \quad \text{in} \quad G \times \{0\}. \tag{4.48}$$

In this formulae we used the following denotations:
$\underline{u} = \underline{u}(\underline{x}, t) \ldots$ vector of displacements, $\omega = \omega(\underline{x}, t) \ldots$ heat flux, $\alpha = \alpha(\underline{x}, t) \ldots$ the unit vector of the outward pointing normal at point x and the time t, $m \ldots$ POISSON number $(m > 2)$ $\kappa := \frac{k}{c} \ldots$ quotient of heat conductivity k and the specific heat c, $\eta := \frac{\gamma}{k} \ldots$ here γ is the temperature extension coefficient, $V \ldots$ volume extension coefficient per time unit.

Furthermore, we have the well-known relations of m and γ to LAMÉ's constants λ and μ, namely $m = 2(1 + \frac{\mu}{\lambda})$ and $\gamma = (3\lambda + 2\mu)V$.
A similar initial-boundary value problem was deduced in [231]. Next we will represent the problem (4.45-4.48) into a hypercomplex notation. For this reason we have to introduce the operators

$$Mu = \frac{m-2}{2(m-1)} u_0 e_0 + \underline{u}, \quad Bu = -(m-2) u_0 e_0 + 2\underline{u}, \quad m > 2.$$

Clearly, the inverse operators M^{-1}, B^{-1} exist and are defined by

$$M^{-1} u =: au = \frac{2(m-1)}{m-2} u_0 e_0 + \underline{u}, \quad B^{-1} u =: bu = \frac{-1}{m-2} u_0 e_0 + \frac{1}{2} \underline{u}.$$

For the moment we assume that ω is known.

Lemma 4.73. *The equations (4.45) and (4.46) are equivalent to*

$$DaD\underline{u} = -2\frac{m+1}{m-2} V D\omega \tag{4.49}$$

$$\partial_{\underline{\alpha}} \underline{u} + \frac{1}{2}(\alpha, \operatorname{rot} \underline{u}) + \underline{\alpha} b D u = \frac{m+1}{m-2} V \omega \underline{\alpha}. \tag{4.50}$$

Proof. Using BOREL–POMPEIU's formulae. #

Proposition 4.74. *[249] The boundary functions $\alpha, \alpha^{-1} \in W_2^{k+\frac{1}{2}}(\Gamma, \mathbb{H})$ can be continued into the domain G with the properties*

$$\mu_1 = D\xi_1 \qquad \text{in } G \quad \text{and } \operatorname{tr}_\Gamma \mu_1 = \alpha$$
$$\mu_2 = \frac{D\xi_2}{|D\xi_2|^2} \qquad \text{in } G \quad \text{and } \operatorname{tr}_\Gamma \mu_2 = \alpha^{-1},$$

where $\xi_1, \xi_2 \in C^k(\overline{G}, \mathbb{H}), \xi_1(\underline{x}), \xi_2(\underline{x}) > 0$ for all $\underline{x} \in G$, $\xi_1(x)$, $\xi_2(x) = 0$ on Γ and $|D\xi_1(x)| = |D\xi_2(x)| = 1$ on Γ.

One can show:

Theorem 4.75. *Let $\underline{u}$ be a weak solution of the problem (4.45,4.46). Then $\underline{u}$ solves the equation*

$$u = F_\Gamma u + T_G P_M B\mu_2\Big(\frac{m+1}{m-2}V\omega\mu_1 - \partial_{\tilde{\alpha}}\underline{u} - \frac{1}{2}(\mu_1, \operatorname{rot}\underline{u})\Big)$$
$$-T_G Q_M M T_G \frac{2(m+1)}{m-2} d_t D\omega, \tag{4.51}$$

where $\partial_{\tilde{\alpha}} := \sum_{i=1}^{3}(\partial_i \xi_1)\partial_i$. ξ_1 has been introduced in Proposition 4.74. On the other hand, for any solution $u \in L_2(W_2^1(G, \mathbb{H}), [0,T])$ of (4.51), $\underline{u}$ is a weak solution of (4.45,4.46).

Proof. The proof is given in [149]. #

Equation (4.51) leads to the following iteration procedure:

$$\begin{aligned} u_n = {} & F_\Gamma u_{n-1} + T_G P_M B\underline{\mu}_2\Big(\tfrac{m+1}{m-2}V\omega\underline{\mu}_1 - \partial_{\tilde{\alpha}} u_{n-1} - \tfrac{1}{2}(\mu_1, \operatorname{rot}\underline{u_{n-1}})\Big) \\ & -T_G Q_M M T_G \tfrac{2(m+1)}{m-2} V D\omega. \end{aligned} \tag{4.52}$$

Let now $u_{n-1} \in \operatorname{im} Q_\Gamma$, then always $u_n \in \operatorname{im} Q_\Gamma$ $(n \in N)$. Consequently, $F_\Gamma u_{n-1}$ in (4.52) vanishes. Next we consider equations (4.47) and (4.48). We write for (4.47)

$$\omega_t = a^2 \Delta\omega + ht \tag{4.53}$$

with $a^2 = \kappa, h = -\eta\omega_0 \operatorname{div} \partial_t \underline{u}$. We need some more abbreviations as follows:

$$N_k := \int_G v_k^2(\underline{y})dy \quad \text{and} \quad s(\underline{x}, \underline{y}, t, \tau) := \sum_{k=1}^{\infty} e^{-\kappa\lambda_k(t-\tau)} \frac{v_k(\underline{x})v_k(\underline{y})}{N_k},$$

where λ_k are the eigenvalues and v_k the eigenfunctions of DIRICHLET's problem. Let

$$X^{(\alpha)} := L_2(W_2^\alpha(G, \mathbb{H}), [0,T]) \qquad (\alpha > 0),$$
$$Y^{(q,\beta)} := W_2^q(W_2^\beta(G, \mathbb{H}), [0,T]) \qquad (\beta \geq 1, q \in \mathbb{N}).$$

Proposition 4.76. [149] *Each solution of the initial boundary value problem (4.47,4.48) permits the representation*

$$\omega(x,t) = \int_0^t \int_G s(\underline{x}, \underline{y}, t, \tau)(-\eta\kappa\omega_0(\underline{y})\partial_\tau u(\underline{y}, \tau))dyd\tau + \\ + \int_G s(\underline{x}, \underline{y}, t, 0)\omega_0(y)dy + \tilde{\psi}, \tag{4.54}$$

where $\tilde{\psi}$ is the $Y^{(1,2)}$-continuation of ψ into G. In order to solve the original coupled problem (4.45-4.48) we have to investigate the convergence of the following algorithm:

$$\begin{aligned}\underline{u}_n = & \operatorname{Vec} T_G P_M B\mu_2\Big(\tfrac{m+1}{m-2}V\omega_{n-1}\mu_1 - \partial_{\tilde{\alpha}}u_{n-1} - \tfrac{1}{2}(\underline{\mu}_1, \operatorname{rot} u_{\underline{n-1}})\Big) \\ & - \operatorname{Vec} T_G Q_M M T_G \tfrac{2(m+1)}{m-2} V D w_{n-1}\end{aligned} \tag{4.55}$$

$$\omega_n = \int_0^t \int_G s(\underline{x}, \underline{y}, \cdot, \tau)(-\eta\kappa\omega_0(y)\partial_\tau u_n(y, \tau)dyd\tau \\ + \int_G s(\cdot, y, \cdot, 0)\omega_0(y)dy + \tilde{\psi}. \tag{4.56}$$

The following theorem is proved in [149].

Theorem 4.77. *Let G be a bounded domain with uniform cone property. Further, let $\underline{g}, \partial_t \underline{g} \in X^{(3/2)}$, $\omega_0 \in W_2^1(G, \mathbb{H})$, $\tilde{\psi} \in Y^{(1,2)}$ and $\underline{u}_0, \partial_t \underline{u}_0 \in X^{(1)} \cap$ $\operatorname{im} Q_\Gamma$. Here $\tilde{\psi}$ denotes an extension of ψ into the domain G. If $||\underline{u}_0||_{X^{(1)}}$, $||\partial_t \underline{u}_0||_{X^{(1)}}$ are sufficiently small then the iteration procedure (4.55,4.56) converges in $X^{(1)} \times X^{(1)} \times Y^{(1,2)}$ to the unique solution $(\underline{u}, \partial_t \underline{u}, \omega)$ of (4.45–4.48).*

4.7 Stokes' equations

At first in this section we will consider STOKES' classical problem. Then we will investigate the linear STOKES boundary value problems of a fluid with variable viscosity and density. Finally, we will generalize these results to a class of non-linear STOKES' problems.

Let again $G \subset \mathbb{R}^n$ be a bounded domain, $u = (u_1, ..., u_n)$ a vector function and p a scalar function. In $\mathbb{R}^3$ u denotes the velocity of the fluid and p the hydrostatic pressure. Furthermore, let f be the function of external forces and η stands for the viscosity of the fluid. Then STOKES' boundary value problem reads as follows:

$$\begin{aligned} -\Delta u + \frac{1}{\eta} \operatorname{grad} p &= f \quad \text{in } G, \\ \operatorname{div} u &= 0 \quad \text{in } G, \\ u &= 0 \quad \text{on } \Gamma. \end{aligned}$$

The hypercomplex notation of this problem is given by

$$DDu + \frac{1}{\eta} Dp = f \quad \text{in } G, \tag{4.57}$$

$$\operatorname{Sc} Du = 0 \quad \text{in } G, \tag{4.58}$$

$$u = 0 \quad \text{on } \Gamma. \tag{4.59}$$

We suppose that $f = \sum_{i=1}^{n} f_i e_i$ and add for $A \notin \{\{1\}, ..., \{n\}\}$ a LAPLACE equation $-\Delta u_A = 0$ with boundary condition $u_A = 0$. Then we have the following lemma.

Lemma 4.78. *Let $f \in L_2(G, C\ell_{0,n})$, $p \in W_2^1(G, \mathbb{R})$. Every $W_2^1(G, C\ell_{0,n})$-solution u of the system (4.57–4.59) permits the representation*

$$u = T_G \mathbf{Q} T_G f - \frac{1}{\eta} T_G \mathbf{Q} p. \tag{4.60}$$

Equation (4.58) may be written as

$$-\operatorname{Sc} (\mathbf{Q} T_G f + \frac{1}{\eta} T_G \mathbf{Q} p) = 0. \tag{4.61}$$

Proof. Applying the operator T_G from the left we get from equation (4.57)

$$T_G DDu + \frac{1}{\eta} T_G \mathbf{Q} p = T_G f.$$

Using BOREL–POMPEIU's formula we immediately have

$$Du - F_\Gamma Du + \frac{1}{\eta} p - \frac{1}{\eta} F_\Gamma Dp = T_G f$$

and

$$\mathbf{Q}Du + \frac{1}{\eta}\mathbf{Q}p = \mathbf{Q}T_G f.$$

Again applying the operator T_G from the the left we get

$$T_G\mathbf{Q}Du + \frac{1}{\eta}T_G\mathbf{Q}p = T_G\mathbf{Q}T_G f.$$

Because of $Du \in \text{im } \mathbf{Q}$ the validity of the equations (4.60) and (4.61) now follows. #

We write for brevity $\|\cdot\|_2 := \|\cdot\|_{L_2(G)}$, and $\|\cdot\|_{2,1} := \|\cdot\|_{W_2^1}$.

Theorem 4.79. *The system of* STOKES' *equations (4.57–4.59) has a unique solution* $\{u, p\}$ *(p is unique up to a constant) which can represented by (4.60) and fulfils the inequality*

$$\left(\frac{\lambda_1}{\lambda_1 + 1}\right)^{1/2} \|u\|_{2,1} + \frac{1}{\eta}\|\mathbf{Q}p\|_2 \leq \sqrt{2}\|T_G f\|_2, \tag{4.62}$$

where λ_1 *is the first eigenvalue of* DIRICHLET'*s eigenvalue problem for the Laplacian.*

Proof. Denote by $M_2^1(G, C\ell_{0,n})$ the space of all $C\ell_{0,n}$-valued functions which have generalized D-derivatives belonging to the space $L_2(G, C\ell_{0,n})$. It can be shown that the subspace $\mathbf{Q}M_2^1(G, C\ell_{0,n})$ permits the orthogonal decomposition

$$\mathbf{Q}M_2^1(G, C\ell_{0,n}) = \ker \text{ div } D(\overset{\circ}{W_2^1}(G, C\ell_{0,n}) \oplus \mathbf{Q}L_2(G, \mathbb{R})$$

with respect to the scalar product

$$[u, v]_2 = \text{Sc} \int_G \overline{u}v \, dG. \tag{4.63}$$

This proof is similar to the quaternionic case in [110].
Now let $z \in \mathbf{Q}M_2^1(G, C\ell_{0,n})$. Then $z = \mathbf{Q}T_G f$. Because of (4.60) a $C\ell_{0,n}$-function $u \in \overset{\circ}{W_2^1}(G, C\ell_{0,n}) \cap \ker \text{div}$ exists and a scalar function $p \in L_2(G, \mathbb{R})$ such that

$$Du + \frac{1}{\eta}\mathbf{Q}p = \mathbf{Q}T_G f. \tag{4.64}$$

Applying BOREL–POMPEIU's formula we obtain

$$u + \frac{1}{\eta}T_G\mathbf{Q}p = T_G\mathbf{Q}T_G f. \tag{4.65}$$

Because of $Du \in \text{im}\,\mathbf{Q}$, $\text{Sc}\,(Du) = 0$ and $\text{Vec}\,p = 0$, we have that $[Du, \mathbf{Q}p]_2 = [Du, p]_2 - [Du, \mathbf{P}p]_2 = 0$. In this way we get by using (4.65)

$$||Du||_2 + \frac{1}{\eta^2}||\mathbf{Q}p||_2 \leq \sqrt{2}||T_G f||_2. \tag{4.66}$$

POINCARE's inequality and the $C\ell_{0,n}$-version to the estimation of $||T||_{[im\mathbf{Q}, W_2^1]}$ (cf. [109]) provides the inequality (4.66). From (4.65) the uniqueness of u immediately follows. $\mathbf{Q}p = 0$ provides $p \in \ker D$. Hence, p is unique up to a real constant. #

Theorem 4.80. *Let $f \in L_2(G, C\ell_{0,n}) \cap W_p^{k,\text{loc}}(G, C\ell_{0,n})$ $(k \in N), 1 < p < \infty$. Then $u \in W_p^{k+2,\text{loc}}(G, C\ell_{0,n})$ and $p \in W_p^{k+1,\text{loc}}(G, \mathbb{R})$.*

Proof. Since $\text{Vec}\,p = 0$ we get $\text{Im}\,\mathbf{P}p = -\ \text{im}\,\mathbf{Q}p \in \ker \Delta(G)$ and therefore, $\text{im}\,\mathbf{Q}\text{p} \in C_{\mathbf{R}}^\infty(G'), \overline{G'} \subset G$. Because of $\text{Sc}\,\mathbf{Q}p = \text{Sc}\,\mathbf{Q}T_G f$ we obtain $\text{Sc}\,\mathbf{Q}p \in W_p^{k+1,\text{loc}}(G, \mathbb{R})$ and hence, $\mathbf{Q}p \in W_p^{k+1,\text{loc}}(G, C\ell_{0,n})$. From (4.66) we have $Du \in W_p^{k+1,\text{loc}}(G, C\ell_{0,n})$ and, finally, $T_G Dp \in W_p^{k+1,\text{loc}}(G, C\ell_{0,n})$. Hence, one gets

$$p = T_G Dp + F_\Gamma p \in W_p^{k+1,\text{loc}}(G, C\ell_{0,n}). \qquad \# \tag{4.67}$$

Of course, we are also interested in global regularity results. For this reason we again study the properties of the boundary operator $\text{tr}_\Gamma T_G F_\Gamma$. Under some additional assumptions concerning the domain of this operator analogously to Theorem 4.22 we can prove the following lemma:

Lemma 4.81. *Let $1 < p < \infty$. Then the operator*

$$\text{tr}_\Gamma T_G F_\Gamma : W_p^{k-\frac{1}{p}}(\Gamma) \cap \text{im}\,P_\Gamma \cap \ker \text{Sc} \mapsto W_p^{k+1-\frac{1}{p}}(\Gamma) \cap \text{im}\,Q_\Gamma \cap \ker \text{tr}_\Gamma \,\text{div}$$

is an isomorphism.

Proof. The regularity properties follow from the mapping properties of the operators F_Γ, T_G, and tr_Γ. We have only to prove the statement about range, domain, and the injectivity.

Let $v \in W_p^{k-\frac{1}{p}}(\Gamma) \cap \text{im}\,P_\Gamma \cap \ker \text{Sc}$ be an arbitrary function. If we suppose that $\text{tr}_\Gamma T_G v = 0$ then we immediately get $v = 0$ because we have here a specialized case of Theorem 4.22.

Let us now assume that $w \in W_p^{k+1-\frac{1}{p}}(\Gamma) \cap \text{im}\,Q_\Gamma \cap \ker \text{tr}_\Gamma \,\text{div}$. In particular, $w \in \text{im}\,Q_\Gamma$ and we have the existence of a function $u \in \ker \Delta$ with $\text{tr}_\Gamma u = w$. From the BOREL–POMPEIU formula we obtain

$$u = F_\Gamma w + T_G Du = T_G v.$$

Here, we substitute $v = Du \in \ker D(G)$. Denoting the boundary values of v by z we get $u = T_G F_\Gamma z$ and, consequently,

$$w = \text{tr}_\Gamma T_G F_\Gamma z.$$

The regularity is ensured because the assumptions imply $v \in W_p^k(G)$ and $T_G v \in W_p^{k+1}(G)$. Finally we have

$$\operatorname{div} T_G v = -\operatorname{Sc} DT_G v = -\operatorname{Sc} v \quad \text{and } \operatorname{tr}_\Gamma \operatorname{div} T_G v = -\operatorname{Sc} v = 0.$$

This implies $\operatorname{Sc} z = 0$ and the proof is complete. #

Now we are able to prove the global regularity result. Additionally, we get a decomposition formula for the solution of the STOKES system. More exactly, we now can write down representation formulas for the velocity or the pressure alone. Clearly, this result also has some importance for numerical methods for the solution of our boundary value problem. But, for the moment we need it only for the regularity theorem.

Theorem 4.82. *Suppose that* $f \in W_p^k(G)$, $k \geq 0$, $1 < p < \infty$. *Then the only solution of (4.57–4.59) can be represented by*

$$u = \frac{\rho}{\eta} T_G \operatorname{Vec} T_G f - \frac{\rho}{\eta} T_G \operatorname{Vec} F_\Gamma (\operatorname{tr}_\Gamma T_G \operatorname{Vec} F_\Gamma)^{-1} \operatorname{tr}_\Gamma T_G \operatorname{Vec} T_G f \tag{4.68}$$

$$p = \rho \operatorname{Sc} T_G f - \rho \operatorname{Sc} F_\Gamma (\operatorname{tr}_\Gamma T_G \operatorname{Vec} F_\Gamma)^{-1} \operatorname{tr}_\Gamma T_G \operatorname{Vec} T_G f. \tag{4.69}$$

For the global regularity we have $u \in W_p^{k+2}(G) \cap \overset{\circ}{W}{}_2^1(G)$ *and* $p \in W_{p,\mathbf{R}}^{k+1}(G)$.

Proof. First we have to prove that the operator $\operatorname{tr}_\Gamma T_G \operatorname{Vec} F_\Gamma$ is correctly defined and that it is invertible.
Clearly, $\operatorname{tr}_\Gamma T_G \operatorname{Vec} T_G f \in \operatorname{im} Q_\Gamma$. Furthermore, we have

$$\operatorname{div} T_G \operatorname{Vec} T_G f = -\operatorname{Sc} DT_G \operatorname{Vec} T_G f = -\operatorname{Sc} \operatorname{Vec} T_G f = 0.$$

Hence,

$$\operatorname{tr}_\Gamma T_G \operatorname{Vec} T_G f \in \ker \operatorname{tr}_\Gamma \operatorname{div}.$$

Assuming now $\operatorname{Sc} z = 0$. If $\operatorname{tr}_\Gamma T_G F_\Gamma z \in \ker \operatorname{tr}_\Gamma \operatorname{div}$ then $\operatorname{tr}_\Gamma \operatorname{Sc} F_\Gamma z = 0$. Because $\operatorname{Sc} F_\Gamma z \in \ker \Delta$ we immediately get that $\operatorname{Sc} F_\Gamma z = 0$ in G. This result implies that $\operatorname{tr}_\Gamma T_G F_\Gamma z$ may be rewritten as $\operatorname{tr}_\Gamma T_G \operatorname{im} F_\Gamma$. Now, using the previous lemma we recognize that the representations (4.68) and (4.69) make sense. Inserting the right-hand sides of (4.68) and (4.69) in (4.57–4.59) we can prove after some computations that the representations (4.68,4.69) describe the solution of the boundary value problem under consideration. It remains the proof of the regularity result. However, this is now an exercise. We have to use only the mapping properties of the occurring operators. #

Now let $\eta = \eta(x)$ be a variable viscosity. With a parameter Λ we get the following system:

$$-\Delta u + \eta^{-1} \operatorname{grad} p = \Lambda f(u) \quad \text{in } G \tag{4.70}$$

$$\operatorname{div} \eta^{-1} u = 0 \quad \text{in } G \tag{4.71}$$

$$u = 0 \quad \text{on } \Gamma. \tag{4.72}$$

Remark 4.83. In case of fluids without sources and sinks the continuity condition provides the following connection between the velocity u and the density ϱ, namely

$$\operatorname{div}(\varrho u) = 0 \quad \text{that means} \quad \frac{1}{\varrho}\operatorname{grad}\varrho \perp u. \tag{4.73}$$

In general case condition (4.73) is to replace by the condition

$$\frac{1}{\varrho}\operatorname{grad}\varrho + \frac{1}{\eta}\operatorname{grad}\eta \perp u.$$

This generalization of STOKES' system (4.57–4.59) may be written in hypercomplex notation

$$DDu + BDp = \Lambda Bf \quad \text{in } G, \tag{4.74}$$

$$\operatorname{Sc} BDu = 0 \quad \text{in } G, \tag{4.75}$$

$$u = 0 \quad \text{on } \Gamma. \tag{4.76}$$

Here B stands for the operator of multiplication by $\eta(x)^{-1}$.

Theorem 4.84. *Let $f \in L_2(G, C\ell_{0,n})$, $p \in W_2^1(G)$, $\eta \in C^\infty(G)$. Then every solution of the system (4.74–4.76) permits the integral representations*

$$u = \Lambda T_G \mathbf{Q} T_G B f - T_G \mathbf{Q} T_G BDp \tag{4.77}$$

$$0 = \operatorname{Sc}\Lambda \mathbf{Q} T_G B f - \operatorname{Sc}\mathbf{Q} T_G BDp. \tag{4.78}$$

Proof. Applying the operator T_G^2 from the left to equation (4.74) we get by the aid of VEKUA's theory

$$u = -T_G^2 BDp + \Lambda T_G^2 Bf + T_G v_2 + v_1 \tag{4.79}$$

with v_2, $v_1 \in \ker D$. Because $\operatorname{tr}_\Gamma u = 0$ and $\operatorname{Vec} Q_\Gamma \cap \operatorname{Vec} P_\Gamma = \{0\}$ BOREL–POMPEIU's formula yields $v_1 \equiv 0$. So we have for u the representation

$$u = -T_G^2 BDp + \Lambda T_G^2 Bf + T_G v_2$$

with $\operatorname{tr}_\Gamma v_2 = (\operatorname{tr}_\Gamma T_G F_\Gamma)^{-1}(\operatorname{tr}_\Gamma T_G^2 BDp - \Lambda \operatorname{tr}_\Gamma T_G^2 Bf)$. In Theorem 4.22. it is verified that $(\operatorname{tr}_\Gamma T_G F_\Gamma)^{-1}$ exists in a certain pair of BANACH spaces. It follows that

$$v_2 = F_\Gamma(\operatorname{tr}_\Gamma T_G F_\Gamma)^{-1}(\operatorname{tr}_\Gamma T_G^2 BDp - \Lambda \operatorname{tr}_\Gamma T_G^2 Bf).$$

Inserting this expression for v_2 in equation (4.79) we obtain after a straightforward computation

$$u = \Lambda T_G \mathbf{Q} T_G Bf - T_G \mathbf{Q} T_G BDp,$$

where $\mathbf{Q} = I - F(\operatorname{tr}_\Gamma T_G F)^{-1}\operatorname{tr}_\Gamma T_G$ is the orthoprojection onto $D\overset{\circ}{W}{}_2^1(G)$. #

Lemma 4.85. *Let* $\{p,u\} \in L_2(G) \times \overset{\circ}{W}{}_2^1(G, C\ell_{0,n})$ *and* $\eta \in C^\infty(G)$ *a solution of system (4.77) and (4.78. Then we have the identity*

$$[Du, \mathbf{Q}T_G(\eta Dp)]_2 = 0.$$

Proof. For the proof we refer to [114]. #

Theorem 4.86. (A-priori estimate) *Let* $\{p,u\} \in L_2(G) \times \overset{\circ}{W}{}_2^1(G, C\ell_{0,n})$ *a solution of the system (4.77,4.78),* $\eta \in C^\infty(G)$. *Then the following inequality is valid:*

$$\sqrt{\frac{\lambda_1}{\lambda_1+1}}\,||u||_{2,1} + ||\mathbf{Q}T_G BDp||_2 \leq \sqrt{2}\,\lambda\,||T_G Bf||_2. \tag{4.80}$$

Proof. Applying the operator D from the left to equation (4.79) we obtain

$$Du + \mathbf{Q}T_G BDp = \Lambda \mathbf{Q} T_G Bf.$$

Hence,

$$\begin{aligned} ||Du + \mathbf{Q}T_G BDp||_2^2 &= |\Lambda|^2 ||\mathbf{Q}T_G Bf||_2^2 \\ ||Du||_2^2 + 2[Du, \mathbf{Q}T_G BDp]_2 + ||\mathbf{Q}T_G BDp||_2^2 &= |\Lambda|^2 ||\mathbf{Q}T_G Bf||_2^2. \end{aligned}$$

Because of ($||\mathbf{Q}||_2 = 1!$), we have

$$||Du||_2 + ||\mathbf{Q}T_G BDp||_2 \leq \sqrt{2}\,|\lambda|\,||\mathbf{Q}T_G Bf||_2.$$

Using the inequality (cf. [109]) $\sqrt{\frac{\lambda_1}{\lambda_1+1}}\,||u||_{2,1} \leq ||Du||_2$ we achieve our assertion. #

Theorem 4.87. *Let* $f \in L_2(G, C\ell_{0,n})$. *Then the system (4.77,4.78) has a unique solution* $\{u,p\}$ *with* $u \in \overset{\circ}{W}{}_2^1(G, C\ell_{0,n})$ *and* $p \in L_2(G)$. *The scalar function* p *is unique up to a real constant.*

Proof. Let $h \in L_2(G)$. Suppose that for all $u \in \overset{\circ}{W}{}_2^1(G, C\ell_{0,n}) \cap \ker(\operatorname{div} B)$ and $p \in L_2(G)$ holds $[Du, \mathbf{Q}T_G h]_2 = [\mathbf{Q}p, \mathbf{Q}T_G h]_2 = 0$. We have to show that h must be 0. One gets $[Du, \mathbf{Q}T_G h]_2 = [u,h]_2 = 0$. Using DE RHAM's theorem [314] we find a scalar function $q \in W_2^1(G)$ with $h = BDq$. Inserting this expression for h in our assumptions then we get

$$[\mathbf{Q}p, \mathbf{Q}T_G h]_2 = [\mathbf{Q}p, T_G BDq]_2 = 0 \qquad (p \in L_2(G)).$$

Therefore, we have $T_G BDq \in \operatorname{im}\mathbf{P}$ and so $DT_G BDq = BDq = h = 0$. In this way $\mathbf{Q}T_G Bf$ can be decomposed with respect to the scalar product $[\cdot,\cdot]_2 = \operatorname{Sc}(\cdot,\cdot)_2$ with a certain $u \in \overset{\circ}{W}{}_2^1(G, C\ell_{0,n}) \cap \ker(\operatorname{div} B)$ and $p \in L_2(G)$. This leads to the equation

$$Du + \mathbf{Q}T_G BDp = \Lambda \mathbf{Q}T_G Bf.$$

Hence, it immediately follows that

$$u + T\mathbf{Q}T_G BDp = \Lambda T\mathbf{Q}T_G Bf.$$

These functions obtained by the orthogonal decomposition satisfy the system (4.77,4.78). In order to prove the uniqueness we assume that two different pairs $\{u_i, p_i\}$ $(i = 1, 2)$ are solutions of system (4.77,4.78). Then we get

$$u_1 - u_2 + T\mathbf{Q}T_G BD(p_1 - p_2) = 0\ .$$

Theorem 4.86 provides that $u_1 = u_2$ and $\|\mathbf{Q}T_G BD(p_1 - p_2)\|_2 = 0$. Hence, $T_G BD(p_1 - p_2) \in \operatorname{Vec}\mathbf{P}$ and also $D(p_1 - p_2) = 0$. The only real-valued $C\ell_{0,n}$-regular functions are constant and so $p_1,\ p_2 = c$. #

Definition 4.88. *The $C\ell_{0,n}$-valued function u is called a weak solution of the system (4.74–4.76), if for all functions $v \in \overset{\circ}{W}{}_2^1(G, C\ell_{0,n}) \cap \ker(\operatorname{div} B)$ the equality*

$$(DDu + BDp, v)_{\mathbf{R}} = \Lambda(Bf, v)_{\mathbf{R}}$$

is valid, where $(u, v)_{\mathbf{R}} = \int_G \sum_{i=1}^n u_i v_i\, dG$.

Theorem 4.89. *Let $u \in \overset{\circ}{W}{}_2^1(G, C\ell_{0,n})$ and $p \in L_2(G)$ solve the system (4.77,4.78). Then $\underline{u} = u$ is a weak solution of the system (4.74–4.76). If $\underline{u}$ is a weak solution of the system (4.74–4.76) then a function $p \in L_2(G)$ will exist such that the pair $\{u, p\}$ with $u = \underline{u}$ solves the system (4.77,4.78).*

Proof. For the proof we again refer to [114]. #

Now we assume that the outer forces depend on the velocity u. The operator function $f(u)$ is acting in the pair of spaces $[W_2^1(G, C\ell_{0,n}), L_2(G, C\ell_{0,n})]$.

Corollary 4.90. *Let $f(u) \in L_2(G, C\ell_{0,n})$, $p \in W_2^1(G, C\ell_{0,n})$. Every solution of the system (4.74–4.76) permits the representation*

$$u = \Lambda T\mathbf{Q}T_G Bf(u) - T\mathbf{Q}T_G BDp, \tag{4.81}$$

$$0 = \operatorname{Sc}\ \Lambda\mathbf{Q}T_G Bf(u) - \operatorname{Sc}\ \mathbf{Q}T_G BDp. \tag{4.82}$$

Proof. Replacing f by $f(u)$ in the proof of Lemma 34.2 the assertion is verified.

Corollary 4.91. (Almost a-priori estimate.)

Let $p \in L_2(G)$ and $u \in \overset{\circ}{W}{}_2^1(G, C\ell_{0,n})$ fulfil the system (4.77,4.78). Then we have

$$\sqrt{\frac{\lambda_1}{\lambda_1 + 1}}\,\|u\|_{2,1} + \|\mathbf{Q}T_G BDp\|_2 \le \sqrt{2}\,\lambda\,\|T_G Bf(u)\|_2.$$

Proof. The proof is similar to the proof of Theorem 4.81. #

In order to prove the existence and uniqueness of the solution of the system (4.77,4.78) the operator function $f(u)$ has to fulfil the conditions

(i) $||f(u) - f(v)||_2 \leq L||u - v||_{2,1}$ for $||u||_{2,1}, ||v||_{2,1} \leq 1$

(ii) $||B||_2 \leq K$,

where $L, K \in \mathbb{R}^+$. The next aim is the construction of an iteration procedure such that we can find a solution of system (4.77,4.78). We fix for the following $T\mathbf{Q}T_G B = R$:

$$\begin{aligned} &u_n = \Lambda R f(u_{n-1}) - RDp_n \\ &\Lambda \,\mathrm{Sc}\, DBRf(u_{n-1}) = \mathrm{Sc}\, DBRDp_n \\ &||u_0||_{2,1} \leq 1 \end{aligned} \tag{4.83}$$

with $u_0 \in \overset{\circ}{W}{}_2^1(G, C\ell_{0,n})$.
In each step a linear problem has to be solved. The existence and uniqueness of linear problems we have shown previously. Now we will formulate conditions for K, L and Λ such that the iteration procedure (4.83) will converge. At first, using the almost-a-priori-estimate for two iteration steps u_n and u_{n-1}, we get

$$\begin{aligned} ||u_n - u_{n-1}||_{2,1} &\leq ||T||_{[\mathrm{im}\mathbf{Q}\cap L_2, L_2]} ||T||_{[L_2, L_2]} K ||f(u_n) - f(u_{n-1})||_2 \\ &\leq ||T||_{[\mathrm{im}\mathbf{Q}\cap L_2, L_2]} ||T||_{[L_2, L_2]} KL ||u_n - u_{n-1}||_{2,1}, \end{aligned}$$

where $||u_k||_{2,1} \leq 1, \;\; k = 0, 1, \ldots$. In the case of

$$\Lambda < \{||T||_{[\mathrm{im}\mathbf{Q}\cap L_2, L_2]} ||T||_{[L_2, L_2]} KL\}^{-1}$$

the iterations u_n cannot leave the unit ball of $W_2^1(G, C\ell_{0,n})$. Applying BANACH's fixed-point theorem we have convergence. In this way we verified:

Theorem 4.92. *System (4.77,4.78) has the unique solution* $\{u, p\} \in L_2(G) \times \overset{\circ}{W}{}_2^1(G, C\ell_{0,n}) \cap \ker(\mathrm{div}\, B)$ *(p is unique up to real additive constant) if the parameter* Λ *satisfies the inequality*

$$\Lambda < \{||T||_{[\mathrm{im}\mathbf{Q}\cap L_2, L_2]} ||T||_{[L_2, L_2]} KL\}^{-1}.$$

The iteration procedure (4.83) will yield the solution of (4.77,4.78).

Remark 4.93. In Section 4.15 we prove that

$$||T||_{[\mathrm{im}\mathbf{Q}, W_2^1]} = \sqrt{1 + \frac{1}{\lambda_1}}, \quad ||T||_{[\mathrm{im}\mathbf{Q}, L_2]} = \frac{1}{\sqrt{\lambda_1}},$$

where λ_1 is the smallest eigenvalue of the problem

$$\{-\Delta u = \lambda u, \;\; \mathrm{tr}_\Gamma\, u = 0 \text{ on } \Gamma\}.$$

4.8 Navier–Stokes equations and related equations

The method explained above to look for the solution of a boundary value problem by the help of hypercomplex integral transformations is applicable in the non-linear case, too. To demonstrate this we have chosen the time-harmonic case of the non-linear stream problem with free convection. The main idea is to transform the system of partial differential equations into an equivalent system of operator-integral equations. In the most cases the properties of the arising integral operators allow to apply classical fixed-point principles for the iterative solution of the non-linear boundary value problem. For considerations of other boundary value problems we refer for instance to [302].

4.8.1 The non-linear boundary value problem

We consider the following boundary value problem:

$$\begin{aligned} -\Delta \underline{u} + a_1(\underline{u}\cdot \mathrm{grad})\underline{u} + f(u) + a_2 \mathrm{grad}\, p + a_3(-e_3)w &= F(x) && \text{in } G \\ \mathrm{div}\, \underline{u} &= 0 && \text{in } G \\ -\Delta w + a_4(\underline{u}\cdot \mathrm{grad})w &= g && \text{in } G \\ \underline{u} &= 0 && \text{on } \Gamma \\ w &= 0 && \text{on } \Gamma. \end{aligned}$$

The coefficients a_i correspond to physical constants. If we denote by ρ the density of the fluid, by η the viscosity, by γ the GRASHOF number, by κ the temperature conductivity, and by m the PRANDTL number, then we have

$$a_1 = \frac{\rho}{\eta},\ a_2 = \frac{1}{\eta},\ a_3 = \frac{\gamma}{\eta},\ a_4 = \frac{m}{\kappa}.$$

As usual p denotes the hydrostatic pressure, $\underline{u}$ stands for the velocity of the fluid, and w is the temperature.

The first step is the translation of this problem into the language of CLIFFORD analysis. For this reason we set $u = u_0 + \underline{u}$. Then, we add to our system the boundary value problem $\{\Delta u_0 = 0 \text{ in } G, u = 0 \text{ on } \Gamma\}$. This is only a formal step to get quaternionic valued functions u. Analogously, we identify the scalar function p with the quaternionic valued function $(p, 0, 0, 0)^T$. Defining $M(u) = a_1(\underline{u}\cdot \mathrm{grad})u + f(u) - F$, we can write our problem as follows:

$$\begin{aligned} D^2 u + M(u) + a_2 \mathrm{grad}\, p + a_3(-e_3)w &= 0 && \text{in } G \\ -\mathrm{Sc}\, Du &= 0 && \text{in } G \\ D^2 w - a_4 \mathrm{Sc}\,(uD)w &= g && \text{in } G \\ u &= 0 && \text{on } \Gamma \\ w &= 0 && \text{on } \Gamma. \end{aligned}$$

Using the same ideas as in the previous section we apply $T\mathbf{Q}T$ and obtain the following system of non-linear operator equations:

$$u = -T\mathbf{Q}T[M(u) - a_3 e_3 w] - a_2 T\mathbf{Q}p \tag{4.84}$$
$$0 = \mathrm{Sc}\,\{\mathbf{Q}T[M(u) - a_3 e_3 w] + a_2 \mathbf{Q}p\} \tag{4.85}$$
$$w = -a_4 T\mathbf{Q}T\mathrm{Sc}\,(uD)w + T\mathbf{Q}Tg. \tag{4.86}$$

We remark that the boundary conditions are satisfied automatically due to the properties of the orthoprojection $\mathbf{Q}$. Indeed, from the definition of $\mathbf{Q}$ we get $\mathbf{Q}u = Dv$ with $v \in \overset{\circ}{W}{}_2^1$. Using BOREL–POMPEIU's formula we obtain

$$\mathrm{tr}_\Gamma\, T\mathbf{Q}u = \mathrm{tr}_\Gamma\, TDv = \mathrm{tr}_\Gamma\,(w - F_\Gamma \mathrm{tr}_\Gamma\, w) = 0.$$

It should be mentioned that special choices of the coefficients a_i lead to special systems with their own importance. If we have $a_3 = 0$ or $a_4 = 0$ then we can decompose the system. In the case $a_3 = 0$ the first two equations describe the stationary NAVIER–STOKES problem. If in addition a_1 vanishes then we have the STOKES problem from Section 4.7. Taking all coefficients a_i to be zero we treat the classical (linear) POISSON equation and the semilinear POISSON equation.

In all that follows we assume that the coefficients a_i are constants. The case of variable coefficients can also be treated with the described methods. For examples we refer to the previous section and [114].

4.8.2 Iteration method

The idea of the following iteration procedure is based on BANACH's fixed-point theorem. Starting with suitably chosen functions u_0 and w_0 we compute for $n = 1, 2, \ldots$

$$u_n = -T\mathbf{Q}T[M(u_{n-1}) - a_3 e_3 w_{n-1}] - a_2 T\mathbf{Q}p_n \tag{4.87}$$
$$0 = \mathrm{Sc}\,\{\mathbf{Q}T[M(u_{n-1}) - a_3 e_3 w_{n-1}] + a_2 \mathbf{Q}p_n\} \tag{4.88}$$
$$w_n = -a_4 T\mathbf{Q}T\,\mathrm{Sc}\,(u_n D)w_n + T\mathbf{Q}Tg. \tag{4.89}$$

For the computation of w_n we use the "inner" iteration:

$$w_n^{(i)} = -a_4 T\mathbf{Q}T\,\mathrm{Sc}\,(u_n D)w_n^{(i-1)} + T\mathbf{Q}Tg \quad (i = 1, 2, \ldots). \tag{4.90}$$

We remark that in each step of this iteration procedure we have to solve only a linear STOKES problem (see Section 4.7) and a POISSON equation (see Section 4.2).

Theorem 4.94.

1. *Let* $u_n \in \overset{\circ}{W}{}_2^1$. *If* $a_4 \neq 4$ *and* $||u_n|| < 1/(a_4KC)$ *where* C *again is the embedding constant from* W_2^1 *in* L_2 *and* $K = a_1||T||_{[\mathrm{im}\mathbf{Q},W_2^1]}||T||_{[L_2,L_2]}$, *then the sequence* $\{w_n^{(i)}\}_{i\in\mathbf{N}}$ *converges in* W_2^1.

2. *Let* $F \in L_2(G), g \in L_2(G), f : W_2^1 \mapsto L_2(G)$ *with* $||f(u) - f(v)||_2 \leq L||u - v||_{W_2^1}$ *for some* $L > 0$, *and* $f(0) = 0$. *Under the additional conditions*

   (i) $a_1||F||_2 + a_3K|d|^{-1}||g||_2 < 1/(16K^2C),\ d = (4 - a_4)\kappa$

   (ii) $||g||_2 < (1 - \frac{1}{\sqrt{2}})a_2^{-1}d^2(1/(32K^3Cm))$

   (iii) $a_4 < 4$

the NEWTON *sequence* $\{u_n, w_n, p_n\}_{n\in\mathbf{N}}$ *converges in* $W_2^1 \times W_2^1 \times L_2$ *to the unique solution*

$$(u, w, p) \in \overset{\circ}{W}{}_2^1 \times \overset{\circ}{W}{}_2^1 \times L_2(G)$$

(p is unique up to a real constant) of the original boundary value problem.

Remark 4.95. The conditions (i) and (ii) can be fulfilled for instance for fluids with a sufficiently large viscosity number. Condition (iii) is easy to check.

For the sake of brevity we will prove this theorem only for the special case of the NAVIER–STOKES equations without the influence of convection. This is on the one hand sufficient for the understanding of the basic ideas and on the other hand, the full problem has been proved already in [110].

In preparation of the proof we have to investigate the non-linear convection term and to prove some necessary norm estimates. It is sufficient to look for the properties of $M^*(u) = \mathrm{Sc}\,(uD)u$ because F does not depend on u and the behaviour of $f(u)$ is fixed by the corresponding assumptions.

Lemma 4.96. *Let* $u \in \overset{\circ}{W}{}_2^1, 1 < p < \frac{3}{2}$. *Then*

$$||M^*(u)||_{L_p} \leq C_1||u||_{W_2^1}^2$$

holds.

Proof. We have

$$||M^*(u)||_{L_p}^p \leq \sum_{i,j=1}^{3} \int_G |u_i\partial_i u_j|^p dG \leq \sum_{i,j=1}^{3} ||u_i||_{L_q}^p ||\partial_i u_j||_{L_2}^p \leq$$

$$\leq \sum_{i,j=1}^{3} C^p \|u_i\|^p_{W_2^1} \|u_j\|^p_{W_2^1} \leq 9C^p \|u\|^{2p}_{W_2^1},$$

where $q < 6, p = \frac{2q}{2+q}$. The last estimate is a consequence of HÖLDER's inequality. The constant C is the embedding constant from $\overset{\circ}{W}{}_2^1$ to L_q and for $9^{\frac{1}{p}}C$ we used the notation C_1. #

To demonstrate that we are able to handle all constants within the theory we will formulate the next lemma about the estimation of C_1.

Lemma 4.97. *The constant C_1 from Lemma 4.96. can be estimated by*

$$C_1 \leq 9^{\frac{1}{p}} \|T_G\|_{[L_2, L_q]}$$

with the same p and q as above.

Proof. Let $u \in \overset{\circ}{W}{}_2^1$. Then we have

$$\|u\|_{L_q} \leq \|TDu\|_{L_q} \leq \|T_G\|_{[L_2,L_q]} \|Du\|_{L_2} \leq \|T_G\|_{[L_2,L_q]} \|u\|_{W_2^1}. \qquad \#$$

Remark 4.98. This idea is already contained in [110]. References for further estimates of $\|T\|$ in L_p-spaces are contained for instance in [98] and [201]. In these papers connections between the norms of T and the smallest eigenvalue (which can be computed approximately by numerical methods) are used. In the new paper [149] other ideas based on algebraical methods are presented. Collecting all these information estimates proved above also seem to be useful for practical computations.

Proof of Theorem 4.94. From Theorem 4.79. we have the existence of a unique solution of the STOKES problems occurring in each step of the iteration procedure. For simplicity we will now omit the norm index in all W_2^1 norms. For L_p-norms we will abbreviate to $\|\cdot\|_p$. In the proof of Theorem 4.62, if we replace the right-hand side f by $M(u)$, then we can repeat the procedure and obtain the inequality

$$\sqrt{\frac{\lambda_1}{1+\lambda_1}} \|u\| + \frac{1}{\eta} \|\mathbf{Q}p\|_2 \leq \sqrt{2} \frac{\rho}{\eta} \|TM(u)\|_2 \tag{4.91}$$

for arbitrary $u \in \overset{\circ}{W}{}_2^1 \cap \ker \operatorname{div}, p \in L_2$. Using inequality (4.91) we get

$$\|u_n - u_{n-1}\| \leq \|T\mathbf{Q}T(M(u_{n-1}) - M(u_{n-2}))\| + \frac{1}{\eta} \|T\mathbf{Q}(p_n - p_{n-1})\| \leq$$

$$\leq 2K \|M(u_{n-1}) - M(u_{n-2})\|_p.$$

As above in Lemma 4.96. we conclude that

$$||M(u_{n-1}) - M(u_{n-2})||_p \leq ||M^*(u_{n-1}) - M^*(u_{n-2})||_p \leq$$

$$\leq ||((u_{n-1} - u_{n-2}) \cdot \text{grad})u_{n-1}||_p + ||(u_{n-2} \cdot \text{grad})(u_{n-1} - u_{n-2})||_p \leq$$

$$\leq C_1||u_{n-1} - u_{n-2}||(||u_{n-1} + u_{n-2})||.$$

We abbreviate $L_n = 2KC_1||(||u_{n-1} + u_{n-2})||$. Then we have

$$||u_n - u_{n-1}|| \leq L_n||u_{n-1} - u_{n-2}||.$$

Now, we assume that $||u_{n-1}|| \leq \frac{1}{4KC_1} - W$ with

$$W = \sqrt{\frac{1}{(4KC_1)^2} - \rho\frac{||F||_p}{\eta C_1}}.$$

Then, it follows that

$$||u_n|| \leq ||T\mathbf{Q}TM(u_{n-1})|| + \frac{1}{\eta}||T\mathbf{Q}p_n|| \leq$$

$$2KC_1||u_{n-1}||^2 + 2K\frac{\rho}{\eta}||F||_p \leq \frac{1}{4KC_1} - W.$$

In a similar way we get from

$$\frac{1}{4KC_1} - W \leq ||u_{n-1}|| \leq \frac{1}{4KC_1} + W$$

that $||u_{n-1}|| \leq ||u_n||$ holds. It is necessary for all these estimates that

$$\frac{\rho}{\eta}||F||_p \leq \frac{1}{16K^2C_1}.$$

Now we know that the NEWTON sequence is completely contained in a ball with fixed radius. If we now use the assumption that $||u_{n-1}|| \leq 1/(4KC_1) - W$ we immediately obtain

$$L_n \leq L = 1 - 4KC_1W < 1.$$

The last thing is to define the radius mentioned above. Let $R = 1/(2KC_1)$. Then we get from $||u_{n-1}|| < R$ that $L_n < 1 - \varepsilon$ with a $\varepsilon > 0$ not depending on n. If we additionally suppose that $||F||_p > (1/(8KC_1)^2)C_1\frac{\rho}{\eta}$ then we can realize $R > 1/(4KC_1) + W$ and no other restrictions occur. All the assumptions necessary for the application of BANACH's fixed-point theorem are now fulfilled.
#

Of course the existence theorem yields an a-priori estimate for u.

Corollary 4.99. *Under the assumptions of Theorem 4.94. specialized as in the proof*

$$||u|| \leq \frac{1}{4KC_1} - W$$

holds.

More interesting for practical applications of the method are error estimates for the NEWTON sequence. From the proof of Theorem 4.94. we immediately get

$$||u_n - u|| \leq L^n ||u_0 - u||$$

and by the help of the a-priori estimate for $u_0 = 0$

$$||u_n - u|| \leq L^n \left(\frac{1}{4KC_1} - W\right).$$

If we have a right-hand side of the differential equation with better regularity as assumed above, our representation formulas may also be used for improving the regularity result of the theorem.

Theorem 4.100. *Let $F \in W_q^k(G), q > \frac{6}{5}, \; f = 0$. Then the solution (u, p) of the uncoupled NAVIER–STOKES system belongs to the space $(W_q^{k+2} \cap \overset{\circ}{W}{}_2^1) \times W_q^{k+1}$.*

Proof. We restrict ourselves to $F \in L_q$. The general case can be treated in the same way using the mapping properties of the integral operators. For the proof we take the solution $u \in \overset{\circ}{W}{}_2^1$ and solve the STOKES problem

$$v + \frac{1}{\eta} T\mathbf{Q}g = \frac{\rho}{\eta} T\mathbf{Q}TM(u)$$

$$\frac{1}{\eta} \operatorname{Sc} \mathbf{Q}g = -\frac{\rho}{\eta} \operatorname{Sc} \mathbf{Q}TM(u).$$

Now, from the regularity results for the STOKES problem we get $v = u \in W_s^2(G)$ and $g = p \in W_s^1(G)$ for $s < 3/2$. Using HÖLDER's inequality and the embedding theorems we obtain $M^*(u) \in L_t(G)$ for all $t < 3$. Now, assuming $q < 3, M(u) \in L_t$ we get by a new study of the same STOKES problem that $u \in W_t^2(G), p \in W_t^1(G)$. Repeating this consideration over and over again we obtain $M^*(u) \in L_r(G) \quad \forall\, r < \infty$. Hence, $M(u) \in L_q(G)$ holds. A final investigation of the STOKES problem used above leads to the statement of the theorem. #

It is easy to see that the full problem with free convection can be treated by the same method. The reason is that all non-linear items have the same quality as $M(u)$ and can be estimated in the same way. If in addition a non-linear right-hand side $f(u)$ occurs then the LIPSCHITZ condition ensures the required properties. Of course, it will be necessary to use stronger assumptions on f for the proof of higher regularity.

4.9 Problems with non-linear right-hand side

A lot of elliptic boundary value problems have on the right-hand side a differential operator of lower order. Here we will consider only operators of the form $f(u)$ which obey a LIPSCHITZ condition.

4.9.1 A general approach

Let $G \subset \mathbb{R}^n$ be a bounded domain with piecewise smooth LIAPUNOV boundary Γ. We assume for the whole section that $f = f(u)$ is a non-linear operator function which fulfils a LIPSCHITZ condition with the LIPSCHITZ constant L. This condition reads as follows: Let $u, v \in W_2^1(G, Cl_{0,n})$, then

$$\|f(u) - f(v)\|_2 < L\|u - v\|_{2,1}, \tag{4.92}$$

where $\|\cdot\|_2 := \|\cdot\|_{L_2(G)}$, and $\|\cdot\|_{2,1} := \|\cdot\|_{W_2^1(G)}$. Furthermore, assume that

$$f(0) = 0. \tag{4.93}$$

These two conditions together yield the inequality

$$\|f(u)\|_2 < L.$$

Let a be an operator (not necessarily linear in the sense of $Cl_{0,n}$) with the following property:

$$au := \sum_{i=0}^{n} a_i u_i e_i + \sum_{\alpha} u_\alpha e_\alpha,$$

where α runs over the all multivectors of the degree greater than two and $a_i = a_i(x) > 0$ are real functions. Hence a has an inverse b, which is of the same form. We have only a_i to replace by a_i^{-1}. We abbreviate again by $B_\varepsilon(x)$ an open ball centred in x with radius ε. Furthermore, D denotes the DIRAC operator. We get the following results:

Theorem 4.101. *The boundary value problem*

$$DbDu = f(u) \quad \text{in} \quad G \tag{4.94}$$

$$u = g \quad on \quad \Gamma \tag{4.95}$$

permits in the neighborhood $B(r(g)) \subset W_2^1(G, Cl_{0,n})$ *of*

$$r(g) := F_\Gamma\, g + T_G a F_\Gamma (\mathrm{tr}_\Gamma T_G a F_\Gamma)^{-1} Q_\Gamma\, g \tag{4.96}$$

a unique solution which can be achieved by the iteration procedure

$$u_n - r(g) = T_G \mathbf{Q}_a a T_G f(u_{n-1}), \quad \mathrm{tr}_\Gamma u_n = g, u_0 = r(g). \tag{4.97}$$

This procedure converges in $W_2^1(G, Cl_{0,n})$ *if the condition*

$$L|G|^{1/n} \max_G |a| < (\sigma_n)^{1/n} \left(\frac{\lambda_1}{\lambda_1 + 1}\right)^{\frac{1}{2}}, \tag{4.98}$$

is fulfilled, where λ_1 *is the first eigenvalue of* DIRICHLET*'s problem over the domain* G.

Proof. Using BOREL–POMPEIU's formula one gets that $F_\Gamma T_G = 0$ in G. With the methods described above we find that the boundary value problem (4.94,4.95) is equivalent to the operator integral equation

$$u = F_\Gamma g + T_G a F_\Gamma (\mathrm{tr}_\Gamma T_G a F_\Gamma)^{-1} Q_\Gamma g + T_G \mathbf{Q}_a a T_G f(u). \tag{4.99}$$

Analogously to Theorem 4.143. (we use it in advance here – but its proof does not depend on the results of the actual section) we get the identity

$$\lambda_1(G) = \frac{1}{||T_G||^2_{[\mathrm{im}\mathbf{Q}_a, W_2^1]} - 1}. \tag{4.100}$$

Here λ_1 is the first eigenvalue of DIRICHLET's problem. The norm of $\mathbf{Q}_a$ is equal to one with respect to the inner product (3.14). An estimation from [107] shows

$$\frac{1}{\sqrt{\lambda_1}} \leq ||T_G||_{L(C)} \leq \left(\frac{|G|}{\sigma_n}\right)^{1/n}. \tag{4.101}$$

The application of BANACH's fixed-point theorem completes the proof. #

Example 4.102. 1. Let $a = I$ then we have the non-linear DIRICHLET problem

$$-\Delta u = f(u) \quad \text{in} \quad G, \qquad u = g \quad \text{on} \quad \Gamma.$$

2. Let $G \subset \mathbb{R}^3$ and $bu := Mu = \frac{m-2}{2(m-1)} u_0 + \underline{u}$. Then arises a non-linear version of LAMÉ's problem. It reads

$$-\Delta \underline{u} - \frac{m}{m-2} \operatorname{grad} \operatorname{div} \underline{u} = f(\underline{u}) \quad \text{in } G, \quad \underline{u} = \underline{g} \quad \text{on } \Gamma.$$

$\underline{u}$ is just the vector of displacements.

3. Let again $G \subset \mathbb{R}^3$ and $bu = (\mu\kappa/\varepsilon)u$ where ε is the dielectrical constant, $\kappa = \kappa(x)$ is the electric conductivity and μ decribes the permeability constant. Denoting by $u := H$ the magnetic field then the problem (4.94,4.95) is just the first boundary value problem for the magnetic field.

4.9.2 Equations of steady oscillation in elasticity

In order to formulate some more applications we call back into mind (cf. Subsection 4.1.2) the operator of the conjugation J defined by $J\,u := u_0 - \underline{u}$. For the following problems it is sufficient to use the algebra of real quaternions $\mathbb{H}$. The DIRAC operator D in this algebra is of the form $D := \sum_{i=1}^{3} e_i \partial_i$.

We consider a boundary value problem of the type

$$D(\alpha + \beta J) D\, u = f(u) \quad \text{in} \quad G, \tag{4.102}$$

$$u = g \quad \text{on} \quad \Gamma, \tag{4.103}$$

where α, β are quaternions. The operator a one has now to identify with the multiplication by $\alpha + \beta J$. Suppose also that $\alpha + \beta J$ has an inverse $\gamma + \delta J$. The $\mathbb{H}$-valued function g on the right-hand side belongs to $W_2^{3/2}(\Gamma, \mathbb{H})$.

Example 4.103. Choosing

$$\alpha = \frac{-m}{4(m-1)} + 1, \ \beta = \frac{-m}{4(m-1)},$$

where m denotes the POISSON number ($2 \leq m \leq 4$). Then the problem (4.94,4.95) with

$$f(\cdot, \underline{u}) = h(\cdot, \underline{u}) - \Theta^2 \rho \underline{u} \quad \text{and} \quad ||h(\cdot, \underline{u}_1) - h(\cdot, \underline{u}_2)||_2 \leq L||\underline{u}_1 - \underline{u}_2||_{2,1}$$

is just the first boundary value problem of the equations of steady oscillation in elasticity. Here Θ denotes the oscillation frequency, $\underline{u}$ the vector of displacements and ρ the density.

Lemma 4.104. *The quaternionic* HILBERT *space* $L_2(G, \mathbb{H})$ *permits the orthogonal decomposition*

$$L_2(G, \mathbb{H}) = (\gamma + \delta J)\, \ker D \cap L_2(G, \mathbb{H}) \oplus_{\alpha+\beta J} D \overset{\circ}{W}{}_2^1 (G, \mathbb{H})$$

relatively to the inner product

$$(u, v)_{\alpha+\beta J} := \int_G \overline{u}(\gamma + \delta J) v dG,$$

where $\gamma + \delta J = (\alpha + \beta J)^{-1}$, $\gamma = \frac{1}{2}[(\alpha + \beta)^{-1} + (\alpha - \beta)^{-1}]$ *and* $\delta = -\frac{1}{2}[(\alpha + \beta)^{-1} - (\alpha - \beta)^{-1}]$.

Corollary 4.105. *The orthoprojections onto the subspaces* $(\gamma + \delta J)\ker D \cap L_2(G, \mathbb{H}), D \overset{\circ}{W}{}_2^1 (G, \mathbb{H})$ *are the operators*

$$\begin{aligned} \mathbf{P}_{\gamma+\delta J} &:= (\gamma + \delta J) F_\Gamma (\mathrm{tr}_\Gamma T_G (\gamma + \delta J) F_\Gamma)^{-1} \mathrm{tr}_\Gamma T_G, \\ \mathbf{Q}_{\gamma+\delta J} &:= I - \mathbf{P}_{\gamma+\delta J}. \end{aligned}$$

Proof. This verifies similarly to Corollary 4.23. #

Theorem 4.106. *The boundary value problem (4.103,4.104) permits in the neighborhood* $B(r(g)) \subset W_2^1(G,H)$ *of* $r(g) := F_\Gamma\, g + T_G \mathcal{P}_{\gamma+\delta J} D\, g^*$, *where* g^* *is a smooth continuation of* g *into the domain* G *a unique solution which can be achieved by the iteration procedure*

$$u_n - r(g) = T_G \mathbf{Q}_{\gamma+\delta J}(\gamma+\delta J) T_G f(u_{n-1}), \;\; \mathrm{tr}_\Gamma u_n = g, \;\; u_0 = r(g).$$

This procedure converges in $W_2^1(G,\mathbb{H})$ *if the condition*

$$L|G|^{1/3} \max_G |\alpha + \beta J| < \left(\frac{4\pi}{3}\right)^{1/3} \left(\frac{\lambda_1}{\lambda_1+1}\right)^{\frac{1}{2}} \tag{4.104}$$

is fulfilled, where λ_1 *is the first eigenvalue of* DIRICHLET*'s problem on the domain* G.

Proof. Using BOREL-POMPEIU's formula and the property $F_\Gamma T_G u = 0$ in G we see that the boundary value problem is equivalent to the operator integral equation

$$u = F_\Gamma\, g + T_G \mathbf{P}_{\gamma+\delta J} D\, \tilde{g} + T_G \mathbf{Q}_{\gamma+\delta J} T_G f(u).$$

with $\tilde{g}$ denoting a smooth extension of g into the domain G. Furthermore, Theorem 4.143. implies

$$\lambda_1(G) = \frac{1}{||T_G||^2_{[\mathrm{im}\mathbf{Q}_{\gamma+\delta J}, \overset{\circ}{W}{}_2^1]} - 1}$$

here λ_1 is the first eigenvalue of DIRICHLET's problem. Obviously, the norms of $\mathbf{P}_{\gamma+\delta J}$ and $\mathbf{Q}_{\gamma+\delta J}$ in $L_2(G,\mathbb{H})$ equal one. It remains only to use the estimation in Theorem 2.3 in [107], namely

$$\frac{1}{\sqrt{\lambda_1}} \le ||T_G||_{L(C)} \le \left(\frac{3}{4\pi}\right)^{1/3} |G|^{1/3}.$$

Then condition (4.104) is necessary in order to prove convergence by BANACH's fixed-point principle. #

4.9.3 Time harmonic equations in moment theory

These results can be transferred to more general situations. For this reason we introduce the matrix operators

$$\mathcal{D} := \begin{pmatrix} D & 0 \\ 0 & D \end{pmatrix}, \quad \mathcal{M} := \begin{pmatrix} \alpha & \beta \\ \gamma & \delta \end{pmatrix},$$

where $\alpha, \beta, \gamma, \delta$ are multiplication operators of the type $\alpha = \alpha_1 + \alpha_2 J, \alpha_i \in \mathbb{C}^{\infty}_{\mathbf{R}}(G)$. Let

$$U := \begin{pmatrix} u_1 \\ u_2 \end{pmatrix}, \quad u_i = \sum_{k=0}^{3} u_{ik} e_k, \quad \Phi(U) := \begin{pmatrix} \phi_1 & \phi_2 \\ \phi_3 & \phi_4 \end{pmatrix} (U),$$

where ϕ_i $(i = 1, 2, 3, 4)$ are differential operators of the first order,

$$H := \begin{pmatrix} h_1 \\ h_2 \end{pmatrix}, \quad g := \begin{pmatrix} g_1 \\ g_2 \end{pmatrix}.$$

Now we consider the non-linear boundary value problem

$$\mathcal{DMD}U = \Phi(U) + H \quad \text{in } G \tag{4.105}$$

$$U = g \qquad \text{on } \Gamma \tag{4.106}$$

with $\det \mathcal{M} \neq 0$.

Corollary 4.107. *Let*

$g \in W_2^{3/2}(\Gamma, \mathbb{H})$ *and* $\|\Phi(U^1) - \Phi(U^2)\|_2 \leq L\|U^1 - U^2\|_{2,1}$.

Then the boundary value problem (4.105,4.106) permits in the neighborhood of

$$R(g) := \mathcal{F}_\Gamma g + \mathcal{V}_n(\mathrm{tr}_\Gamma \mathcal{V}_n)^{-1}\mathcal{Q}_\Gamma g$$

with

$$\mathcal{F}_\Gamma = \begin{pmatrix} F_\Gamma & 0 \\ 0 & F_\Gamma \end{pmatrix}, \qquad \mathcal{Q}_\Gamma = \begin{pmatrix} Q_\Gamma & 0 \\ 0 & Q_\Gamma \end{pmatrix},$$

$$\mathcal{V} = \begin{pmatrix} V & 0 \\ 0 & V \end{pmatrix}, \qquad (V\,u)(x) = \frac{1}{4\pi} \int_\Gamma \frac{\alpha(y)}{|x-y|} u(y) d\Gamma_y.$$

a unique solution in $W_2^1(G, \mathbb{H})$ *which can be obtained by the iteration procedure*

$$U_n - R(g) = \mathcal{T}_G \mathbf{Q} \mathcal{M}^{-1} \mathcal{T}_G \mathcal{H}(U_{n-1}) \tag{4.107}$$

$$U_0 = R(g) \quad (n = 1, 2, \ldots), \tag{4.108}$$

where $\mathcal{H}(U) = \Phi(U) + H$ *and* $\mathcal{T}_G = \begin{pmatrix} T_G & 0 \\ 0 & T_G \end{pmatrix}$.

This procedure converges in $W_2^1(G, \mathbb{H})$, *if*

$$L |\, G\,|^{1/3} \|\mathcal{M}^{-1}\|_{C(G)} < \frac{1}{2} \left(\frac{4\pi}{3}\right)^{1/3} \left(\frac{\lambda_1}{\lambda_1 + 1}\right)^{\frac{1}{2}}.$$

Example 4.108. (cf. [154]) The time harmonic equations in the moment theory in linear elasticity which are defined over a domain G read as follows:

$$D(\mu+\gamma)D\underline{u} + \operatorname{grad}(\lambda+\mu-\gamma)\operatorname{div}\underline{u} + 2\,\Lambda\operatorname{rot}\underline{w} = h_1,$$
$$D(\nu+\Theta)D\underline{w} + \operatorname{grad}(\epsilon-\nu-\Theta)\operatorname{div}\underline{w} + 2\Lambda\operatorname{rot}\underline{u} - 4\Lambda\underline{w} = h_2,$$

On the boundary Γ we have DIRICHLET conditions, i.e., $\underline{u} = g_1$ and $\underline{w} = g_2$. All occurring constants have a well-defined physical meaning, and have for this reason to satisfy the conditions $\mu > 0, 3\lambda + 2\mu > 0, \Lambda > 0, \varepsilon > 0, 3\varepsilon + 2\nu > 0, \Theta > 0$. We should mention that $\underline{u}$ is the vector of the displacements and with $\underline{w}$ we denote the rotation vector. The entries of the matrix $\mathcal{M}$ introduced above we have to choose now in the following way:

$$\beta = \gamma = 0,$$
$$\alpha = \alpha_1 + \alpha_2 J = \frac{\lambda+3\mu+\gamma}{2} + \frac{\lambda+\mu-\gamma}{2}J,$$
$$\delta = \delta_1 + \delta_2 J = \frac{\varepsilon+3\nu+\Theta}{2} + \frac{\varepsilon+\nu-\Theta}{2}J.$$

Furthermore, we have to set

$$\phi_1 = 0, \quad \phi_2 = \phi_3 = (-DJ + JD), \quad \phi_4 = -4.$$

Note that

$$\mathcal{M}^{-1} = \left\{ \begin{vmatrix} \alpha_1 & \alpha_2 \\ \delta_1 & \delta_2 \end{vmatrix} + \begin{vmatrix} \alpha_1 & -\alpha_2 \\ \delta_1 & \delta_2 \end{vmatrix} J \right\}^{-1} \begin{pmatrix} \delta & 0 \\ 0 & \alpha \end{pmatrix}.$$

In this example we have $H \equiv 0$. Under the conditions formulated above we can find a solution for our problem by the iteration procedure (4.107).

4.9.4 Method of harmonic approximation

We will mention here the so-called *harmonic approximation* (cf. [299]). We consider in a bounded domain $G \subset \mathbb{R}^n$ the following non-linear problem:

$$D\,u = f(u) \text{ in } G \tag{4.109}$$

$$P_\Gamma u = g \text{ on } \Gamma. \tag{4.110}$$

We assume that the operator $f : W_2^1(G,\mathbb{H}) \longmapsto W_2^2(G,\mathbb{H})$ satisfies the condition $\|f(u)\|_{W_2^1(G,\mathbf{H})} \leq M$ and $g \in W_2^{3/2}(\Gamma,\mathbb{H}) \cap \operatorname{im} P_\Gamma$, where P_Γ is the HARDY projection, which maps onto all $\mathbb{H}$-valued functions with an $\mathbb{H}$-regular extension into the bounded domain G.

Theorem 4.109. *The solution u^* (in case of existence!) can be approximated by the harmonic $\mathbb{H}$-valued function u_h*

$$u_h = F_{\Gamma_0} g + T_{G_0} F_{\Gamma_0} f(\tilde{g}),$$

in the domain G_0, where $\tilde{g}$ is a smooth continuation of the function g into the domain G up to a parallel surface Γ_1 and G_0 is just the arising annulus domain between both smooth surfaces Γ_0 and Γ_1. Furthermore, the following error estimate is valid:

$$||u_h - u^*||_{C(G,\mathbf{H})} \leq C|G_0|^{1/3}.$$

The constant C is bounded by

$$M\left(\frac{3}{4\pi}\right)^{1/3}(|G|^{1/3} + ||P_\Gamma||)$$

with $\Gamma = \Gamma_0 + \Gamma_1$.

Remark 4.110. This method can be easily generalized to equations which contain polynomials of the DIRAC operator (cf. [299]).

4.10 On strong non-linear problems

Numerous problems in engineering lead to non-linear boundary value problems of partial differential equations where the non-linearity appears in the highest order. We focus our attention on a special class of such type of problems. Assume that the non-linearity depends on partial derivatives of the unknown function. One such problem is the so-called distortion problem of an elastic–plastic material. This problem reads as follows:

$$-\mathrm{div}\ \varphi(|\mathrm{grad}\ u_0|)\mathrm{grad}\ u_0 = f_0 \quad \text{in} \quad G \tag{4.111}$$

$$u_0 = 0 \quad \text{on} \quad \Gamma, \tag{4.112}$$

where $G \subset \mathbb{R}^2$ is a simple-connected domain with the piecewise smooth boundary Γ and f_0 is just a measure of the distortion of the material (cf. [120], [200]). In [302] another problem of this type

$$D\varphi(|Du|)Du = f(u) \quad \text{in} \quad G \tag{4.113}$$

$$u = g \quad \text{on} \quad \Gamma \tag{4.114}$$

is discussed. Here D is the DIRAC operator and $\varphi : \mathbb{R}^+ \to \mathbb{R}^+$ a sufficiently smooth function. Note, that for $n = 2$, $u = u_0$, $f = f_0$, $g = 0$ problem (4.111,4.112) arise. We will be able to construct with the help of a suitable attachment a "fundamental" system of solutions of this non-linear boundary value problem with well-defined but not too restrictive conditions for

$$\varphi : \mathbb{R}^+ \to \mathbb{R}^+_c = \{x \in \mathbb{R}^+ : x > c\},\ f \quad \text{and} \quad g.$$

We use the abbreviations of Section 4.9.

4.10.1 A boundary value problem with strong non-linear potential.

We now consider the following boundary value problem for $\mathbb{H}$-valued functions u:

$$\begin{aligned} -\Delta u + D|Du|^k &= f \quad \text{in } G \subset \mathbb{R}^3 \\ u &= g \quad \text{on } \Gamma. \end{aligned}$$

Applying again BOREL–POMPEIU's formula we get

$$Du + |Du|^k = T_G f + \phi\ ,\ \phi \in \ker D.$$

The square of the quaternionic equation provides

$$|Du|^2 + |Du|^{2k} - 2\ \mathrm{div}\ \underline{u}|Du|^k = |T_G f + \phi|^2 = |\mathcal{F}|^2$$

and so

$$|Du|^2 + |Du|^{2k} + 2\mathcal{F}_0|Du|^k - 2|Du|^{2k} = |\mathcal{F}|^2.$$

Setting now $|Du| = p$ we obtain

$$-p^{2k} + 2\mathcal{F}_0 p^k + p^2 = |\mathcal{F}|^2.$$

In addition the projection onto the scalar part leads to

$$-\operatorname{div}\underline{u} + |Du|^k = \operatorname{Sc} T_G f + \phi_0 = \mathcal{F}_0.$$

Let p^* be a solution of this equation and we obtain

$$\begin{aligned} -\Delta u &= f - D|p^*|^k = f^*, \\ u &= g, \end{aligned}$$

which can be solved explicitly. Using the representation formulas from Section 4.2 (and the following sections) we get

$$u = F_\Gamma g + TF_\Gamma(\operatorname{tr}_\Gamma TF)^{-1} Q_\Gamma g + T_G \mathbf{Q}\left[T_G f - (I - F_\Gamma)|p^*|^k\right].$$

Special cases:

$$\begin{aligned} k = 1: &\qquad p^* = \frac{|\mathcal{F}|^2}{2\mathcal{F}_0}, \\ k = 2: &\qquad -p^4 + (2\mathcal{F}_0 + 1)p^2 = |\mathcal{F}|^2, \\ &\qquad\quad p^2 =: q \Rightarrow \quad -q^2 + (2\mathcal{F}_0 + 1)q = |\mathcal{F}|^2, \\ &\qquad\quad q^2 - (2\mathcal{F}_0 + 1)q + |\mathcal{F}|^2 = 0, \end{aligned}$$

$$q_{\frac{1}{2}} = \frac{2\mathcal{F}_0 + 1}{2} \pm \sqrt{\left(\frac{2\mathcal{F}_0 + 1}{2}\right)^2 - |\mathcal{F}|^2}.$$

For

$$\left(\frac{2\mathcal{F}_0 + 1}{2}\right) - |\mathcal{F}|^2 > 0 \quad \text{it follows that} \quad p^* = +\sqrt{q_{\frac{1}{2}}}.$$

4.10.2 A non-linear problem with integral boundary conditions

Let $G \subset \mathbb{R}^3$ and u a $\mathbb{H}$-valued function. Further, let be $\psi : (u, Du) \to \psi(u, Du) \in \mathbb{R}$. We consider the boundary value problem

$$-\Delta u + D\psi(u, Du) = f \quad \text{in} \quad G,$$

$$\operatorname{tr}_\Gamma \int_G e(x-y)u(y)dy = g \quad \text{on} \quad \Gamma,$$

where f and g are given functions.

Using BOREL–POMPEIU's formula we get immediately

$$Du + \psi(u, Du) = T_G f + \phi, \quad (\phi \in \ker D).$$

In addition we have

$-\mathrm{div}\,\underline{u} + \psi(u, Du) = \mathrm{Sc}\,T_G f + \phi_0$

where $\phi_0 = \mathrm{Sc}\,\phi$ is harmonic in G. Sc is just the projection onto the scalar part. This leads to

$-\Delta u + D(\mathrm{Sc}\,T_G f + \phi_0) + \mathrm{grad}\,\mathrm{div}\,\underline{u} = f$

and so

$D(I - \mathrm{Sc}\,)Du = f - D\mathrm{Sc}\,T_G f - D\phi_0.$

Now denote $I - \mathrm{Sc} =: \mathrm{Vec}$ the projection onto the vector part and we get

$D\mathrm{Vec}\,Du = D\mathrm{Vec}\,T_G f + \phi_1, \quad \phi_1 \in \ker D.$

Let $u \in L_2(G, \mathbb{H}) \cap W_2^2(G, \mathbb{H})$. Then we have for the representation

$u = \phi^* + Dw$

with $w \in \overset{\circ}{W}{}_2^1(G, \mathbb{H}) \cap W_3^1(G, \mathbb{H})$.
This delivers the equation

$\mathrm{Vec}\,\Delta w = \Delta\mathrm{Vec}\,w = \mathrm{Vec}\,T_G f + T_G\phi_1 + \phi_2 \quad (\phi_k \in \ker D;\ k = 1, 2).$

Now adding the equation

$\Delta\mathrm{Sc}\,w = 0 \Leftrightarrow w_0 = 0$

we obtain

$\Delta w = \mathrm{Vec}\,T_G f + T_G\phi_1 + \phi_2 \quad \text{in} \quad G, \quad w = 0 \quad \text{on} \quad \Gamma.$

Then we get the solution with $T_G\phi_1 + \phi_2 = H$

$w = T_G\mathbf{Q}T_G\mathrm{Vec}\,T_G f + T_G\mathbf{Q}T_G H$

and for u

$u = \phi^* + \mathbf{Q}T_G\mathrm{Vec}\,T_G f + \mathbf{Q}T_G H,$

where H is a harmonic quaternion. Realizing now the boundary condition

$g = \mathrm{tr}_\Gamma T_G u = \mathrm{tr}_\Gamma T_G F\phi^* + \mathrm{tr}_\Gamma T_G\mathbf{Q}T_G(\mathrm{Vec}\,T_G f + H).$

We gain

$$\mathrm{tr}_\Gamma\phi^* = (\mathrm{tr}_\Gamma TF)^{-1}\mathrm{tr}_\Gamma T_G u = (\mathrm{tr}_\Gamma TF)^{-1}g \quad (\text{cf.}\,[110])$$

$$\Rightarrow \quad u = F_\Gamma(\mathrm{tr}_\Gamma TF)^{-1}g + \mathbf{Q}T_G(\mathrm{Vec}\,T_G f + H).$$

Therefore we have the theorem:

Theorem 4.111. *Let $f \in L_2(G, \mathbb{H}), g \in W_2^{\frac{3}{2}}(\Gamma, \mathbb{H}) \cap \operatorname{im} Q_\Gamma$. Then a general solution of the boundary value problem*

$$\begin{aligned} -\Delta u + D\psi(u, Du) = f \quad &in \quad G \\ \operatorname{tr}_\Gamma T_G u = g \quad &on \quad \Gamma \end{aligned}$$

can be given by

$$u = F_\Gamma(\operatorname{tr}_\Gamma TF)^{-1} g + \mathbf{Q} T_G(\operatorname{Vec} T_G f + H)$$

with an arbitrary harmonic function H.

4.10.3 A generalized problem of Yanushauskas

A. YANUSHAUSKAS considered in [337] the following system of partial differential equations of second order:

$$\begin{aligned} &-\Delta u_i + \partial_i(\lambda_0 \operatorname{div} \underline{u} + \underline{\lambda} \cdot \operatorname{rot} \underline{u}) = f \quad \text{in} \quad G \\ &(i = 1, 2, 3) \end{aligned}$$

with the boundary condition

$$\underline{u} = \underline{g} \quad \text{on} \quad \Gamma.$$

Let $G \subset \mathbb{R}^3$ be compact with a piecewise smooth LIAPUNOV boundary Γ. By addition of the equation

$$-\Delta \underline{u} + \operatorname{grad}(\underline{\lambda} \cdot \operatorname{grad}) \mathrm{u}_0 = 0$$

we obtain with $\lambda = \lambda_0 + \underline{\lambda}$

$$\begin{aligned} -\Delta u + \operatorname{grad}(\lambda \cdot \mathrm{Du}) &= f \quad \text{in} \quad G \subset \mathbb{R}^3 \\ u &= g \quad \text{on} \quad \Gamma. \end{aligned}$$

Here the DIRAC operator is again defined by

$$D := \sum_{i=1}^{3} e_i \partial_i.$$

Assume that $\underline{\lambda} = \underline{\lambda}(u)$, $f = f(u)$, and $f : W_2^1(G, Cl_{0,n}) \longmapsto L_2(G, Cl_{0,n})$ is non-linear and satisfies the conditions:

(i) $||f(\cdot, u)||_2 \leq M,\ M > 0,$ (4.115)

(ii) $||f(\cdot, u_1) - f(\cdot, u_2)||_2 \leq L||u_1 - u_2||_{2,1}\ , L > 0.$ (4.116)

For λ we assume the same conditions. Already J.C. MAXWELL used the decomposition of the Laplacian into a product of two DIRAC operators, i.e. $\Delta = -DD$. Then we get the new compact formulation

$$\begin{aligned} DDu + D(\lambda \cdot Du) &= f \quad \text{in} \quad G \\ u &= g \quad \text{on} \quad \Gamma. \end{aligned}$$

Using BOREL–POMPEIU's formula it follows that

$$Du + (\lambda \cdot Du) = T_G f + \Phi \quad \text{with} \quad \Phi \in \ker D.$$

Furthermore, it follows with $\lambda_0 \neq -1$ that

$$Du \cdot \lambda = (1+\lambda_0)^{-1}(T_G f(u) + \Phi) \cdot \lambda(u).$$

After resubstitution we get

$$\begin{aligned} DDu &= f(u) - D(1+\lambda_0)^{-1}(T_G f(u) + \Phi) \cdot \lambda(u) =: \mathcal{F}(u) \quad \text{in} \quad G \\ u &= g \quad \text{on} \quad \Gamma. \end{aligned}$$

Using the representation formulae from Section 4.2 we obtain

$$\begin{aligned} &u = F_\Gamma + T_G F_\Gamma (\text{tr}_\Gamma T_G F_\Gamma)^{-1} Q_\Gamma g + T_G \mathbf{Q} T_G f(u) \\ &-T_G \mathbf{Q}(1+\lambda_0)^{-1}(T_G f(u) + \Phi) \cdot \lambda(u) \end{aligned}$$

A class of solutions can be obtained by application of Theorem 4.101 on the problem with the right-hand side $\mathcal{F}(u)$.

4.11 Riemann–Hilbert problems

The history of complex RIEMANN–HILBERT problems can be traced back to the origins of complex function theory. This problem was first mentioned in a paper by B. RIEMANN as he was studying conditions which uniquely define a function holomorphic in a domain $G \subset \mathbb{R}^2$. B. RIEMANN himself treated the problem only heuristically. A first method appropriate for solving the linear problem was suggested by D. HILBERT. He introduced the singular integral operator and transformed the problem to a FREDHOLM integral equation. Up to now, complex RIEMANN–HILBERT problems, also referred as HILBERT problems are a central object of research in complex analysis. Thus, it is more than natural to search for CLIFFORD analytic generalizations of this kind of problems. In two papers, [268], [269], the Ukrainian mathematicians M. SHAPIRO and V. VASILEVSKI gave a quaternionic formulation of RIEMANN–HILBERT problems and transferred basic results. Furthermore, we refer a former paper by D. PASCALI [223] and investigations of XU ZHENYAN [334]. We mention also the fundamental paper of A. DZURAEV [73], where among other considerations RIEMANN–HILBERT problems were discussed. We follow here the arguments of S. BERNSTEIN.

The simplest version of a RIEMANN boundary value problem reads as follows:

Let Γ be a closed LIAPUNOV *surface in $\mathbb{R}^3$ bounding a domain $G = G^+$. Given $g \in C^{0,\alpha}(\Gamma, Cl_{0,3})$ $0 < \alpha \leq 1$ and a, $H \in Cl_{0,3}$, to find a pair of functions*

$$u^\pm \in (\ker D_a)(G^\pm), \quad D_a := D + a, \quad G^- := \mathbb{R}^3 \setminus \overline{G^+}$$

which fulfils the boundary condition

$$u^+(x) = u^-(x)H + g(x), \qquad x \in \Gamma.$$

In [269] the FREDHOLM theory of different generalizations of the RIEMANN boundary value problem in the case of $a = 0$ has been developed. Non-commutativity of the quaternionic multiplication resulted two direct analogues of RIEMANN–HILBERT problems. Such we have the *"right"* RIEMANN *boundary value problem.* This reads as follows:

To find a pair of functions $u^\pm$ from $(\ker D)(G^\pm)$ *satisfying the following boundary condition on Γ:*

$$u^+a = u^-b + g,$$

where g is an element of a suitable function space on Γ while a and b are multiplicators.

If changing the boundary condition into

$$au^+ = bu^- + g$$

we obtain the *"left"* RIEMANN *boundary value problem.*

Having applied PLEMELJ–SOKHOTZKI's formulae the problems reduce in a classical way to a singular integral equation with the operators

$$M^c P_\Gamma + M^d Q_\Gamma \qquad \text{or} \qquad {}^c M P_\Gamma +^d M Q_\Gamma.$$

In this denotation M^r, ${}^r M$ mean a multiplication by r from the right or from the left, respectively. Here r is a quaternion. P_Γ and Q_Γ are the HARDY projections introduced in Section 4.1.

Remark 4.112. The denotations of the multiplication operators were introduced by M.V. SHAPIRO and N.V. VASILEVSKI.

For an arbitrary $a \in Cl_{0,n}$ we consider the operator D_a. We can state the same types of boundary value problems with the requirement to find $u^\pm$ from $(\ker D_a)(G^\pm)$. Then these problems reduce in the same manner to singular integral equations on Γ defined by the operators

$$M^c P_a + M^d Q_a \qquad \text{or} \qquad {}^c M P_a +^d M Q_a,$$

respectively. The operators P_a, Q_a are defined by

$$P_a := \frac{I + S_a}{2}, \qquad Q_a := \frac{I - S_a}{2}$$

and S_a is just the singular integral operator

$$(S_a u)(x) := 2 \int_\Gamma e_a(x - y)\alpha(y)u(y)d\Gamma_y$$

with

$$e_a(x) := e^{-(\underline{a},x)} e_{a_0}(x),$$

where again

$$e_{a_0}(x) := \frac{a_0}{(2\pi)^{n/2}} \frac{(a_0|x|)^{n/2-1}}{|x|^{n-2}} \left(\frac{a_0\, x}{|x|} K_{n/2}(a_0|x|) + K_{n/2-1}(a_0|x|) \right)$$

and $K_\nu(x)$ denotes the so-called MACDONALD function which we have already introduced. The FREDHOLM theory for these operators, as well as the description of the algebra generated by them and by multiplication operators, the reader can find in detail in [268] and [269]. We only mention here that the FREDHOLM theory for operators of the type $M^c P_a + M^d Q_a, {}^c M P_a +^d M Q_a$ coincides substantially with that for the operators $M^c P_\Gamma + M^d Q_\Gamma, {}^c M P_\Gamma +^d M Q_\Gamma$.

This is caused by the fact that the operator S_a coincides with S_Γ up to a compact operator if the boundary Γ of the domain G is compact. Unfortunately, FREDHOLM theory gives no information on the uniqueness of the solution.

We now follow the investigations of S. BERNSTEIN in ([22]). We will observe in more detail the special case of the lower half space $\mathbb{R}^n_-$ the relations between RIEMANN–HILBERT problems and singular integral equations over the boundary Γ.

We consider the following RIEMANN–HILBERT problem:

$$u^+ = Hu^- + h$$

with functions $u^\pm \in \ker \mathrm{D}(\mathbb{R}^n_\pm)$. The corresponding singular integral equation over the boundary $\mathbb{R}^{n-1}$ is just

$$P_\Gamma u + HQ_\Gamma u = \frac{1}{2}(1+H)u + \frac{1}{2}(1-H)S_\Gamma u = h.$$

It is necessary to give the following statement:

Lemma 4.113. *The singular integral operator S_Γ is self-adjoint in the* HILBERT *module $L_2(Cl_{0,n}, \mathbb{R}^{n-1})$ with the inner product*

$$(u,v) := \int_{\mathbf{R}^n} \widetilde{u(x)} v(x) dx.$$

($\widetilde{u(x)}$ has been introduced in Section 1.4).

Proof. We have

$$\begin{aligned}
(S_\Gamma u, v) &= \int_{\mathbf{R}^{n-1}} 2\Big[\int_{\mathbf{R}^{n-1}} \widetilde{\Big(e(x-y)e_n u(y)\Big)} dy\Big] v(x) dx \\
&= \int_{\mathbf{R}^{n-1}} 2 \int_{R^{n-1}} \widetilde{u(y)}(-e_n)\overline{e(x-y)} dy v(x) dx \\
&= \int_{\mathbf{R}^{n-1}} \widetilde{u(y)}(-e_n) 2 \int_{\mathbf{R}^{n-1}} \overline{e(x-y)} v(x) dx dy \\
&= \int_{\mathbf{R}^{n-1}} \widetilde{u(y)} 2 \int_{R^{n-1}} e(y-x) e_n v(x) dx dy \\
&= (u, S_\Gamma v)
\end{aligned}$$

and thus $S_\Gamma^* = S_\Gamma$. #

Remark 4.114. Using the stereographic projection this result may be transferred to the unit sphere S^{n-1}.

The following theorem can be proved:

Theorem 4.115. *Let*

(i) $H(x) = \sum_\beta H_\beta(x) e_\beta$ *and all* $H_\beta(x) \in L_\infty(\mathbb{R}^{n-1})$ *are real-valued.*

(ii) $(1 + H(x))\overline{(1 + H(x))} \in \mathbb{R}$ *and* $H(x)\overline{H(x)} \in \mathbb{R}$ *for all* $x \in \mathbb{R}^{n-1}$.

(iii) *There exists a number* $\varepsilon > 0$ *such that* $H_0(x) > \varepsilon$ *for all* $x \in \mathbb{R}^{n-1}$.

Then the singular integral equation

$$\frac{1}{2}(1 + H)u + \frac{1}{2}(1 - H)S_\Gamma u = h \quad \text{on} \quad \mathbb{R}^{n-1}$$

associated with the RIEMANN–HILBERT *problem described above*

$$(P_\Gamma u)(x) + H(x)(Q_\Gamma u)(x) = h(x) \quad \text{on} \quad \mathbb{R}^{n-1}$$

is uniquely solvable in $L_2(Cl_{0,n}, \mathbb{R}^{n-1})$ *and the successive approximation*

$$u_n := 2(1 + H)^{-1}h - (1 + H)^{-1}(1 - H)S_\Gamma u_{n-1}$$

with an arbitrary $u_0 \in L_2(Cl_{0,n}, \mathbb{R}^{n-1})$ *converges to the unique solution.*

Proof. From Lemma 4.113 we know that $\|S_\Gamma\|_{L_2} = 1$. Now if there exists a number $\varepsilon > 0$ such that

$$|(1 + H(x))^{-1}(1 - H(x))| < 1 - \varepsilon \quad \text{on} \quad \mathbb{R}^{n-1}$$

the theorem follows from BANACH's fixed-point theorem. From $H(x)\overline{H(x)} \in \mathbb{R}$ we get

$$(1 - H(x))\overline{(1 - H(x))} = 1 - 2H_0(x) + |H(x)|^2 \in \mathbb{R}.$$

Due to (ii) we can write

$$(1 + H(x)^{-1}) = \frac{1 + \overline{H(x)}}{|1 + H(x)|^2}.$$

Moreover,

$$(1 + H(x))^{-1}(1 - H(x))\overline{(1 + H(x)^{-1}(1 - H(x))}$$

$$\frac{(1 + \overline{H(x)})(1 - H(x))(1 - \overline{H(x)})(1 + H(x))}{|1 + H(x)|^4} \in \mathbb{R}$$

on $\mathbb{R}^{n-1}$. Finally,

$$
\begin{aligned}
|(1+H(x))^{-1}(1-H(x))|^2
&= \left|\frac{(1+\overline{H(x)})(1-H(x))}{(1+H(x))(1+\overline{H(x)})}\right|^2 \\
&= \left|\frac{1-2[H(x)-H_0(x)]-|H(x)|^2}{1+2H_0(x)+|H(x)|^2}\right|^2 \\
&= \frac{(1-|H(x)|^2-2[H(x)-H_0(x)])(1-|H(x)|^2+2[H(x)-H_0(x)])}{(1+2H_0(x)+|H(x)|^2)^2} \\
&= \frac{(1-|H(x)|^2)^2+4[H(x)-H_0(x)]\overline{H(x)-H_0(x)}}{(1+|H(x)|^2+2H_0(x))^2}.
\end{aligned}
$$

A simple computation shows that the inequality

$$\frac{(1-|H(x)|^2)^2+4[H(x)-H_0(x)]\overline{H(x)-H_0(x)}}{(1+|H(x)|^2+2H_0(x))^2} < 1-\varepsilon$$

with $\varepsilon > 0$ is equivalent to

$$\varepsilon(1+|H(x)|^2+2H_0(x))^2 < 4H_0(x)|1+H(x)|^2.$$

Now setting

$$\sigma := \varepsilon \operatorname{ess\,sup}(1+|H(x)|^2+2H_0(x)).$$

Thus

$$0 < \sigma < 4H_0(x)|(1+H(x))|^2$$

is only fulfilled if $\inf\limits_{x\in\mathbb{R}^{n-1}} H_0(x)$ is a positive number. #

4.12 Domain perturbation problems

We will compare the solutions of a class of first boundary value problems over different basic domains with common intersection. Such problems one can often find in practice (obstacles in a streaming fluid, cracked elastic materials). It will be considered the influence of a subdomain $G' \subset G \subset \mathbb{R}^n$ on the solution of elliptic boundary value problems in the remaining domain $G \backslash \overline{G}' \subset \mathbb{R}^n$ with suitable corresponding comparable boundary conditions. Much work is done in the field of "small" changes of a given domain ([309]). Almost every numerical procedure contains an approximation of the given domain with "complicated" geometry by a suitable sequence of "more simple" domains. We recall, for instance, the approximation of boundaries with irregular points by regular boundaries which is called WIENER procedure [266]. Since each procedure breaks off after a finite number of steps one has to compare solutions over "different" domains under comparable boundary conditions. We will deal with boundary value problems of the type (4.17,4.18) and STOKES' boundary value problem.

4.12.1 On perturbed Lamé problems

Let be G_i $(i = 1,2)$ domains in $\mathbb{R}^n$ which are are bounded by piecewise LIAPUNOV surfaces Γ_i $(i = 1,2)$. Furthermore, let $G_2 \subset G_1$ and set $\overline{G}' := \overline{G}_1 \setminus G_2$ as well as $\partial G' = \Gamma'$.

We consider the boundary value problem (4.17,4.18)

$$\begin{aligned} DbDu^{(i)} &= f^{(i)} \quad \text{in} \quad G_i \\ u^{(i)} &= g^{(i)} \quad \text{on} \quad \Gamma_i \end{aligned}$$

for $i = 1,2$. Furthermore, we suppose on the common boundary $\tilde{\Gamma} = \Gamma_1 \cap \Gamma_2$ that $g^{(1)} = g^{(2)} = \tilde{g}$. We know from Section 4.4 that the unique solution of this problem permits the representation

$$Du = \mathbf{P}_a Dh + \mathbf{Q}_a T_G f.$$

Here h was just a sufficient smooth extension of the boundary data into the domain G.

Consequently we obtain for our domains G_1 and G_2 $(G_1 \supset G_2)$ the equations

$$\begin{aligned} Du^{(1)} &= \mathbf{P}_a^{(1)} Dh^{(1)} + \mathbf{Q}_a^{(1)} T_{G_1} f^{(1)} \\ Du^{(2)} &= \mathbf{P}_a^{(2)} Dh^{(2)} + \mathbf{Q}_a^{(2)} T_{G_2} f^{(2)}. \end{aligned}$$

Write in brief $d := u^{(1)} - u^{(2)}$ in G_2 then

$$Dd = (\mathbf{P}_a^{(1)} Dh^{(1)} - \mathbf{P}_a^{(2)} Dh^{(2)}) + (\mathbf{Q}_a^{(1)} T_{G_1} f^{(1)} - \mathbf{Q}_a^{(1)} T_{G_2} f^{(2)}). \qquad (4.117)$$

We abbreviate the first term by A_1 and the second by A_2. Under the assumption $f^{(1)}|_{G_2} = f^{(2)}$ we can easily get:

$$A_2 = \mathbf{Q}_a^{(1)} T_{G'} f^{(1)} + (\mathbf{Q}_a^{(1)} - \mathbf{Q}_a^{(2)}) T_{G_2} f^{(2)}.$$

Lemma 4.112. *Let $u \in W_2^k(G, Cl_{0,n})$ $(k \in \mathbb{N})$. Then the following relations are valid:*

(i) $\mathbf{Q}_a^{(2)} \mathbf{P}_a^{(1)} u = 0$; (ii) $\mathbf{Q}_a^{(2)} \mathbf{Q}_a^{(1)} u = \mathbf{Q}_a^{(2)} u$;
(iii) $\mathbf{P}_a^{(2)} \mathbf{P}_a^{(1)} u = \mathbf{P}_a^{(1)} u$; (iv) $\mathrm{im} \mathbf{P}_a^{(1)} \subset \mathrm{im} \mathbf{P}_a^{(2)}$;
(v) $\mathrm{im} \mathbf{Q}_a^{(2)} \subset \mathrm{im} \mathbf{Q}_a^{(1)}$.

Proof. First we will prove the identity

$$\mathbf{Q}_a^{(2)} a F_{\Gamma_2} b \mathbf{P}_a^{(1)} u = 0.$$

Then, we have

$$\begin{aligned} \mathbf{Q}_a^{(2)} F_{\Gamma_2} b \mathbf{P}_a^{(1)} u &= a F_{\Gamma_2} b \mathbf{P}_a^{(1)} u - \mathbf{P}_a^{(2)} F_{\Gamma_2} b \mathbf{P}_a^{(1)} u \\ &= a F_{\Gamma_2} b \mathbf{P}_a^{(1)} u - a F_{\Gamma_2} (\mathrm{tr}_{\Gamma_2} T_{g_2} a F_{\Gamma_2})^{-1} \mathrm{tr}_{\Gamma_2} T_{G_2} a F_{\Gamma_2} b \mathbf{P}_a^{(1)} = 0. \end{aligned}$$

Applying Borel–Pompeiu's formula and using that $b\mathbf{P}_a^{(1)} u \in \ker D(G_1)$ we obtain

$$\begin{aligned} \mathbf{Q}_a^{(2)} \mathbf{P}_a^{(1)} u &= \mathbf{Q}_a^{(2)} a F_{\Gamma_1} b \mathbf{P}_a^{(1)} u - \mathbf{Q}_a^{(2)} a F_{\Gamma_2} b \mathbf{P}_a^{(1)} u \\ &= \mathbf{Q}_a^{(2)} a (F_{\Gamma_1} - F_{\Gamma_2}) b \mathbf{P}_a^{(1)} u = \mathbf{Q}_a^{(2)} a T_{G'} D b \mathbf{P}_a^{(1)} u = 0. \end{aligned}$$

Relation(ii) follows immediately from (i), as $\mathbf{Q}_a^{(2)} \mathbf{Q}_a^{(1)} u = \mathbf{Q}_a^{(2)} (I - \mathbf{P}_a^{(1)}) u = \mathbf{Q}_a^{(2)} u$. Relation (iii) follows from (ii) and relations (iv) and (v) can be obtained by straightforward calculation. #

Corollary 4.113. *Let $u \in W_2^k(G, Cl_{0,n})$, $k \in \mathbb{N}$. Then*

$$(\mathbf{Q}_a^{(1)} - \mathbf{Q}_a^{(2)}) u = \mathbf{P}_a^{(2)} \mathbf{Q}_a^{(1)} u.$$

Proof. The application of Lemma 4.112. leads to

$$\mathbf{Q}_a^{(1)} u - \mathbf{Q}_a^{(2)} u = (\mathbf{Q}_a^{(1)} - \mathbf{Q}_a^{(2)}) \mathbf{Q}_a^{(1)} u = (\mathbf{P}_a^{(2)} - \mathbf{P}_a^{(1)}) \mathbf{Q}_a^{(1)} u = \mathbf{P}_a^{(2)} \mathbf{Q}_a^{(1)} u. \quad \#$$

Remark 4.114. The operators $\mathbf{P}_a^{(1)} \mathbf{P}_a^{(2)}$ and $\mathbf{Q}_a^{(1)} \mathbf{Q}_a^{(2)}$ are projections. Indeed, we have by using of Lemma 4.112. (iii) and (ii)

$$\begin{aligned} \mathbf{P}_a^{(1)} \mathbf{P}_a^{(2)} \mathbf{P}_a^{(1)} \mathbf{P}_a^{(2)} &= \mathbf{P}_a^{(1)} \mathbf{P}_a^{(1)} \mathbf{P}_a^{(2)} = \mathbf{P}_a^{(1)} \mathbf{P}_a^{(2)}, \\ \mathbf{Q}_a^{(1)} \mathbf{Q}_a^{(2)} \mathbf{Q}_a^{(1)} \mathbf{Q}_a^{(2)} &= \mathbf{Q}_a^{(1)} \mathbf{Q}_a^{(2)} \mathbf{Q}_a^{(2)} = \mathbf{Q}_a^{(1)} \mathbf{Q}_a^{(2)}. \end{aligned}$$

In addition one can prove $\mathbf{P}_a^{(2)} \mathbf{P}_a^{(1)} \mathbf{Q}_a^{(2)} \mathbf{Q}_a^{(1)} = 0$.

Lemma 4.115. *Let $u \in \mathrm{im}(\mathbf{Q}_a^{(1)} - \mathbf{Q}_a^{(2)})$. Then u permits the representation $u = D\,z$ with $z \in \ker DbD$ and vanishes on Γ_1.*

Proof. If $u \in \operatorname{im}(\mathbf{Q}_a^{(1)} - \mathbf{Q}_a^{(2)})$, then for $\operatorname{im}\mathbf{Q}_a^{(2)} \subset \operatorname{im}\mathbf{Q}_a^{(1)}$ we obtain $u \in \operatorname{im}\mathbf{Q}_a^{(1)}$. Hence there exists a function $z \in D(\overset{\circ}{W}{}_2^1(G, Cl_{0,n}))$ with $u = Dz$. On the other hand $u \in \operatorname{im}\mathbf{P}_a^{(2)}$, i.e., $u = az$ with $z \in \ker D(G_2)$. This leads to $Dbu = 0$ on G_2 and so $DbDz = 0$ in G_2. #

Corollary 4.116. *Let $\mathcal{M}$ be the set of all $Cl_{0,n}$-valued functions $u \in L_2(G_1)$, which vanish on $\Gamma_1 = \partial G_1$ and are harmonic in the subdomain G_2. Then*

$$\mathcal{M} = T_{G_1}(\operatorname{im}(\mathbf{Q}_1^{(1)} - \mathbf{Q}_1^{(2)})).$$

Theorem 4.117. *Assume that $f^{(1)} \in W_2^k(G_1, Cl_{0,n})$, $g^{(1)} \in W_2^{k+\frac{3}{2}}(\Gamma_1, Cl_{0,n})$, $g^{(2)} \in W_2^{k+\frac{3}{2}}(\Gamma_2, Cl_{0,n})$, $g^{(1)}|_{\Gamma_2} = g^{(2)}$ and $f^{(1)}|_{G_2} = f^{(2)}$. Then we have on G_2 for Dd the representation*

$$Dd = \mathbf{Q}_a^{(1)} T_{G'} f^{(1)} + \mathbf{P}_a^{(2)} \mathbf{Q}_a^{(1)} (T_{G_2} f^{(2)} - Dh^{(1)}),$$

where $h^{(1)}$ is such a smooth extension of the boundary data $g^{(1)}$ into the domain G_1 such that $h^{(1)}|_{\Gamma_2} = g^{(2)}$.

Proof. The proof is a consequence of (4.117) and property (iii) of Lemma 4.112. as well as its Corollary 4.113. #

A special choice of the assumptions leads to:

Corollary 4.118. *For $g^{(1)} = 0, g^{(2)} = 0$ we obtain*

$$Dd = \mathbf{P}_a^{(2)} \mathbf{Q}_a^{(1)} T_{G_2} f^{(2)} + \mathbf{Q}_a^{(1)} T_{G'} f^{(1)}.$$

Corollary 4.119. *Let $f^{(1)} \in L_2(G_1, Cl_{0,n})$. Then the following estimate is valid:*

$$\|u^{(1)} - u^{(2)}\|_F = \|d\|_F \leq \sqrt{\frac{1 + \lambda_1^{(2)}}{\lambda_1^{(2)}}} \operatorname{diam} G' \|f^{(1)}|_{G'}\|_2,$$

where $\|\cdot\|_F$ denotes the norm in the factor space $W_2^1(G_2, Cl_{0,n}) / \ker DbD(G_2)$ and $\lambda^{(2)}$ the first eigenvalue of DIRICHLET'S *problem on G_2.*

Proof. As already shown one can find a function z which is harmonic in G_2 such that the representation

$$Dz = \mathbf{P}_a^{(2)} \mathbf{Q}_a^{(1)} (T_{G_2} f^{(1)} - Dh^{(1)})$$

holds, with $z \in \ker DbD(G_2)$. Theorem 4.117. now yields

$$D(d - z) = \mathbf{Q}_a^{(1)} T_{G'} f^{(1)}.$$

Let $\lambda_1^{(2)}$ be the first eigenvalue of DIRICHLET's problem on G_2. Then we have the estimate (cf. [107])

$$\|u\|_{2,1} \leq \sqrt{\frac{1+\lambda_1^{(2)}}{\lambda_1^{(2)}}}\|Du\|_2.$$

Therefore,

$$\|d-z\|_{2,1} \leq \sqrt{\frac{1+\lambda_1^{(2)}}{\lambda_1^{(2)}}}\, \|\mathbf{Q}_a^{(1)}\|_{L(L_2(G_1))}\|T_{G'}\|_{L(L_2(G'))}\|f^{(1)}\|_{L_2(G')}$$

and by using Theorem 3.14. we obtain our statement. #

4.12.2 On a perturbed Stokes' problem

As further example of domain perturbation problems we will consider STOKES' equations. In domains G_i $(i=1,2)$ $G_2 \subset G_1$ we have

$$\begin{aligned} -\Delta \underline{u}^{(i)} + \eta^{-1}\operatorname{grad} p^{(i)} &= \underline{f}^{(i)} && \text{in } G_i \\ \operatorname{div} \underline{u}^{(i)} &= f_0^{(i)} && \text{in } G_i \\ \underline{u}^{(i)} &= \underline{g}^{(i)} && \text{on } \Gamma_i. \end{aligned}$$

Using results of Section 4.9 we get for the solution

$$u^{(i)} = T_{G_i}\mathbf{Q}_0^{(i)}\operatorname{Vec} T_{G_i} f^{(i)} \qquad (i=1,2),$$

where $\mathbf{Q}_0^{(i)} = I - \mathbf{P}_0^{(i)}$ and $\mathbf{P}_0^{(i)} := \operatorname{Vec} F_{\Gamma_i}(\operatorname{tr}_{\Gamma_i} T_{G_i} \operatorname{Vec} F_{\Gamma_i})^{-1}\operatorname{tr}_{\Gamma_i} T_{G_i}$. Besides assume $f^{(2)} = f^{(1)}|_{G_2}$. It is easy to see that the operators $\mathbf{P}_0^{(i)}$ and $\mathbf{Q}_0^{(i)}$ are orthoprojections onto the subspaces $\operatorname{Vec}(\ker D)(G_i)$ and $\operatorname{Vec} D \overset{\circ}{W}{}_2^1 (G_i, Cl_{0,n})$, respectively. Instead of the domain G_1 we consider the new domain G_2 with $G_1 \supset G_2$ and $\underline{u}^{(2)} = 0$ on $\Gamma_2 \setminus \Gamma_1$. We obtain

$$D(u^{(1)} - u^{(2)}) = \left[(\mathbf{Q}_0^{(1)} - \mathbf{Q}_0^{(2)})\operatorname{Vec} T_{G_1} f^{(1)} + \mathbf{Q}_0^{(2)}\operatorname{Vec} T_{G'} f^{(1)}\right]$$

and therefore (Corollary 4.113.)

$$D(u^{(1)} - u^{(2)}) = [\mathbf{P}_0^{(2)}\mathbf{Q}_0^{(1)}\operatorname{Vec} T_{G_1} f^{(1)} + \mathbf{Q}_0^{(2)}\operatorname{Vec} T_{G'} f^{(1)}]$$

BOREL–POMPEIU's formula now provides :

Theorem 4.120. *The solution $u^{(2)}$ over the domain G_2 of* STOKES' *problem is given by*

$$u^{(2)} = u^{(1)} + F_{\Gamma_2\setminus\Gamma_1} u^{(1)} + T_{G_2}\left[\mathbf{Q}_0^{(2)}\operatorname{Vec} T_{G'} f^{(1)} + \mathbf{P}_0^{(2)}\mathbf{Q}_0^{(1)}\operatorname{Vec} T_{G_1} f^{(1)}\right],$$

where $F_{\Gamma_2\setminus\Gamma_1}u^{(1)}$ *denotes the* CAUCHY*-type integral over the open surface* $\Gamma_2 \setminus \Gamma_1$. *Furthermore, we have the estimation*

$$\|u^{(1)} - u^{(2)}\|_{W_2^1(G,Cl_{0,n})} \leq \sqrt{\frac{\lambda_1^{(2)} + 1}{\lambda_1^{(2)}}}\left[\|\mathbf{Q}_0^{(1)}\mathrm{Vec}\, T_{G_1} f^{(1)}\|_2 + \|\mathrm{Vec}\, T_{G'} f^{(1)}\|_2\right],$$

where $\lambda_1^{(2)}$ *denotes the first eigenvalue of* DIRICHLET*'s problem over* G_2.

4.13 Transmission problems

A transmission problem will be considered in a bounded domain $G \subset \mathbb{R}^3$ with a smooth boundary and k inclusions G_i $(i = 1, .., k)$:

$$\begin{aligned} \Delta^* u^{(i)} &= 0 \quad \text{in } G_i, \qquad u^{(0)} = 0 \quad \text{on } \Gamma_0, \\ u^{(i)} &= u^{(0)} \quad \text{on } \Gamma_i, \ \tau_\alpha u^{(i)} - \tau_\alpha u^{(0)} = h_i \quad \text{on } \Gamma_i, \end{aligned} \tag{4.118}$$

where $\bar{G}_i \subset G$, $G_0 = G \setminus \cup_{i=1}^k \bar{G}_i$, $\bar{G}_i \cap \bar{G}_j = \emptyset$ for $i \neq j, \Gamma_i = \partial G_i$, $\Gamma_0 = \partial G$ and the operator Δ^* stands for the LAMÉ operator $\Delta + \frac{m}{m-2}\text{grad div}$ using the number $m = 1/\nu$, where ν is the POISSON ratio. The operator τ_α is the operator of the normal stresses on Γ_i.

Such transmission problems have been well studied. In the plane case it is usual to solve the problem using complex function theory (see e.g. [209], [317] and further references therein). For spatial problems boundary integral equation methods will be used in most cases. These methods require complicated investigations of systems of singular (weakly singular or in some cases hypersingular) integral equations. It is the aim of this section to show how (and how easy) quaternionic analysis can be applied to solve system (4.118). For simplicity we start with the case of $m = 0$ ($\Delta^* = \Delta$ and $\tau_\alpha = \partial_\alpha u$). We give a description of the orthogonal complement of the subspace of all $\mathbb{H}$-regular functions in $L_2(G_0)$ and then we can prove criterions which allow us to answer the question if a given set of functions $u_0, ..., u_k$ defined on $\Gamma_0, ..., \Gamma_k$ has a $\mathbb{H}$-regular extension into the domain G_0. Using these results an explicit representation formula for the solution of the transmission problem will be given. Considering the boundary values and the jump relations on Γ_i we obtain a system of integral equations for the unknown density $(u_1, ..., u_k)$ on $(\Gamma_1 \times \ldots \times \Gamma_k)$. We discuss uniqueness and existence of the solution of the integral equation system in a constructive way. Finally we will give some information on the general case $m \neq 0$.

4.13.1 Boundary value problems in multiply connected domains

Let G be a sufficiently smooth bounded domain in $\mathbb{R}^3$ with the boundary Γ. For translating the LAMÉ equation into hypercomplex language we rewrite the LAMÉ-operator $\Delta^* = DMD$ in G_i with the multiplication operator M ($Mu = \frac{2m-2}{m-2}\text{Sc}\, u + \text{Vec}\, u$). Clearly, all results from previous sections on our integral operators can be applied to operators which are connected with G_0, G_i, Γ_0 and Γ_i. The generalization of BOREL–POMPEIU's formula to the case of the multiply connected domain G_0 reads as follows:

$$u = F_{\Gamma_0}u - \sum_{i=1}^{k} F_{\Gamma_i}u + T_{G_0}Du \qquad \forall\, u \in W_2^1(G_0). \tag{4.119}$$

We denote by tr_j the *trace operator* on Γ_j, where Γ_j is considered as part of ∂G_0. Applying tr_j we get from (4.119)

$$u_j := \mathrm{tr}_j\, u = \mathrm{tr}_j\, F_{\Gamma_0}u_0 - \sum_{i=1}^{k} \mathrm{tr}_j\, F_{\Gamma_i}u_i + \mathrm{tr}_j\, T_{G_0}Du$$

and for functions $u \in \ker D(G_0)$ we have

$$u_j = \mathrm{tr}_j F_{\Gamma_0}u_0 - \sum_{i=1}^{k} \mathrm{tr}_j F_{\Gamma_i}u_i \qquad (j = 0, ..., k).$$

If we collect these equations we get the following system:

$$\begin{pmatrix} u_0 \\ u_1 \\ \vdots \\ u_k \end{pmatrix} = \begin{pmatrix} P_{\Gamma_0} & -\mathrm{tr}_0 F_{\Gamma_1} & \dots & -\mathrm{tr}_0 F_{\Gamma_k} \\ \mathrm{tr}_1 F_{\Gamma_0} & Q_{\Gamma_1} & \dots & -\mathrm{tr}_1 F_{\Gamma_k} \\ \vdots & \vdots & \ddots & \vdots \\ \mathrm{tr}_k F_{\Gamma_0} & -\mathrm{tr}_k F_{\Gamma_1} & \dots & Q_{\Gamma_k} \end{pmatrix} \begin{pmatrix} u_0 \\ u_1 \\ \vdots \\ u_k \end{pmatrix}.$$

We denote this matrix operator by P_Γ and set $\vec{\mathbf{u}} := (u_0, ..., u_k)^T$. Now we can prove the following theorems using the same ideas as the previous sections for simply connected domains.

Theorem 4.125. *The operator P_Γ is a projection. It holds for $j = 1, ..., k$*

$$(u_0, \dots, u_k)^T \in \mathrm{im}\, P_\Gamma \cap W_2^{1/2}(\partial G_0) \Longleftrightarrow \exists u \in \ker D \cap \overset{o}{W}{}_2^1(G_0) : \mathrm{tr}_j\, u = u_j$$

$L_2(G_0)$ admits the following orthogonal decomposition $L_2(G_0) = L_2(G_0) \cap \ker D(G_0) \oplus D(\overset{o}{W}{}_2^1(G_0))$. The corresponding orthoprojections onto $\ker D(G_0)$ and $\overset{o}{W}{}_2^1(G_0)$ are denoted by $\mathbf{P}_{G_0}$ ($\mathbf{P}_{G_0} : L_2(G_0) \longrightarrow \ker D(G_0)$) and $\mathbf{Q}_{G_0}$($\mathbf{Q}_{G_0} : L_2(G_0) \longrightarrow D(\overset{o}{W}{}_2^1(G_0))$), respectively.

We will now investigate the formal matrix operator

$$\mathrm{tr}_\Gamma\ T_{G_0}\vec{F_\Gamma} := \begin{pmatrix} \mathrm{tr}_0 \\ \mathrm{tr}_1 \\ \vdots \\ \mathrm{tr}_k \end{pmatrix} T_{G_0}\vec{F_\Gamma}$$

where $\vec{F_\Gamma} := (F_{\Gamma_0}, -F_{\Gamma_1}, \dots, -F_k)$. The operator T_{G_0} acts component-wise on $\vec{F_\Gamma}$. In general this operator is not invertible, but if we use some suitable restrictions we get the following result.

Theorem 4.126. *The operator* $\mathrm{tr}_\Gamma T_{G_0}\vec{F_\Gamma} : \mathrm{im}\, P_\Gamma \cap W_2^{1/2}(\partial G_0) \longrightarrow \mathrm{im}\, Q_\Gamma \cap W_2^{3/2}(\partial G_0)$ *is invertible.*

Proof. Using the orthoprojections $\mathbf{P}_{G_0}$ and $\mathbf{Q}_{G_0}$ we can show that the boundary value problem $\{\Delta u = 0 \text{ in } G_0, \mathrm{tr}_i\, u = g_i \text{ on } \Gamma_i\ (i = 1, ..., k)\}$ has a unique solution for arbitrarily given functions u_i $(i = 1, ..., k)$. The function

$$u = F_{\Gamma_0} u_0 - \sum_{i=1}^{k} F_{\Gamma_i} u_i + T_{G_0}\mathbf{P}_{G_0} Dh$$

is a solution for an arbitrary extension (with correct regularity) h of the boundary values u_i into G_0. We obtain the uniqueness with the help of BOREL–POMPEIU's formula for multiply connected domains. If we now have that $\mathrm{tr}_\Gamma\, T_{G_0}\vec{F_\Gamma}\vec{g} = 0$ with $\vec{g} \in \mathrm{im} P_\Gamma$, then it follows that $T_{G_0}\vec{F_\Gamma}\vec{g} = 0$, $\vec{F_\Gamma}\vec{g} = 0$ and $\vec{g} = 0$. Conversely, let $\vec{g} \in \mathrm{im} Q_\Gamma$. We solve the problem $\{\Delta u = 0, \mathrm{tr}\, u = \vec{g}\}$ and then we obtain

$$u = T_{G_0}\mathbf{P}_{G_0} Dh = T_{G_0}\vec{F_\Gamma}\Phi,\ \ \mathrm{tr}_\Gamma\, \Phi \in \mathrm{im}\mathbf{P}_\Gamma.$$

Therefore, $\mathrm{tr}_\Gamma\ u = \vec{g} = \mathrm{tr}_\Gamma\, T_{G_0}\vec{F_\Gamma}\mathrm{tr}_\Gamma\, \Phi$ and the proof is complete. #

Using Theorem 4.126 we can describe the orthoprojection $\mathbf{P}_{G_0}$ explicitly:

Theorem 4.127. *We have*

$$\mathbf{P}_{G_0} = \vec{F_\Gamma}\{\mathrm{tr}_\Gamma T_{G_0}\vec{F_\Gamma}\}^{-1}\mathrm{tr}_\Gamma T_{G_0}$$

A representation for the solution of the DIRICHLET *problem in the multiply connected domain* G_0 *is given by*

$$u^{(0)} = \vec{F_\Gamma}\vec{g} + T_{G_0}\vec{F_\Gamma}\{\mathrm{tr}_\Gamma T_{G_0}\vec{F_\Gamma}\}^{-1}\mathrm{tr}_\Gamma T_{G_0} D\vec{h}$$

In this formula $\vec{g}$ *stands for the vector* $(g_0, g_1, \ldots, g_n)^T$ *and the function* $\vec{h}$ *is an arbitrarily chosen extension of* $\vec{g}$ *into the domain* G_0.

By the help of BOREL–POMPEIU's formula we have $\mathrm{tr}_\Gamma T_{G_0} D\vec{h} = Q_\Gamma\vec{g}$ and we can prove the following statement:

Corollary 4.128. *The solution of* DIRCHLET*'s problem reads as follows* $u^{(0)} = \vec{F_\Gamma}\vec{g} + T_{G_0}\vec{F_\Gamma}\{\mathrm{tr}_\Gamma T_{G_0}\vec{F_\Gamma}\}^{-1}Q_\Gamma\vec{g}$

4.13.2 Solution of the transmission problem

In each subdomain G_i we can use the known representation formula [110]

$$u^{(i)} = F_{\Gamma_i} u_i + T_{G_i} F_{\Gamma_i}(\mathrm{tr}_i\, T_{G_i} F_{\Gamma_i})^{-1} Q_{\Gamma_i} u_i.$$

From this we have a formal solution in G

$$u = \begin{cases} \vec{F_\Gamma}\vec{u} + T_{G_0}\vec{F_\Gamma}\{\mathrm{tr}_\Gamma T_{G_0}\vec{F_\Gamma}\}^{-1}Q_\Gamma\vec{u} & \text{in } G_0 \\ F_{\Gamma_i} u_i + T_{G_i} F_{\Gamma_i}(\mathrm{tr}_i\, T_{G_i} F_{\Gamma_i})^{-1} Q_{\Gamma_i} u_i & \text{in } G_i \quad (i = 1, ..., k). \end{cases} \quad (4.120)$$

Here we used the notation $\vec{u} := (0, u_1, \ldots, u_n)^T$. This representation formula can be simplified. From the representation in G_i we see that

$$T_{G_i} F_{\Gamma_i} (\mathrm{tr}_\Gamma T_{G_i} F_{\Gamma_i})^{-1} Q_{\Gamma_i} u_i \in \ker D(\mathrm{co}\, \overline{G_i}).$$

Furthermore, it holds that $\mathrm{tr}_i\, F_{\Gamma_i} u_i = -Q_{\Gamma_i} u_i$ and $-F_{\Gamma_i} u_i \in \ker D(\mathrm{co}\, \overline{G_i})$. Using the identity theorem for $\mathbb{H}$-regular functions [299] we get

$$-F_{\Gamma_i} u_i = T_{G_i} F_{\Gamma_i} (\mathrm{tr}_\Gamma T_{G_i} F_{\Gamma_i})^{-1} Q_{\Gamma_i} u_i \quad \text{in} \quad G_0 \qquad (\forall\, i \neq 0). \tag{4.121}$$

In a similar way we can show that

$$\sum_{s=1}^{k} F_{\Gamma_s} u_s = T_{G_0} \vec{F_\Gamma} \{\mathrm{tr}_\Gamma T_{G_0} \vec{F_\Gamma}\}^{-1} Q_\Gamma \vec{u}. \tag{4.122}$$

By the help of the equations (4.121) and (4.122) we get a unique representation of the solution in the whole domain G.

$$u = T_{G_0} \vec{F_\Gamma} \{\mathrm{tr}_\Gamma T_{G_0} \vec{F_\Gamma}\}^{-1} Q_\Gamma \vec{u} + \sum_{s=1}^{k} T_{G_s} F_{\Gamma_s} (\mathrm{tr}_s T_{G_s} F_{\Gamma_s})^{-1} Q_{\Gamma_s} u_s. \tag{4.123}$$

This function u fulfils the first three equations of system (4.118). The remaining problem is also to satisfy the fourth equation by controlling the unknown densities u_i on the boundaries Γ_i. For this purpose we use the single layer potential V. (In our former denotation we wrote V_α.) In the case of $\phi \in \ker D(G)$ we have from [110] that $V_\Gamma \mathrm{tr}_\Gamma \phi = T_G \phi$. In our situation it follows that

$$T_{G_i} F_{\Gamma_i} (\mathrm{tr}_i T_{G_i} F_{\Gamma_i})^{-1} Q_{\Gamma_i} u_i = V_{\Gamma_i} (\mathrm{tr}_i T_{G_i} F_{\Gamma_i})^{-1} Q_{\Gamma_i} u_i \quad \text{in} \quad \mathbb{R}^3 \tag{4.124}$$

using again the identity theorem for $\mathbb{H}$-regular functions. We abbreviate as follows:

$$\vec{\psi} := (\psi_0, \ldots, \psi_n)^T := \{\mathrm{tr}_\Gamma T_{G_0} \vec{F_\Gamma}\}^{-1} Q_\Gamma \vec{u} \in \mathrm{im} P_\Gamma$$

and start the investigation of $T_{G_0} \vec{F_\Gamma} \vec{\psi}$. We find that

$$T_{G_0} \vec{F_\Gamma} \vec{\psi} = V_{\Gamma_0} \psi_0 - \sum_{i=1}^{k} V_{\Gamma_i} \psi_i.$$

Now, we recall the jump formula

$$[\partial_{\alpha_{\mathrm{ext}}} - \partial_{\alpha_{\mathrm{int}}}] V_{\Gamma_i} \phi = -\alpha_i(y) \phi_i(y),$$

where α_i denotes the unit normal on Γ_i directed into G_0. Applying this formula to equation (4.124) we obtain

$$\alpha_i \psi_i - \alpha_i (\mathrm{tr}_i T_{G_i} F_{\Gamma_i})^{-1} Q_{\Gamma_i} u_i = h_i$$

or

$$(\mathrm{tr}_i T_{G_i} F_{\Gamma_i})^{-1} Q_{\Gamma_i} u_i - \psi_i = \alpha_i h_i \qquad (i = 1, \ldots, k). \tag{4.125}$$

We have to prove now that this system of equations determines the unknown functions u_i. First, we are looking for the uniqueness and we assume that there are two different solutions. Then we get for the difference, which we will denote by ψ again, the following properties.

Proposition 4.129. *We have*

(i) $\psi_i \in \operatorname{im} P_{\Gamma_i}$ $(i = 1, ..., k)$ $\Rightarrow \psi_0 \in P_{\Gamma_0}$.

(ii) *From (4.124), (4.125) and the definition of* $\vec{\psi}$ *we have*

$u = V_{\Gamma_0}\psi_0$ *and so* $\mathrm{tr}_0 u = \mathrm{tr}_0 V_{\Gamma_0}\psi_0 = 0 \Rightarrow V_{\Gamma_0}\psi_0 = 0 \Rightarrow$
$DV_{\Gamma_0}\psi_0 = F_{\Gamma_0}\psi_0 = 0 \Rightarrow \psi_0 = 0.$

(iii) *From* $\psi_i \in \operatorname{im} P_{\Gamma_i}$ *and*

$$\psi_i = -\sum_{s=1, s\neq i}^{k} \mathrm{tr}_s\, F_{\Gamma_s}\psi_s + Q_{\Gamma_i}\psi_i = 0$$

it follows that $\vec{\psi} = 0$.

$$\Rightarrow Q_{\Gamma_i} u_i = 0 \quad i = 1, ...k \Rightarrow u_i \in \mathrm{im} P_{\Gamma_i}, u_0 = 0$$
$$\Rightarrow u = T_{G_0}\vec{F}_\Gamma\{\mathrm{tr}_\Gamma T_{G_0}\vec{F}_\Gamma\}^{-1}Q_\Gamma\vec{u} \Rightarrow \mathrm{tr}_\Gamma u = Q_\Gamma\vec{u} \Rightarrow P_\Gamma\vec{u} = 0.$$

(iv) *Further, we have that with* $\vec{\psi} = 0$ *this is also valid* $\{\mathrm{tr}_\Gamma T_{G_0}\vec{F}_\Gamma\}^{-1}Q_\Gamma\vec{u} = 0$ *and this means* $Q_\Gamma\vec{u} = 0 \Rightarrow u_i = 0 \quad \forall\, i$.

In this way we proved

Theorem 4.130. *If the system*

$$(\mathrm{tr}_i T_{G_i} F_{\Gamma_i})^{-1} Q_{\Gamma_i} u_i - \psi_i = \alpha_i h_i \qquad (i = 1, ..., k)$$

has a solution it is unique.

Applying D to representation (4.121) of the solution we get

$$Du = \begin{cases} \vec{F}_\Gamma\{\mathrm{tr}_\Gamma T_{G_0}\vec{F}_\Gamma\}^{-1}Q_\Gamma\vec{u} & \text{in } G_0 \\ F_{\Gamma_j}(\mathrm{tr}_j\, T_{G_j}F_{\Gamma_j})^{-1}Q_{\Gamma_j}u_j & \text{in } G_j, \quad (j = 1, ..., k). \end{cases}$$

A consideration of the (interior and exterior, respectively) boundary values of Du leads to $\mathrm{tr}_{j,\mathrm{ext}}Du = \psi_j$ and $\mathrm{tr}_{j,\mathrm{int}}Du = (\mathrm{tr}_j\, T_{G_j}F_{\Gamma_j})^{-1}Q_{\Gamma_j}u_j$. It follows that

$$(\mathrm{tr}_{j,\mathrm{ext}} - \mathrm{tr}_{j,\mathrm{int}})(\alpha_j Du) = \alpha_j\psi_j - \alpha_j(\mathrm{tr}_j\, T_{G_j}F_{\Gamma_j})^{-1}Q_{\Gamma_j}u_j = h_j$$

and that means that the functions $\alpha_j Du$ and $\partial\alpha_j u$ have the same jump if we are crossing the boundary Γ_j. Obviously the equation $\mathrm{tr}_0 Du = \psi_0$ holds and our problem (4.118) can be transformed into the following generalized RIEMANN–HILBERT *transmission problem* for the function $v = Du$.

$$\begin{aligned} Dv &= 0 \quad \text{in} \quad G_i, \\ (\mathrm{tr}_{j,\mathrm{ext}} - \mathrm{tr}_{j,\mathrm{int}})v &= -\alpha_j h_j \quad \text{on} \quad \Gamma_j \quad (j = 1, ..., k), \\ \mathrm{tr}_0 v &= \psi_0 \quad \text{on} \quad \Gamma_0. \end{aligned}$$

We try to find a solution of this problem in the following way:

$$v = \sum_{s=1}^{k} F_{\Gamma_s}(\alpha_s h_s) - F_{\Gamma_0}\theta_0.$$

Using this expression for computing $\mathrm{tr}_0 v$ and $\mathrm{tr}_{i,\mathrm{ext}} v$ we come to

$$\begin{pmatrix} \psi_0 \\ \vdots \\ \psi_j \\ \vdots \end{pmatrix} = \begin{pmatrix} -P_{\Gamma_0}\theta_0 + \sum_{i=1}^{k} \mathrm{tr}_0 F_{\Gamma_i}(\alpha_i h_i) \\ \vdots \\ -\mathrm{tr}_j F_{\Gamma_0}\theta_0 + \sum_{i=1, i\neq j}^{k} \mathrm{tr}_j F_{\Gamma_i}(\alpha_i h_i) - Q_{\Gamma_j}(\alpha_j h_j) \\ \vdots \end{pmatrix}.$$

With the help of BOREL–POMPEIU's formula for $u^{(i)}$ in G_i and $v = Du$ we obtain

$$Q_{\Gamma_i} u_i = \sum_{s=1}^{k} \mathrm{tr}_i T_{G_i} F_{\Gamma_s}(\alpha_{s h_s}) - \mathrm{tr}_i T_{G_i} F_{\Gamma_0}\theta_0 \qquad (i = 1, ..., k).$$

From (4.121) we have

$$P_{\Gamma_i} u_i = \mathrm{tr}_i T_{G_0}(F_{\Gamma_0}\psi_0 - \sum_{s=1}^{k} F_{\Gamma_s}\psi_s) + \mathrm{tr}_i \sum_{s=1, s\neq i}^{k} F_{\Gamma_s} Q_{\Gamma_s} u_s$$

and by inserting the already known functions $\psi_i, Q_{\Gamma_i} u_i$ we obtain, after some computation, the following result:

$$u_i = -\mathrm{tr}_i T_G F_{\Gamma_0}\theta_0 + \mathrm{tr}_i T_G \sum_{s=1}^{k} F_{\Gamma_s}(\alpha_s h_s).$$

Finally, we get by inserting in equation (4.123)

$$\mathrm{tr}_0 T_G F_{\Gamma_0}\theta_0 = \mathrm{tr}_0 T_G N(h) \quad \text{or} \quad \theta_0 = (\mathrm{tr}_0 T_G F_{\Gamma_0})^{-1} \mathrm{tr}_0 T_G N(h),$$

where we used the abbreviation $N(h) = \sum_{s=1}^{k} F_{\Gamma_s}(\alpha_s h_s)$. The above considerations prove our main theorem in this section.

Theorem 4.131. *The transmission problem (4.118) has the solution*

$$u = T_G(I - F_{\Gamma_0}(\mathrm{tr}_0 T_G F_{\Gamma_0})^{-1} \mathrm{tr}_0 T_G) N(h) \tag{4.126}$$

for arbitrarily given functions $h_i \in W_2^{1/2}(\Gamma_i)$.

4.13.3 Transmission Problems for the Lamé system

We will give now some information what we have to do in the case of the LAMÉ system ($m > 2$). For the representation of a solution u of (4.118) in G_i we can use the following result:

$$u^{(i)} = F_{\Gamma_i} u_i + T_{G_i} M^{-1} F_{\Gamma_i} (\mathrm{tr}_i T_{G_i} M^{-1} F_{\Gamma_i})^{-1} Q_{\Gamma_i} u_i$$

Similar to the previous section we have to consider the matrix operator:

$$\{\mathrm{tr}_\Gamma T_{G_0} M^{-1} \vec{F_\Gamma}\}$$

and we obtain the following representation formula:

$$u = T_{G_0} M^{-1} \vec{F_\Gamma} \{\mathrm{tr}_\Gamma T_{G_0} M^{-1} \vec{F_\Gamma}\}^{-1} Q_\Gamma \vec{u} + \sum_{s=1}^{k} F_{\Gamma_s} u_s$$

For a bounded domain Ω with the sufficiently smooth boundary γ we have now to investigate the operator $K_\Omega D$, where K_Ω is the *volume potential operator* of the linear elasticity and we get the following result:

$$V_\gamma^{\mathrm{elast}} u = \frac{m-2}{2m-2} M^{-1} T_\Omega M^{-1} u \qquad (\forall\, u \in \ker D),$$

where the operator $V_\gamma^{\mathrm{elast}}$ is of single layer type defined by

$$V_\gamma^{\mathrm{elast}} u(x) = -\frac{1}{4\pi} \int_\gamma \sum_{i=1}^{4} H_i(x-y)[\alpha\ u]_i(y) d\Gamma_y$$

with

$$H_0(x) = \frac{1}{|x|},\ H_i(x) = \frac{3m-4}{4m-4} \frac{1}{|x|} e_i - \frac{m}{4m-4} \sum_{j=1}^{3} \frac{x_i x_j}{|x|^3} e_j .$$

We can prove the jump formula

$$(\tau_{\alpha,\mathrm{ext}} - \tau_{\alpha,\mathrm{int}}) V_\gamma^{\mathrm{elast}} u = \alpha\ u$$

and then our further investigations can be done in the same way as above.

Remark 4.132. The representation formulas in the form of equation (4.126) allow the analytical properties of the solution of the transmission problem to be studied in a convenient (and explicit) way. We can look for stability and regularity results (as above in Section 4.4), but it is also possible to study the behaviour of the solution if we disturb the boundaries Γ_i (see Section 4.12). On the other hand, the presented method is open for numerical applications (see Section 5.4).

4.14 Hypercomplex Beltrami equation

By the help of functional analytical methods it is possible to generalize many results of the classical function theory to classes of functions containing essentially more than holomorphic functions. Often the transformation of partial differential equations to integral equations by means of singular integral operators is the starting point for such methods. We will illustrate this with a simple example. Let us consider the two-dimensional system of linear first-order partial differential equations:

$$\begin{aligned} a_{11}\frac{\partial u}{\partial x} + a_{12}\frac{\partial u}{\partial y} + b_{11}\frac{\partial v}{\partial x} + b_{12}\frac{\partial v}{\partial y} + a_1 u + b_1 v &= f_1, \\ a_{21}\frac{\partial u}{\partial x} + a_{22}\frac{\partial u}{\partial y} + b_{21}\frac{\partial v}{\partial x} + b_{22}\frac{\partial v}{\partial y} + a_2 u + b_2 v &= f_2. \end{aligned}$$

This system of partial differential equations can be transformed into a generalized CAUCHY–RIEMANN equation:

$$\frac{\partial w}{\partial \bar{\zeta}} = a(\zeta)\, w + b(\zeta)\, \bar{w},$$

where the coordinate transform has to satisfy a so-called BELTRAMI equation

$$\frac{\partial \zeta}{\partial \bar{z}} = q(z)\frac{\partial \zeta}{\partial z}.$$

Here $q(z)$ depends on the coefficients a_{ij} and b_{ij}. For the advantages of this transformation and the general theory of generalized analytic functions we refer to [319]. In general the BELTRAMI equation

$$\frac{\partial w}{\partial \bar{z}} = q(z)\frac{\partial w}{\partial z}$$

is the complex form of the first-order elliptic system

$$\begin{aligned} v_y &= a u_x + b u_y \\ -v_x &= b u_x + d u_y \end{aligned}$$

where $w = u + iv$, $z = x + iy$, $ad - b^2 = 1$, $a > 0$, and $b > 0$. Using the attachment

$$w(z) = z + T_G h$$

with

$$T_G h(z) = \frac{1}{2\pi i}\int_G \frac{h(\xi)}{\xi - z} d\xi_1 d\xi_2$$

we get the singular integral equation:

$$h = q(z)(1 + \Pi_G h)$$

where

$$\Pi_G h(z) = \frac{1}{2\pi i} \int_G \frac{h(\xi)}{(\xi - z)^2} d\xi_1 d\xi_2$$

is the complex Π-operator, defined as the derivative of the T_G-operator.

In multidimensional complex analysis many of these methods and efforts are lost. One of the most important reasons is the fact that in the theory of several complex variables there are no "good enough" analogies of the complex differential operators $\partial/\partial z$ and $\partial/\partial \bar{z}$. That means these analogies have not the property to factorize the Laplacian. But, as we have seen earlier in this book, "good" elliptic analogies of $\partial/\partial z$ and $\partial/\partial \bar{z}$ exist in another generalization of one-dimensional complex analysis, namely, in the CLIFFORD analysis. The hypercomplex CAUCHY–FUETER operator has most of the properties of its complex origin.

But, up to now only one of the two integral operators used in our above example — the T-operator — was applied in generalized form. Here we want to generalize the second operator the so-called Π-operator, too.

4.14.1 Definition of hypercomplex Π-operators

On the first view this definition should be an easy task. In the complex plane we have defined the Π-operator as the derivative of the complex T-operator, so why we shall not go in the same way? But here a problem arises. A. S. MEHLIKHZON proved in the 1940's that a quaternionic derivative in the usual sense exists at most for linear functions [198], which means we have differentiability and derivatives only in the case of the CLIFFORD-algebra $C\ell_{0,1}$. So we have the big problem how we generalize the derivative in a way that preserves most of its useful properties. For this we will keep in mind that we have in $\mathbb{C}$ the formula

$$df = \frac{\partial f}{\partial z} dz + \frac{\partial f}{\partial \bar{z}} d\bar{z}. \tag{4.127}$$

For holomorphic functions we have $\partial f / \partial \bar{z} = 0$ and then for the derivative $f' = \partial f / \partial z$ holds. If we can "generalize" this formula to the higher dimensional case in some sense, then it will give us a good idea what a generalization of a derivative means. One way is to start with the usual formula of the total differential in $\mathbb{R}^{n+1}$:

$$df(x) = \frac{\partial f}{\partial x_0} dx_0 + \ldots + \frac{\partial f}{\partial x_n} dx_n$$

and introduce a new structure of the pre-image set (cf. Sec. 2.2, 3.9, and [187])

$$z_0 = x_0, z_k = x_k - x_0 e_k \qquad (k = 1, \ldots, n).$$

With this new structure one gets the formula

$$df(x) = dz_0 Df + dz_1 \frac{\partial f}{\partial x_1} + \ldots + dz_n \frac{\partial f}{\partial x_n},$$

which can be viewed as a hypercomplex analogue to our formula (4.127). Using this formula we find that each monogenic function is differentiable (it is also true the other way around) and the derivative is given by the vector $(\partial f/\partial x_1, \ldots, \partial f/\partial x_n)$. We have to mention that this differential operator has another quality as the complex Π-operator. It maps a usual CLIFFORD-valued function into a vector function, whose components are CLIFFORD-valued functions.

Another way is to use the $\overline{D}$-operator. I.N. MITELMAN and M.V. SHAPIRO have shown that this operator can be considered as the derivative-operator (hyperderivative called) in some sense [206]. In fact, it was done again by a "generalization" of our complex formula (4.127), but in a really non-trivial way. Nevertheless, we get the advantage of an "one-dimensional" CLIFFORD-valued derivative.
For all that follows we have to keep in mind that we have two different ways to define a derivative of a monogenic function, which have both their advantages and their weaknesses.
Using the $\overline{D}$-operator as the derivative we can define our Π-operator in the following way:

Definition 4.121. *The operator* Π, *defined by*

$$\Pi f = \overline{D} T f$$

is called the generalized Π*-operator.*

This operator acts from $\mathcal{C}^\alpha(\overline{G})$ to $\mathcal{C}^\alpha(\overline{G})$ $(0 < \alpha \leq 1)$ [298]. We remark that this operator is a direct generalization of the complex Π-operator. That means in the case of $C\ell_{0,1} = \mathbb{C}$ our generalized Π-operator coincides with the usual complex Π-operator up to the factor two.
Historically we remark that this operator was first introduced in 1978 [298] and then again in a more general way (and another context) by M.V. SHAPIRO and N.V. VASILEVSKI in 1992 [267].
Another way to generalize the complex Π-operator is to use the idea of a derivative in the sense of H.R. MALONEK [189].

Definition 4.122. *The operator* $\vec{\Pi}$, *defined by*

$$\vec{\Pi} f = \nabla_{\vec{x}} T f,$$

where $\nabla_{\vec{x}} = (\partial/\partial x_1, \ldots, \partial/\partial x_n)$ *is called the generalized* $\vec{\Pi}$*-operator.*

Corresponding to the properties of the derivative defined by the gradient, this operator maps a usual CLIFFORD-valued function into a vector function, which components are CLIFFORD-valued functions. But, nevertheless, it is also a direct generalization of the usual complex Π-operator.

4.14.2 Properties of hypercomplex Π-operators

In the following we will give some important properties of the generalized Π-operator. For the proofs of these properties and more information about this generalization of the complex Π-operator we refer to [104] and [105]. Note that we are interested in properties of Π acting on functions defined in a bounded domain. In the case of $G = \mathbb{R}^{n+1}$ these properties are studied in [298] and [104]. In particular, Π is a unitary operator in this case. From our above-mentioned problems it is clear that here we need some information on the invertibility of Π in the case of a bounded domain. Let us mention that this problem is also not completely solved in the complex case.
First we find that the generalized Π-operator acts from $W_p^k(G, Cl_{0,n})$ to $W_p^k(G, Cl_{0,n})$, $p > 1, k \in \mathbb{N} \cup \{0\}$. Moreover, this operator has an integral representation of the form

$$(\Pi f)(x) = -\frac{1}{\sigma_{n+1}} \int_G \frac{(n-1) + (n+1)\dfrac{\overline{(y-x)}^2}{|y-x|^2}}{|y-x|^{n+1}} f(y)\, dG_y + \frac{1-n}{1+n} f(x),$$

Due to the strong singularity of the kernel function the integral has to be understood in the sense of CAUCHY*'s principal value*. Under the same assumptions as before we have for the conjugate operator

$$\overline{\Pi} f = D\overline{T} f.$$

Lemma 4.123. *We have*

$\Pi : \ker \overline{D}(G) \cap L_2(G, Cl_{0,n}) \mapsto \ker D(G) \cap L_2(G, Cl_{0,n})$.

In terms of the orthoprojections we obtained

$\Pi : \operatorname{im} \mathbf{P}^{\overline{D}} \mapsto \operatorname{im} \mathbf{P}^{D}$.

We remember that $\mathbf{P}^A$ denotes the orthoprojection onto $\ker A \cap L_2(G, Cl_{0,n})$ and $\mathbf{Q}^A = I - \mathbf{P}^A$.

Lemma 4.124. *We have that*

$\Pi : D(\overset{\circ}{W}{}_2^1(\Omega)) \mapsto \overline{D}(\overset{\circ}{W}{}_2^1(\Omega))$.

Lemma 4.124. means that $\Pi : \operatorname{im} \mathbf{Q}^{\overline{D}} \mapsto \operatorname{im} \mathbf{Q}^{D}$.

Remark 4.125. *Again investigating the above mapping properties for the conjugate operator* $\overline{\Pi}$ *we get*

$$\overline{\Pi}: \quad \operatorname{im}\mathbf{P}^D \mapsto \operatorname{im}\mathbf{P}^{\overline{D}},$$
$$\overline{\Pi}: \quad \operatorname{im}\mathbf{Q}^D \mapsto \operatorname{im}\mathbf{Q}^{\overline{D}}.$$

Theorem 4.126. *Suppose* $f \in W_p^k(G, Cl_{0,n}),\ (1 < p < \infty), k \in \mathbb{N}$, *then we have*

$$\overline{\Pi}\Pi f = D\overline{T}\,\overline{D}Tf = D(I - \overline{F}_\Gamma)Tf = f - D\overline{F}_\Gamma Tf$$

$$\Pi\overline{\Pi}f = \overline{D}TD\overline{T}f = \overline{D}(I - F_\Gamma)\overline{T}f = f - \overline{D}F_\Gamma\overline{T}f.$$

From this theorem we derive a sufficent condition for the invertibility of Π of the form in the case $p = 2$:

Proposition 4.127. *We have*

$$\overline{\Pi}\Pi f = f \quad \forall f \in \operatorname{im}\mathbf{Q}^{\overline{D}}$$

$$\Pi\overline{\Pi}f = f \quad \forall f \in \operatorname{im}\mathbf{Q}^{D}.$$

From our proposition we are able to give a necessary and sufficient criterion for the invertibility of Π. We have to remark that this is also a necessary and sufficient condition for the invertibility of the usual complex Π-operator. Up to now this was an unsolved problem.

Theorem 4.128. *The operator*

$$\Pi\ : \operatorname{im}\mathbf{Q}^D \cap \operatorname{im}\mathbf{Q}^{\overline{D}} \longrightarrow \operatorname{im}\mathbf{Q}^D \cap \operatorname{im}\mathbf{Q}^{\overline{D}}$$

is invertible.

For the proof we refer to [105]. This criterion is of special importance in investigating the singular integral equation

$$\Pi u = g,$$

which has, according to our theorem, only in the case of $g \in \operatorname{im}\mathbf{Q}^D \cap \operatorname{im}\mathbf{Q}^{\overline{D}}$ a unique solution. We remark that the set $\operatorname{im}\mathbf{Q}^D \cap \operatorname{im}\mathbf{Q}^{\overline{D}}$ is non-empty. We mentioned that in the case of $L_2(\mathbb{R}^{n+1})$ the Π-operator is a unitary operator [104].
With the help of the theory of Calderon and Zygmund (cf. [202]) one can also estimate the norm of the generalized Π-operator in the following form:

$$||\Pi||_{[(L_2(G)),(L_2(G))]} \leq \frac{n-1}{n+1} + \frac{2n\sqrt{c_4}}{\sigma_{n+1}}, \tag{4.128}$$

where

$$c_4 = \left(\int_S | \ln \frac{1}{|\cos \gamma|} + \frac{i\pi}{2} \text{sign} \cos \gamma |^4 dS_{\theta'} \right)^{\frac{1}{2}}$$

is a constant [104].

Next we give some important properties of the $\vec{\Pi}$-operator. For the $\vec{\Pi}$-operator we get the representation formula:

$$\vec{\Pi} f = (\Pi_1 f, \ldots, \Pi_n f) = -\frac{1}{\omega} \int_G \vec{K}(\xi, z) f(\xi)\, dG + \frac{\vec{i}}{n+1} f$$

with

$$\vec{i} = (-e_1, \ldots, -e_n) \text{ and } \vec{K}(y, x) = \nabla_{\vec{x}} \left(\frac{\overline{(y-x)}}{|y-x|^{n+1}} \right).$$

Theorem 4.129. *Suppose that* $1 < p < \infty, k \in \mathbb{N} \cup \{0\}$. *Then, we have*

$$\vec{\Pi} : W_p^k(G) \mapsto W_p^k(G), \quad (k \geq 0, p > 1).$$

Moreover, we can estimate the norm of this operator by

$$||\vec{\Pi}||_{[L_2(G), L_2(G)]} \leq \frac{(n+2)\sqrt{n \cdot c_4}}{\sigma_{n+1}},$$

where c_4 is the same constant as in the above norm estimate of the Π-operator. At this point of our considerations the question about the connections between these two generalizations of the complex Π-operator arises. First we have the problem that these two operators are not only defined in a very different manner but also act in a very different way. Nevertheless, we are able to prove the relationships:

$$\Pi f = f + 2 \sum_{k=1}^{n} \overline{e}_k \Pi_k f,$$

with $\vec{\Pi} f = (\Pi_k f)_{k=1}^n$ and

$$\vec{\Pi} f = \nabla_{\vec{x}} \overline{T} \Pi f + \nabla_{\vec{x}} \overline{F}_\Gamma T f.$$

It may be observed that we need not only Πf or $\vec{\Pi} f$, respectively, but also the knowledge of f, except some special cases like $L_2(\mathbb{R}^{n+1})$, where we have the invertibility of the Π-operator. Up to now it is an open problem to get relationships without explicit knowledge of the values of f. Nevertheless, these relationships are important connections between $\vec{\Pi}$ and Π because it gives us the possibility to get the analytical properties of $\vec{\Pi}$ by the study of the Π-operator, which is easier to deal with from the analytical point of view.

4.14.3 The Clifford–Beltrami equation

Now we will demonstrate the possibilities of the application of our Π-operator by the example of a CLIFFORD–BELTRAMI equation. For further applications see [104], [144], and [189].
Let $G \subset \mathbb{R}^{n+1}$ be a bounded, simply connected domain with a sufficiently smooth boundary, and $q : G \mapsto C\ell_{0,n}$ a measurable function. Moreover, let $w : G \mapsto C\ell_{0,n}$ be a sufficiently smooth function. Then we call the equation

$$Dw = q\,\overline{D}w \tag{4.129}$$

CLIFFORD–BELTRAMI *equation.* Applying the attachment

$$w = \phi + Th, \tag{4.130}$$

where ϕ is an arbitrary left-monogenic function, we transform the equation (4.129) into the integral equation

$$h = q(\overline{D}\phi + \Pi h). \tag{4.131}$$

Obviously, w is a solution of equation (4.129) if h is a solution of equation (4.131). On the other hand, each solution of equation (4.129) can be represented by (4.131). Investigating the norm of the operator $q\Pi$ we have that in the case of

$$||q|| \leq \frac{1}{||\Pi||}$$

this operator is contractive. That means we can get a solution of equation (4.131) using BANACH's fixed-point theorem (the other conditions of BANACH's fixed-point theorem can easily be verified). Applying our norm estimate (4.128) for the generalized Π-operator we get the condition

$$||q||_{[L_2(G),L_2(G)]} \leq \frac{1}{(n-1)/(n+1) + (2n\sqrt{c_4})/(\sigma_{n+1})}$$

being sufficient for the existence of a solution of equation (4.127). We remark only that in the case of the complex BELTRAMI equation we have the condition $||q|| \leq ||q_0|| < 1$.

4.15 Lower bounds for eigenvalues

In this section we will investigate the first eigenvalue of elliptic boundary value problems with DIRICHLET boundary condition. It will be shown that the quaterionic calculus can be used for the proof of lower bounds for the first eigenvalues have been studied very extensively. We will mention here only a few results. Let $|B_1|$ be the volume of the unit ball in $\mathbb{R}^n$ and j_k be the first (positive) zero of BESSEL's function J_k. Then, the well-known FABER–KRAHN inequality states that

$$\lambda_1(G) \geq |K_1|^{\frac{2}{n}} |G|^{-\frac{2}{n}} \, j^2_{\frac{(n-1)}{2}},$$

where λ_1 stands for the first eigenvalue of the Laplacian under DIRICHLET's condition. In particular, for $n = 3$

$$\lambda_1(G) \geq \left(\frac{4\pi}{3}\right)^{\frac{2}{3}} |G|^{-\frac{2}{3}} \, \pi^2$$

holds. This fundamental result has been improved many times using additional assumptions concerning geometric properties of the domain G. We will mention here only one estimate proved by H. LEVINE and M.H. PROTTER in 1985.

$$\Lambda_1(G) \geq \left(\frac{4\pi}{3}\right)^{\frac{2}{3}} |G|^{-\frac{2}{3}} \frac{3}{5} \left(\frac{3\pi}{2}\right)^{\frac{2}{3}} \tag{4.132}$$

The estimates presented here do not need such assumptions. We only need the usual properties of G and ∂G necessary for the definition of our integral operators. This is an advantage because the same method can be used for all operators under consideration. The method also works after discretization (see Sec. 5.5–6.2) for finite difference operators in the same way. The disadvantage is that in cases where we can compare several estimates our results are not the best.

In passing we will recognize that the norms of the integral operators used in this book are strongly connected to the eigenvalues of the investigated boundary value problems. In some cases there are equalities. This opens the possibility of using eigenvalue estimates for the estimation of operator norms and vice versa. Of course, if we want to have an operator norm we can use all the other known results on eigenvalues and numerical approaches too(see e.g. [89]).

In the subsection 4.15.2 we show that the method can be extended to some semi-linear boundary value problems, and in the subsection 4.15.3 we will see that using representation formulas for the solutions of boundary value problems and also parameter dependence of eigenvalues can be studied in a convenient way

4.15.1 The first eigenvalue of the Laplacian

For simplicity we abbreviate $\|\cdot\|_2 := \|\cdot\|_{L_2(G,\mathbb{H})}$, $\|\cdot\|_{2,1} := \|\cdot\|_{W_2^1(G,\mathbb{H})}$.

Lemma 4.142. *The operator* $D : \overset{\circ}{W}{}_2^1 (G,\mathbb{H}) \to D(\overset{\circ}{W}{}_2^1 (G,\mathbb{H}))$ *is invertible and we have* $D^{-1} = T_G$.

The proof is an exercise. The operator D is continuous, surjective by definition, and from $Du = 0$, $u \in \overset{\circ}{W}{}_2^1 (G,\mathbb{H})$ with the help of BOREL–POMPEIU's formula it follows that $u = T_G Du = 0$. Clearly, $D^{-1} = T_G$, because T_G is a right-inverse of D. #

We already know that $D \overset{\circ}{W}{}_2^1 = \mathrm{im}\,\mathbf{Q}$. This lemma implies that

$\|T_G v\|_2 \le c\|v\|_2 \quad (\forall\, v \in \mathrm{im}\,\mathbf{Q}\,).$

Now, we are able to prove the representation of $\lambda_1(G)$ with the help of the norm of T_G.

Theorem 4.143. *Let* $\lambda_1(G)$ *be the first eigenvalue of* $\{-\Delta, \mathrm{tr}_\Gamma\}$. *Then,*

$$\lambda_1(G) = \frac{1}{\|T_G\|^2_{[\mathrm{im}\mathbf{Q},L_2]}} \tag{4.133}$$

and

$$\lambda_1(G) = \frac{1}{\|T_G\|^2_{[\mathrm{im}\mathbf{Q},\overset{\circ}{W}{}_2^1(G)]} - 1}. \tag{4.134}$$

Proof. From the invertibility of D we get $\|Du\|_2 \ge c^{-1}\|u\|_2$. Using

$(u, \Delta u)_2 = (Du, Du)_2 = \|Du\|_2^2$

we obtain from the known scalar result

$$(Du, Du)_2 \ge \lambda_1(G)\|u\|_2^2 \Longrightarrow \|T_G\|_{[\mathrm{im}\mathbf{Q},L_2]} \le \frac{1}{\sqrt{\lambda_1}}.$$

Now, let u_1 be an eigenfunction of $-\Delta$ with respect to λ_1. Starting with $w = Du_1$ we get

$$\|T_G w\|_2^2 = \|T_G Du\|_2^2 = \|u\|_2^2 = (-\Delta u, u)_2 \frac{1}{\lambda_1} = \frac{1}{\lambda_1}\|Du\|_2^2 = \frac{1}{\lambda_1}\|w\|_2^2.$$

The W_2^1-statement can be obtained as follows:

$$\begin{aligned}\|T_G\|^2_{[\mathrm{im}\mathbf{Q},\overset{\circ}{W}{}_2^1(G)]} &= \sup_{u\in\mathrm{im}\mathbf{Q}} \frac{\|T_G u\|_{2,1}^2}{\|u\|_2^2} = \sup_{v\in\overset{\circ}{W}{}_2^1} \frac{\|T_G Dv\|_{2,1}^2}{\|Dv\|_2^2} \\ &= \sup_{v\in\overset{\circ}{W}{}_2^1} \frac{\|v\|_{2,1}^2}{\|Dv\|_2^2} \le \sup_{v\in\overset{\circ}{W}{}_2^1} \frac{\|Dv\|_2^2}{\|Dv\|_2^2} + \sup_{v\in\overset{\circ}{W}{}_2^1} \frac{\|v\|_2^2}{\|Dv\|_2^2} = 1 + \frac{1}{\lambda_1}.\end{aligned}$$

Finally, we have to take again $w = Du_1$. #

Corollary 4.144. *The eigenvalue λ_1 can be estimated by*

$\lambda_1(G) \geq (4\pi)^{\frac{2}{3}}|G|^{-\frac{2}{3}}.$

Proof. F. KIPPIG proved in [149] that

$$||T_G||_{[L_2,L_2]} \leq \sqrt[3]{\frac{1}{4\pi}}\sqrt[3]{|G|} \quad .$$

Now, we have only to use $||T_G||_{[L_2,L_2]} \geq 1/\sqrt{\lambda_1}$. This immediately follows from Theorem 4.133. #

Assuming $\partial G \in C^\infty$, it can be proved that an arbitrary eigenfunction of $\{-\Delta, \mathrm{tr}_\Gamma\}$ belongs to $C^\infty(G)$. For this we have to use the representation formula $u = \lambda T_G \mathbf{Q} T_G u$ for the solution of the eigenvalue problem $\{-\Delta u = \lambda u, \mathrm{tr}_\Gamma u = 0\}$. From the mapping properties we get $u \in W_2^2(G, \mathbb{H}) \cap \overset{\circ}{W}{}_2^1 (G, \mathbb{H})$. Repeating this consideration we obtain $u \in W_2^{2k}(G, \mathbb{H}) \cap \overset{\circ}{W}{}_2^1 (G, \mathbb{H})$ $(\forall\, k \in \mathbf{N})$.

Now we will look for non-linear eigenvalue problems and we recognize that the above method works successfully. As an example we will deal with a non-linear STOKES eigenvalue problem.

4.15.2 A non-linear Stokes eigenvalue problem

Let $f : W_2^1(G, \mathbb{H}) \to L_2(G, \mathbb{H})$ be a non-linear operator function. We assume that f fulfils a local LIPSCHITZ condition:

$$||f(u) - f(v)||_2 \leq L||u - v||_{2,1} \quad \forall\, u, v : ||u||_{2,1} \leq 1, ||v||_{2,1} \leq 1.$$

Consider the following non-linear boundary value problem

$$\begin{aligned} -\Delta\underline{u} + \frac{1}{\eta}\mathrm{grad}\, p &= \Lambda f(\underline{u}) \quad \text{in} \quad G \\ \mathrm{div}(\underline{u}) &= 0 \quad \text{in} \quad G \\ \underline{u} &= 0 \quad \text{on} \quad \Gamma, \end{aligned}$$

where all notations are used as in Section 4.7. The number Λ is called the *eigenvalue parameter*. In quaternionic language we have

$$\begin{aligned} D^2 u + \frac{1}{\eta}Dp &= \Lambda f(u) \quad \text{in} \quad G \\ \mathrm{Sc}\, Du &= 0 \quad \text{in} \quad G \\ u &= 0 \quad \text{on} \quad \Gamma \end{aligned} \tag{4.135}$$

with $u = u_0 e_0 + \underline{u}$, $f(u) = f(\underline{u})$. Applying $T_G \mathbf{Q} T_G$ we obtain the equivalent problem

$$\begin{aligned} u + \frac{1}{\eta} T_G \mathbf{Q} p &= \Lambda T_G \mathbf{Q} T_G f(u) \\ \frac{1}{\eta} \mathrm{Sc} \mathbf{Q} p &= \Lambda \mathrm{Sc} \mathbf{Q} T_G f(u). \end{aligned}$$

We have now to explain more exactly the meaning of Λ.

Definition 4.145. *If for a parameter Λ functions $u \neq 0, p \neq 0$ exist which solve the problem (4.135) then Λ is called a point of the spectrum σ of (4.135). If u has to satisfy $||u||_{2,1} \leq r$ then we mark this by $\sigma_{\mathbf{r}}$.*

Theorem 4.146. *Let $\Lambda \in \sigma_{\mathbf{1}}, f(0) = 0$. Then,*

$$|\Lambda| \geq (L ||T_G||_1 ||T_G||_2)^{-1}$$

Proof. Differentiating the representation formula (4.136) we get

$$Du + \frac{1}{\eta} \mathbf{Q} p = \Lambda \mathbf{Q} T_G f(u).$$

As in the case of the linear STOKES system we have

$$||Du||_2^2 + \frac{1}{\eta} ||\mathbf{Q} p||_2^2 = |\Lambda|^2 ||\mathbf{Q} T_G f(u)||_2^2 \tag{4.136}$$

and this implies by the help of Theorem 4.79 that

$$\sqrt{\frac{\lambda_1}{1+\lambda_1}} ||u||_{2,1} \leq |\lambda| \, ||T_G||_2 ||f(u)||_2 \leq |\lambda| \, ||T_G||_2 L.$$

Here, we used that $||f(u)||_2 \leq L$ for $||u||_{W_2^1} \leq 1$ is a consequence of the local LIPSCHITZ condition. Hence,

$$||u||_{2,1} \leq |\lambda| \, ||T_G||_1 ||T_G||_2 L.$$

Now, we look to the iteration procedure

$$\begin{aligned} u_0 &\in \overset{\circ}{W_2^1}(G, \mathbb{H}) \cap \ker \operatorname{div}, \quad ||u_0|| \leq 1 \\ u_n &= \lambda T_G \mathbf{Q} T_G f(u_{n-1}) - \frac{1}{\eta} T_G \mathbf{Q} p_n \quad (n = 1, 2, \ldots). \end{aligned}$$

At each step a linear STOKES problem has to be solved. Following the same path as in Section 4.1, we end up with the estimate

$$||u_n - u_{n-1}||_{2,1} \leq |\Lambda| \, ||T_G||_1 \, ||T_G||_2 L ||u_{n-1} - u_{n-2}||_{2,1}$$

Assuming $|\Lambda| \, ||T_G||_1 ||T_G||_2 L \leq 1$ we conclude the contractivity of the non-linear mapping. With the help of (4.136) it is ensured that $||u_n||_{W_2^1} \leq 1$ $(\forall\, u \in \mathbf{N})$.

Now, we can apply BANACH's fixed-point theorem and we obtain that $\{u_n, p_n\}$ converges to the unique solution. Therefore, Λ with

$$|\Lambda| \le \frac{1}{L\|T_G\|_1\|T_G\|_2}$$

cannot belong to $\boldsymbol{\sigma_1}$. #

What happens when we omit the restrictive condition $f(0) = 0$? Then, it cannot be ensured that $u_n \to 0$, although the proof of convergence goes through. If we assume that $f : \overset{\circ}{W}{}_2^1 (G, \mathbb{H}) \to L_p(G, \mathbb{H})$ $(\frac{6}{5} < p < \frac{3}{2})$ and

$$\|f(u) - f(v)\|_p \le L\|u - v\|_{2,1},$$

$$\forall\, u, v \in W_2^1(G, \mathbb{H}) : \ \|u\|_{2,1} \le 1,\ \|v\|_{2,1} \le 1.$$

Then, we can prove that

$$\inf_{\Lambda \in \sigma_1} |\Lambda| \ge [L\,\|T_G\|_1\|T_G\|_{[L_p,L_2]}]^{-1}.$$

This is a consequence of

$$\sqrt{\frac{\lambda_1}{1+\lambda_1}}\|u\|_{2,1} \le |\lambda|\,\|T_G\|_{[L_p,L_2]}L$$

which can be concluded from (4.136). The assumption $p > \frac{6}{5}$ guarantees that $T_G : L_p \to L_2$ is bounded. It is also possible to change the LIPSCHITZ condition. Under the assumption

$$\|f(u) - f(v)\|_2 \le L\|u - v\|_2, \quad \forall\, u, v : \|u\|_2 \le 1, \|v\|_2 \le 1$$

we can prove the following theorem.

Theorem 4.147. *Let $\Lambda \in \boldsymbol{\sigma_1}$, $f(0) = 0$. Then, we have that*

$$\inf_{\Lambda \in \sigma_1} |\Lambda| \ge \lambda_1 L^{-1}.$$

Proof. We already know $\|Du\|_2 \ge \sqrt{\lambda_1}\|u\|_2$ $(\forall\, u \in \overset{\circ}{W}{}_2^1)$. Therefore, from (4.136)

$$\lambda_1\|u\|_2^2 \le |\lambda|^2\|T_G\|_2^2\|f(u)\|_2^2 \le |\Lambda|^2\|T_G\|_2^2L^2\|u\|_2^2$$

follows. Using $\|T_G\|_2 \le 1/\sqrt{\lambda_1}$ the proof is complete. #

The LIPSCHITZ condition in L_2 is more useful than the condition between L_2 and W_2^1. But we have to pay for this convenience. Now we have another eigenvalue in our final result. Nevertheless, all estimates are constructive, because λ_1 as well T_G-norms can be explicitly estimated by properties of G. At the end of this section we will study the behaviour of eigenvalues if the boundary value problem contains a varying parameter (coefficient).

4.15.3 Eigenvalues of the Lamé system

We consider the eigenvalue problem

$$\begin{aligned} DMDu &= \Lambda u \quad \text{in} \quad G, \\ u &= 0 \quad \text{on} \quad \Gamma, \end{aligned} \tag{4.137}$$

with $\Lambda \in \mathbb{R}$, $\mathrm{Sc}\, u = 0$.
System (4.137) is equivalent to

$$\begin{aligned} (I - \Lambda T_G \mathbf{Q}_M M^{-1} T_G)u = 0 \quad &\text{in} \quad G \quad , \\ u = 0 \quad &\text{on} \quad \Gamma \quad . \end{aligned} \tag{4.138}$$

This system has only a trivial solution if

$$|\Lambda| \| T_G \mathbf{Q}_M M^{-1} T_G \|_{[L_2, L_2]} < 1,$$

and by using

$$\| \mathbf{Q}_M M^{-1} \|_{[L_2, L_2]} \leq 1$$

we obtain an estimation of the first eigenvalue $\Lambda_1(m, G)$.

$$\Lambda_1(m, G) \geq \| T_G \mathbf{Q}_M M^{-1} T_G \|^{-1}_{[L_2, L_2]} \geq \| T_G \|^{-2}_{[L_2, L_2]}.$$

Using the estimate

$$\| T_G \|_{[L_2, L_2]} \leq \left(\frac{3}{4\pi} \right)^{1/3} |G|^{1/3}$$

obtained in [110], we get

$$\Lambda_1(m, G) \geq \left(\frac{4\pi}{3} \right)^{2/3} |G|^{-2/3}. \tag{4.139}$$

The quality of the estimation which is contained in [146], based on the result (4.132), cannot be improved.
Now, we consider the behaviour of $\Lambda_1(m, G)$ in dependence on $m \in (2, \infty)$. If A denotes the LAMÉ operator, then the first eigenvalue under DIRICHLET's condition is characterized by

$$\begin{aligned} \Lambda_1(m, G) &= \inf\{(DMDu, u)/(u, u), u \in \overset{\circ}{W}{}^1_2(G), \mathrm{Sc}\, u = 0, u \neq 0\} \\ &= \inf\{(MDu, Du)/(u, u), u \in \overset{\circ}{W}{}^1_2(G), \mathrm{Sc}\, u = 0, u \neq 0\} \\ &\leq \inf\{(MDu, Du)/(u, u), \hat{u} \in \overset{\circ}{W}{}^1_2(G) \cap \ker\ \mathrm{div}, \mathrm{Sc}\, u = 0, u \neq 0\} \\ &= \inf\{(Du, Du)/(u, u), \hat{u} \in \overset{\circ}{W}{}^1_2(G) \cap \ker\ \mathrm{div}, \mathrm{Sc}\, u = 0, u \neq 0\} \\ &= \Lambda_* . \end{aligned}$$

Now we can state the following:

Lemma 4.148. *We have*

$\Lambda_1(m,G) \geq \Lambda_* < \infty \quad (\forall\, m > 2).$

In the following we investigate the function $\Lambda_1(m,G)$:

$$\Lambda_1(m,G) = \inf\left\{ \left(\|Du\|_2^2 + \frac{m}{m-2}\|\mathrm{Sc}\, Du\|_2^2 \right) / \|u\|_2^2,\ \hat{u} \in \overset{\circ}{W}{}_2^1, \mathrm{Sc}\, u = 0, u \neq 0 \right\}.$$

We now get for every $u \in \overset{\circ}{W}{}_2^1 (G,\mathbb{H})$ and $m' \leq m$ that the inequality

$$\|Du\|^2 + \frac{m}{m-2}\|\mathrm{Sc}\, Du\|^2 \leq \|Du\|^2 + \frac{m'}{m'-2}\|\mathrm{Sc}\, Du\|^2$$

holds and we obtain that $\Lambda_1(m,G)$ is a monotonic decreasing function in $(2,\infty)$.

Lemma 4.149. *Here Λ_{**} exists with*

$$\lim_{m\to 2+0} \Lambda_1(m,G) = \Lambda_{**} \leq \Lambda_*.$$

Proof. Monotony and boundedness of $\lambda_1(m,G)$ provides the proof. #

In the following we denote by $u_m(x)$ a normed (in $\overset{\circ}{W}{}_2^1 (G,\mathbb{H})$) eigenfunction with respect to the eigenvalue $\Lambda_1(m,G)$. We consider the case $m \to 2+0$. Using the condition $\|u_m\|_{2,1} = 1$ we obtain the existence of a subsequence $\{u_{m'}\}$ which converges weakly in $W_2^1(G,\mathbb{H})$. For simplicity we will denote this subsequence again by $\{u_m\}$

$$\begin{aligned} u_m &\rightharpoonup u_* \quad \text{in} \quad \overset{\circ}{W}{}_2^1 (G,\mathbb{H}) \\ \Rightarrow \quad u_m &\to u_* \quad \text{in} \quad L_2(G,\mathbb{H}). \end{aligned} \tag{4.140}$$

Further, we get the following estimations of $\mathrm{Sc}\, Du_m$:

$$\begin{aligned} \|\mathrm{Sc}\, Du_m\| &\leq \Lambda_* \|\mathrm{Sc}\, M^{-1}\mathbf{Q}_m T_G u_m\| \leq \Lambda_* \sqrt{\frac{m-2}{2m-2}}\|T_G u_m\|_2 \\ &\leq \Lambda_* \sqrt{\frac{m-2}{2m-2}}\|T_G\|_2\, \|u_m\|_2 \to 0 \text{ for } m \to 2. \end{aligned} \tag{4.141}$$

Further, it is valid that

$$\begin{aligned} Du_m &\rightharpoonup Du_* \quad \text{in} \quad L_2(G,\mathbb{H}), \\ \mathrm{Sc}\, Du_m &\rightharpoonup \mathrm{Sc}\, Du_* \quad \text{in} \quad L_2(G,\mathbb{H}). \end{aligned}$$

From (4.140) if follows that

$$\mathrm{Sc}\, Du_* = 0 \quad \text{and} \quad \mathrm{Sc}\, u_* = 0.$$

$$\Rightarrow u_* \in \overset{\circ}{W}{}_2^1 (G,\mathbb{H}) \cap \ker \operatorname{div}. \tag{4.142}$$

We now consider the sequences $\{\Lambda_1 = \Lambda_1(m,G)\}$ and $\{u_m\}$ for $m \to 2+0$, and we investigate properties of Λ_{**}. Note that $Mu \to \underline{u}$ for $m \to 2$. We have

$$\begin{aligned}\Lambda_1 T_G \mathbf{Q}_M M^{-1} T_G u_m \quad = \quad & \Lambda_1(m,G)(T_G \mathbf{Q}_M M^{-1} T_G u_m - T_G M^{-1} \mathbf{Q}_M T_G u_*) \\ & + \Lambda_1(m,G)(T_G \mathbf{Q}_M M^{-1} T_G u_* - T_G \mathbf{Q} T_G u_*) \\ & + (\Lambda_1(m,G) - \Lambda_{**}) T_G \mathbf{Q} T_G u_* \\ & + (\Lambda_1(m,G) - \Lambda_{**}) \frac{1}{\eta} T_G \mathbf{Q} p_* \qquad (4.143) \\ & + (\Lambda_{**} - \Lambda_1(m,G)) \frac{1}{\eta} T_G \mathbf{Q} p_* + \Lambda_{**} T_G \mathbf{Q} T_G u_*,\end{aligned}$$

where p_* (with a suitable η) is defined by

$$u_* + \frac{1}{\eta} T_G \mathbf{Q} p_* = T_G \mathbf{Q} T_G u_*, \qquad (4.144)$$

$$u_* \in \ker \operatorname{div} \cap \overset{\circ}{W}{}_2^1 (G, \mathbb{H}).$$

We observe that $\mathrm{Sc}\, u_* = 0$. Further, for $m \to 2$

$$\|\Lambda_1(m,G)(T_G \mathbf{Q}_M M^{-1} T_G u_m - T_G \mathbf{Q}_M M^{-1} T_G u_*\| \leq$$

$$\leq \Lambda_* \|T_G\|_2^2 \|u_m - u_*\|_2 \to 0$$

and

$$\|\Lambda_1(m,G)(T_G \mathbf{Q}_M M^{-1} T_G u_* - T_G \mathbf{Q} T_G u_* + \frac{1}{\eta} T_G \mathbf{Q} p_*\|_2 \leq$$

$$\leq \Lambda_* \|u_m - u_*\|_2 \to 0$$

hold, where $u_m = T_G M^{-1} \mathbf{Q}_m T_G u_*$, and u_* is the solution to STOKES' problem (4.144). We used the result $u_m \to u_*$ (in $W_2^1(G, \mathbb{H})$) which has been proved in [96]. From Lemma 4.149. it follows that

$$\|\Lambda_1(m,G)) \frac{1}{\eta} T_G \mathbf{Q} T_G u_*\| \to 0 \quad \text{for } m \to 2,$$

$$\|\Lambda_{**} - \Lambda_1(m,G)) \frac{1}{\eta} T_G \mathbf{Q} T_G p_*\| \to 0 \quad \text{for } m \to 2.$$

Hence,

$$\Lambda_1(m,G) T_G M^{-1} \mathbf{Q}_M T_G u_m \longrightarrow \Lambda_{**} T_G \mathbf{Q} T_G u_* - \Lambda_{**} \frac{1}{\eta} T_G \mathbf{Q} p_*$$

and $u_m \longrightarrow u_*$. Therefore, (with $p^* = \Lambda_{**} p_*$) $\{u_*, p^*\}$ solve the system

$$u_* = \Lambda_{**} T_G \mathbf{Q} T_G u_* - \frac{1}{\eta} T_G \mathbf{Q} p^*, \quad u_* \in \overset{\circ}{W}{}_2^1 (G) \cap \ker \operatorname{div}. \qquad (4.145)$$

If we investigate the relations between Λ_* and Λ_{**} we can state $\Lambda_{**} \leq \Lambda_*$ because of Lemma 4.149. On the other hand we have

$$u_* = \Lambda_{**} T_G \mathbf{Q} T_G u_* - \frac{1}{\eta} T_G \mathbf{Q} p^*$$

and by application of D from the left we get

$$Du_* = \Lambda_{**} T_G \mathbf{Q} T_G u_* - \frac{1}{\eta} T_G \mathbf{Q} p^*. \tag{4.146}$$

By partial integration and by using the properties $DT_G = I$ and $D\mathbf{Q} = D$ we obtain

$$\begin{aligned} \|Du_*\|_2^2 &= \Lambda_{**}(\mathbf{Q} T_G u_*, Du_*)_2 - \frac{1}{\eta}(\mathbf{Q} p^*, u_*)_2 \\ &= \Lambda_{**}(D\mathbf{Q} T_G u_*, u_*)_2 - \frac{1}{\eta}(D\mathbf{Q} p^*, u_*)_2 \\ &= \Lambda_{**}(u_*, u_*)_2 - \frac{1}{\eta}(Dp^*, u_*)_2 = \Lambda_{**}\|u_*\|_2^2. \end{aligned}$$

Hence,

$$\Lambda_{**} = \frac{\|Du_*\|_2^2}{\|u_*\|_2^2} \geq \inf\left\{ \frac{(Du, Du)_2}{(u, u)_2} \right\},$$

where $\hat{u} \in \overset{\circ}{W}{}_2^1 (G, \mathbb{H}) \cap \ker \operatorname{div}$, $\operatorname{Sc} u = 0, u \neq 0\} = \Lambda_*$. The last inequality proves $\Lambda_* = \Lambda_{**}$. From 4.145 we have that Λ_* is an eigenvalue of STOKES' equation under DIRICHLET's condition. By repeating the above estimations for any eigenvalue of STOKES' problem we have obtained that Λ_* must be the smallest eigenvalue of STOKES' problem. If we collect all results obtained and the quoted estimates we can state the following theorem.

Theorem 4.150. *Let Λ_* be the first eigenvalue of* STOKES *equation under* DIRICHLET*'s condition and $m > 2$. Then we have*

$$\left(\frac{4\pi}{3}\right)^{2/3} |G|^{-2/3}\pi^2 \leq \lambda_1(G) < \Lambda_1(m, G) \leq \Lambda_*(G). \tag{4.147}$$

$$\Lambda_1(m, G) \longrightarrow \Lambda_*(G) \quad \textit{for} \quad m \to 2.$$

The estimation 4.147 gives an upper bound of $\Lambda_1(m, G)$ independently of $m > 2$. We also proved that $\Lambda_*(G)$ is the best uniform upper bound for $\Lambda_1(m, G)$. Of course, we have obtained at the same time lower bounds for the first eigenvalue $\Lambda_*(G)$ of STOKES' problem under DIRICHLET's condition.

4.16 Kernel functions

It is known from the complex function theory that kernel functions play an important role. One of the most remarkable kernels is the so-called reproducing kernel, which is defined as follows:

Let Y be a HILBERT space of functions defined on X. Then, a function $k : X \times X \mapsto C\ell_{0,n}$ is called reproducing kernel of Y if for any $y \in Y$

1. $k(\cdot, y) \in Y$
2. $f(y) = (k(\cdot, y, f)_Y \quad (\forall f \in Y)$.

Nowadays, reproducing kernels are a well-known tool in the theory of functions of one or several complex variables, in CLIFFORD analysis, and also in abstract HILBERT spaces. Reproducing kernels are very useful for

- the investigation of conformal mappings (in complex analysis)
- the explicit description of a linear functional by its representatives
- the construction of complete function systems in Y
- the proof of projection methods
- the solution of elliptic boundary value problems.

Some of the sources are the papers of N. ARONSZAJN [11] from 1950 and of S. BERGMAN [17]. The topic is well investigated from the theoretical point of view (for a review see also [36] and [48]). For more practical applications it is necessary to know the reproducing kernel explicitly. Such a description is possible only in some very special cases. In [48] one can find the BERGMAN kernel for a ball and some other special domains and relationships between BERGMAN and SZEGÖ kernel. M.V. SHAPIRO/N.V. VASILEVSKI/R.M. PORTER [236] studied representations of the BERGMAN kernel by the help of the GREEN function for the Laplacian under DIRICHLET conditions.

They also tried to find out an integral representation of the BERGMAN projection. But, also in this paper the result is a powerful theoretical tool restricted in its practical application to domains with known GREEN function.
In Section 3.5 we have described the BERGMAN projection in another context (the orthoprojection $\mathbf{P}$) and and in Section 4.2 we have represented this projection by the help of $\mathrm{tr}_\Gamma, T_G, F_\Gamma$ (and adjoint operators). We will now look for connections with the results cited above. This should be possible because the introduction of $\mathbf{P}$ used the solution of some special boundary value problems for the

Laplacian. The BERGMAN projection will be represented by the help of GREEN's function or using the fundamental solution of D. Finally, the BERGMAN kernel function can be represented. Initially, it seems to be again only a theoretical result which connects some already known results. But, in Section 5.3 we will repeat the considerations in the same way at the level of discretization methods and then we will get representation formulas which can be directly used in numerical procedures.

4.16.1 Some known results

Without proof we will refer here to some basic results on kernel functions, which have been known for a long time.

Theorem 4.151. *Let Y be a unitary* HILBERT *space of functions defined on X. Then Y has a reproducing kernel iff for all $x \in X$ there exists a constant $C(\xi)$ such that*

$$|f(\xi)| \leq C(\xi)\|f\|_Y \qquad (\forall f \in Y)$$

Proof. We refer to [36]. #

Obviously, due to HARNACK's theorem (see also Proposition 3.73.) the right HILBERT space $L_2(G, C\ell_{0,n}) \cap \ker D$ satisfies the conditions of the theorem. We will denote the BERGMAN kernel by $B(x,\xi)$. Then, we assume for all what follows that $Y = L_2(G, C\ell_{0,n}) \cap \ker D$, and $X = G$. Now we can choose a dense subset of points $\{x_i\}_{i\in\mathbb{N}} \subset G$ and define functions ψ_i by

$$\psi_i(x) = B(x, x_i) \quad (i \in \mathbb{N}).$$

Then, it can be shown that $\{\psi_i\}_{i\in\mathbb{N}}$ is complete in $L_2(G, C\ell_{0,n}) \cap \ker D$. This is a consequence of the implication

$$(\psi_i, u) = 0 \, \forall i \in \mathbb{N} \Longrightarrow (B(x, x_i), u) = 0 \Longrightarrow u(x_i) = 0 \Longrightarrow u(x) = 0 \, \forall x \in G.$$

If we now take an orthonormal complete system of functions ϕ_i then we can prove a FOURIER series expansion for all functions $f \in L_2(G, C\ell_{0,n}) \cap \ker D$

$$f(x) = \sum_{k=1}^{\infty} \phi_k(x)(\phi_k, f)$$

and analogously to the complex case

$$K(x,\xi) = \sum_{j=1}^{\infty} \overline{\phi_j(x)}\phi_j(\xi).$$

All the results mentioned underline that the CLIFFORD analysis and one variable complex analysis are closely connected.

4.16.2 Representations of the Bergman kernel

The GREEN function of $\{\Delta, \mathrm{tr}_\Gamma\}$ is defined by

$$\mathcal{G}(x,y) = E(x,y) + h(x,y),$$

where $(x,y) \in G \times G \setminus \{(x,x)\ x \in G\}$, E is the fundamental solution of Δ and h is the solution of the boundary value problem

$$\begin{aligned} \Delta_y h(x,y) &= 0 && \text{in } G \\ \mathrm{tr}_\Gamma h(x,y) &= -\mathrm{tr}_\Gamma\, E(x,y) && \text{on } \Gamma. \end{aligned}$$

Then, using ideas from [110], we have

$$h(x,y) = -\overline{F}_\Gamma \mathrm{tr}_\Gamma E_x - \overline{T}_G F_\Gamma (\mathrm{tr}_\Gamma \overline{T}_G F_\Gamma)^{-1} \overline{Q_\Gamma} E_x$$

and

$$\begin{aligned} G(x,y) &= E_x(y) + h_x(y) = E - \overline{F}_\Gamma E - \overline{T}_G F_\Gamma (\mathrm{tr}_\Gamma \overline{T}_G F_\Gamma)^{-1} \mathrm{tr}_\Gamma (E - \overline{F}_\Gamma E) \\ &= \overline{T}_G \overline{D} E - \overline{T}_G \mathbf{P} \overline{D} E = (\overline{T}_G \mathbf{Q} e_x)(y). \end{aligned}$$

As above, $e = \bar{D}E$ stands for the fundamental solution of the CAUCHY-FUETER operator D. Now, we have a description of the GREEN function G by the help of $\mathbf{Q}$ and $\overline{T}_G$. Note that this representation of G is advantageous in cases where norm estimates of G are required. Obviously, $\|\mathbf{Q}\|_{[L_2(G,C\ell_{0,n}),\mathcal{L}_2(G,C\ell_{0,n})]} = 1$ and the norm of $\overline{T}_G\ : L_2 \cap \operatorname{im} \mathbf{Q} \mapsto \overset{\circ}{W}{}_2^1(G, C\ell_{0,n})$ is known exactly using the first eigenvalue of $\{-\Delta, \mathrm{tr}_\Gamma\}$ (see e.g. [110]).

In the following we study a similar boundary value problem:

$$\begin{aligned} \Delta_y h_1(x,y) &= 0 && \text{in } G \\ \mathrm{tr}_\Gamma\, h_1(x,y) &= -\mathrm{tr}_\Gamma \overline{e_x}(y) && \text{on } \Gamma. \end{aligned}$$

Using again representation formulas from [110] we get

$$\begin{aligned} h_1(x,y) &= -\overline{F_\Gamma}\, \overline{e_x}(y) - \overline{T}_G\, F_\Gamma (\mathrm{tr}_\Gamma \overline{T}_G\, F_\Gamma)^{-1} \overline{Q_\Gamma} \overline{e_x}(y) \\ &= \overline{T}_G\, F_\Gamma (\mathrm{tr}_\Gamma \overline{T}_G\, F_\Gamma)^{-1} \overline{Q_\Gamma} \overline{e_x}(y). \end{aligned}$$

Applying the conjugate CAUCHY–FUETER operator we obtain

$$b(x,y) := \overline{D}_y h_1(x,y) = -F_\Gamma (\mathrm{tr}_\Gamma \overline{T}_G\, F_\Gamma)^{-1} \overline{Q_\Gamma} \overline{e_x}(y)$$

It is easy to see that $b(x,y) \in \ker D_y \cap L_2(G, C\ell_{0,n})$. Let $\phi \in \ker D \cap W_2^1(G, C\ell_{0,n})$. Then we have

$$(\overline{D_y} h_1, \phi) = -(h_1, D\phi) + \int_\Gamma \overline{h_1}(x,y) \alpha(y) \phi(y)\, d\Gamma_y = F_\Gamma \phi = \phi.$$

If we assume that $\phi \in \operatorname{im} \mathbf{Q}, \phi = \overline{D}u, u \in \overset{\circ}{W}{}_2^1(G, C\ell_{0,n})$, then we immediately obtain

$$(\overline{D}_y h_1, \phi) = (\overline{D}_y h_1, \overline{D}u) = (\Delta h_1, u) = 0.$$

These considerations prove that $b(x, y)$ has the reproducing property in the subspace of monogenic functions. Together with the projection property we obtain $b(x, y)$ as the BERGMAN kernel function.

Chapter 5

Numerical Clifford analysis

5.1 Discrete quaternionic calculus

It is easy to see that our (analytical) representation formulas are not suitable for the explicit computation of solutions. It is necessary to study numerical approximations of the representation formulas. This idea is not new. Because we also have similar problems in the complex case these problems have been under consideration for a long time.

The first (and simplest) idea was the direct discretization of the representation formulas by quadrature formulas. This approach fails in almost all cases, because without additional investigation the matrix representation of the approximation of tr TF_Γ is not invertible. A second idea comes from complex considerations. This approach is based on the idea of translating the whole function theory into the language of discretization methods, for example into the language of finite difference approximations, and this procedure ends with a discrete function theory (for lattice functions). This work was initiated in the 1940s and 1950s ([59], [62], [69], [70], [72], [75], [122]). We will briefly repeat a part of this work here which will enable us to recognize the difficulties and also the new possibilities of hypercomplex approaches to numerical methods. It should be remarked that these references include information about discrete function theories on the level of research articles. To get some information on the teaching level we refer to the textbooks [265] for the general theory of finite difference methods and [250] for discrete potential theory.

5.1.1 Some problems in complex discrete function theory

Complex-valued analytic functions can be defined as the set of solutions of the CAUCHY–RIEMANN system. An essential property of analytic functions is also to be harmonic functions. The first idea in consideration of discretizations is to approximate the partial derivatives $\frac{\partial}{\partial x}$ and $\frac{\partial}{\partial y}$ by finite differences. To do so we need some definitions:

$$\mathbb{R}_h^n = \{(i_1 h, i_2 h, ..., i_n h), h > 0, i_1, ..., i_n \in \mathbf{Z}\}$$

$\mathbb{R}_h^n$ describes an equidistant lattice in $\mathbb{R}^n$ with mesh width h. We consider functions defined on $\mathbb{R}_h^n$. Sometimes we abbreviate $u(i_1 h, i_2 h, ..., i_n h)$ by $u_{i_1,i_2,...,i_n}$. The discretization of $\frac{\partial}{\partial x_k}$ is defined by

$$(D_{k,h}^{\pm} u)_{i_1,...,i_n} = \pm\frac{1}{h}(u_{i_1,...,i_k\pm 1,...,i_n} - u_{i_1,...,i_k,...,i_n})$$

A usual approximation of the Laplacian can be defined by

$$\Delta_h u = \sum_{k=1}^{n} D_{k,h}^{+} D_{k,h}^{-} u.$$

The first problem arises now from the fact that not all of these partial difference operators commute in a certain sense. We study for $n = 2$ the action of $D^+_{1,h}D^-_{2,h}$ and $D^+_{2,h}D^-_{1,h}$.

$$(D^+_{1,h}D^-_{2,h}u)_{k,j} = \frac{1}{h^2}[u_{k+1,j} - u_{k,j} - u_{k+1,j-1} + u_{k,j-1}],$$

$$(D^+_{2,h}D^-_{1,h}u)_{k,j} = \frac{1}{h^2}[u_{k,j+1} - u_{k,j} - u_{k-1,j+1} + u_{k-1,j}].$$

Obviously, both approximations are different. If we now approximate the CAUCHY–RIEMANN operator and its adjoint operator by forward and backward differences, respectively, then we obtain

$$(D^+_{1,h}+iD^+_{2,h})(D^-_{1,h}+iD^-_{2,h}) = \underbrace{D^+_{1,h}D^-_{1,h} + D^+_{2,h}D^-_{2,h}}_{-\Delta_h} + i\underbrace{(D^+_{2,h}D^-_{1,h} - D^+_{1,h}D^-_{2,h})}_{\neq 0}$$

That means we have no factorization of Δ_h by the standard approximations defined above.

Exercise: Try other combinations of forward and backward differences to factorize the Laplacian.

To overcome these fundamental difficulties in [70] other approximations were studied. The idea was on the one hand to change the approximation of Δ and on the other hand to approximate the CAUCHY–RIEMANN operator directly without approximations of single partial derivatives.
We now approximate the Laplacian by

$$(\widetilde{\Delta}_h u)_{k,j} = \frac{1}{4h^2}[4u_{k,j} - u_{k-1,j-1} - u_{k+1,j-1} - u_{k-1,j+1} - u_{k+1,j+1}]$$

and we define

$$(D^+_h u)_{k,j} = \frac{1-i}{2h}[u_{k+1,j+1} - u_{k,j}] + \frac{1+i}{2h}[u_{k,j+1} - u_{k+1,j}]$$

$$(D^-_h u)_{k,j} = \frac{1+i}{2h}[u_{k,j} - u_{k-1,j-1}] + \frac{1-i}{2h}[u_{k-1,j} - u_{k,j-1}]$$

Then, D^+_h approximates $(-i)(\frac{\partial}{\partial x} + i\frac{\partial}{\partial y})$ and D^-_h approximates $i(\frac{\partial}{\partial x} - i\frac{\partial}{\partial y})$ for $h \to 0$. Clearly, we have now the factorization $D^+_h D^-_h = -\widetilde{\Delta}_h$.
One of the main problems in this approach to a discrete function theory is that from $f \in \ker D^+_h$, $g \in \ker D^+_h$ in general it does not follow that $fg \in \ker D^+_h$. The set of so-called discrete analytic functions is not a commutative algebra. That means we have no analytic powers and so on, and many of the known tools from complex function theory are not applicable. The way out of this was by a change of the product. The pointwise product of functions was replaced by a convolutional product (see e.g. [59], [72]). As a result a commutative algebra of

analytic functions was obtained. But now an analogue of the BOREL–POMPEIU formula was missed. Nowadays, looking to the results of CLIFFORD analysis, we know that the commutativity of the product is not necessary for the foundation of a function theory and its applications.

5.1.2 A general theorem

Omitting the commutativity of the product, we may hope to be successful in establishing a discrete function theory. It may be recalled that there is a general theorem from the theory of right-invertible operators (see e.g. [250]).

Theorem 5.1. *We assume that*

(1) $L : X \to Y$ *is a right-invertible operator and* G *an arbitrary right inverse of* L

(2) $Tr : X \to Z$ *is a generalized trace-operator*

(3) $\ker L \cap \ker Tr = \{0\}$,

where X, Y, Z *are* BANACH *spaces. We define*

$$\begin{aligned} P &= I - GL, \\ P_r &= Tr\ P. \end{aligned}$$

Then, we have

(1) $\ker L \cap \ker Tr = \{0\} \Leftrightarrow \ker \mathrm{Tr} \subset \ker P$,

(2) $Pu = P(\mathrm{Tr}\, u) = P\xi \qquad \Rightarrow P : Z \to X$,

(3) $P_r : Z \to Z$,

(4) $P^2 = P, P_r^2 = P_r$,

(5) $P_r\xi = \xi \Leftrightarrow \xi = Tr\ u, \quad u \in \ker L$,

(6) $u = P\xi + GLu$,
$\xi = P_r\xi + \mathrm{Tr}\, GLu$.

Example 5.2. $L = D, G = T, X = W_2^1(G), Y = L_2(G), Z = W_2^{1/2}(\Gamma)$, $\mathrm{Tr} = \mathrm{tr},\ P = I - TD, P_\gamma = \mathrm{tr}\ P$.

Using the BOREL–POMPEIU formula we see that $P = I - TD = F_\Gamma$. That means the operator P may be redefined as an operator acting only on the boundary data of the functions from X and this is the statement of (2). Property (5) fixes necessary and sufficient conditions for the solvability of the "boundary value problems".

$$\begin{aligned} Lu &= 0, u \in X, \\ \mathrm{Tr}\, u &= \xi,\ \xi \in Z. \end{aligned}$$

The equations (6) are analogous to the BOREL–POMPEIU formula. For $u \in \ker L$ we have $u = P\xi$, an analogue of CAUCHY's integral formula. Theorem 5.1. can be seen as a "schedule" for our desired numerical approach. We have to define our finite difference operators and the corresponding function spaces in such a way that at least the assumptions (1)—(3) from Theorem 5.1. are fulfilled.

5.1.3 Definition of discrete operators

We restrict our considerations to the dimension $n = 3$ and to approximations of the DIRAC operator. Let f, g be defined on $\mathbb{R}_h^3$. Then, we introduce the inner product

$$\langle f, g\rangle_2 = \sum_{x \in \mathbb{R}_h^3} \overline{f(x)} g(x) h^3.$$

Using this inner product we define the HILBERT space $l_2(\mathbb{R}_h^3)$.

$$f \in l_2(\mathbb{R}_h^3) \Leftrightarrow \|f\|_{l_2}^2 = \sum_{x \in \mathbb{R}_h^3} |f(x)|^2 h^3 < \infty.$$

The DIRAC operator will be approximated by

$$(D_h^{\pm} u)(x) = \sum_{i=1}^{3} e_i (D_{i,h}^{\pm} u)(x)$$

and the Laplacian analogously to the ideas of [69] by

$$(-\tilde{\Delta}_h)(x) = (D_h^+ D_h^- u)(x).$$

The lattice function

$$\delta_h(x) = \begin{cases} \dfrac{1}{h^3} & x = 0 \\ 0 & x \in \mathbb{R}_h^3 \setminus \{0\} \end{cases}$$

is called *discrete Delta function.* We should mention that δ_h in our theory is a function which, of course, belongs to all l_p-spaces. Then, we study the difference equation

$$(A_h E_h)(x) = \delta_h(x) \qquad \forall\, x \in \mathbb{R}_h^3,$$

where A_h stands for an arbitrary difference operator. If a solution E_h exists it is called *fundamental solution* of A_h. We have to prove existence, uniqueness, representations, and algorithms for the computation of these fundamental solutions.

For simplicity we skip these problems here. In the next section we will investigate the problem of fundamental solutions in more detail. A collection of already computed fundamental solutions is contained in Appendix B. Furthermore, we refer to the literature:

- [34], [309], [315] for existence (general results)
- [34], [69], [96], [103] special fundamental solutions, including uniqueness results, representations and algorithms for computation
- [110], [99] discussion of the above-mentioned problems for $D_h^{\pm}, \widetilde{\Delta}_h, \Delta_h$.

Thus, we can assume at this moment that all necessary knowledge on fundamental solutions is available.
Let $G \subset \mathbb{R}^3$ be a bounded domain. Then, we define $G_h := G \cap \mathbb{R}_h^3$ and call it discrete domain. A point x belongs to the discrete boundary ∂G_h if at least one of its neighbours (in the lattice) belongs to $\mathbb{R}_h^3 \backslash G_h$. Discrete analogues of the integral operators used above will be defined by finite sums.

$$(T_h^{\pm} f)(x) = \sum_{y \in \widetilde{G}_h} e_h^{\pm}(x-y) f(y) h^3 \quad \text{discrete (TEODORESCU transformation).}$$

Here $e_h^{\pm}$ stands for the fundamental solution of $D_h^{\pm}$.
Application of $D_h^{\pm}$ to these finite sums proves

$$(D_h^{\pm} T_h^{\pm} f)(x) = f(x) \qquad \forall\, x \in \widetilde{G_h}$$

If we denote by $\operatorname{int} G_h$ the set $G_h \backslash \partial G_h$ then it is natural to use $\widetilde{G_h}$ with the property $\operatorname{int} G_h \subseteq \widetilde{G_h} \subseteq G_h$.
In this way it is no problem to satisfy (1) and (2) from Theorem 5.1.. For the definition of a restriction operator tr it seems to be natural to define it by the help of the values of the functions at boundary points. But, in this new context we have also to look for assumption (3) of Theorem 5.1. Therefore, it is required that the boundary may be fixed more prcisely.
Further, we have to explain what happens with the normal vector $\alpha(y)$ at corners and edges. This procedure is very time-consuming and we refer to the Appendix for the explicit description of a suitable $\gamma_h \subset \partial G_h$ with $\ker \operatorname{tr}_{\gamma_h} \cap \ker L = \{0\}$ and to [110] for the proof of the description mentioned. This description is not needed in the discussion of the main ideas in discrete function theories.
Following our "schedule" from Theorem 5.1. we define

$$(F_h^{\pm} f)(x) = f(x) - (T_h^{\pm} D_h^{\pm} f)(x), \qquad \forall\, x \in G_h$$

and call $F_h^{\pm}$ to be a discrete CAUCHY integral. One can prove (Theorem 5.1., see also [110] and the Appendix)

$$(F_h^{\pm} f)(x) = \sum_{y \in \gamma_h^{\pm}} e_h^{\pm}(x-y) \alpha(y) f(y) h^2.$$

Using Theorem 5.1. we have the possibility of finding the points of γ_h without the explicit description from [110]. We know that

$$\begin{aligned} & f(x) - (T_h^{\pm} D_h^{\pm} f)(x) : \mathrm{Z} \longrightarrow \mathrm{X} \\ \Rightarrow \quad & f(x) - (T_h^{\pm} D_h^{\pm} f)(x) = 0, \qquad \forall\, f : \operatorname{supp} f \cap \gamma_h^{\pm} = \emptyset. \end{aligned}$$

Then we can check with test functions $\delta_h(x-y)$, whether or not $x \in \gamma_h^+$. This can be done by a computer program and is sufficient for the numerical application. For theoretical investigations, e.g., for the proof of approximation $F_h^{\pm} \to F_\Gamma$ for $h \to 0$, we explicitly need $\gamma_h^{\pm}$.

Theorem 5.3. *We have*

$$\begin{aligned}
F_h^+ f &\in \ker D_h^+ \quad (\operatorname{int} G_h), \\
((F_h^+)^2 f)(x) &= (F_h^+ f)(x) \quad \forall x \in \operatorname{int} G_h, \\
(F_h^+ f)(x) &= \begin{cases} f(x) & \forall\, f \in \ker D_h^+(\operatorname{int} G_h), x \in \operatorname{int} G_h \\ 0 & \forall\, f \in \ker D_h^+(\operatorname{int} G_h), x \in \operatorname{co} G_h \end{cases}, \\
T_h^+ f &\in \ker D_h^+(\operatorname{co} G_h), \\
(T_h^+ D_h^+ f)(x) &= -(F_h^+ f)(x), \quad x \in \operatorname{co} G_h.
\end{aligned}$$

The proof is an exercise. We do not need all these properties in the following, but it is interesting to see a lot of equivalent properties in comparison with the continuous theory already at this stage of investigation.
The following theorem plays a central role in the theory of discrete monogenic functions.

5.1.4 Discrete orthoprojections

Theorem 5.4. (orthogonal decomposition)*:*

$$L_{2,h} = \ker D_h^+(\operatorname{int} G_h) \oplus_{\langle,\rangle_h} D_h^+(\overset{\circ}{W}{}^1_{2,h}(G_h)).$$

Proof. We assume that $\langle f, g\rangle_h = 0 \quad \forall\, g \in D_h^-(\overset{\circ}{W}{}^1_{2,h}(G_h))$ or in other words $\sum_{x\in G_h} \overline{f(x)} g(x) h^3 = 0$. Then we have $\forall\, s \in \overset{\circ}{W}{}^1_{2,h}(G_h)$

$$\begin{aligned}
&\sum_{x\in G_h} \overline{f(x)} (D_h^- s)(x) h^3 = 0 \\
\Rightarrow \quad &\sum_{x\in \operatorname{int} G_h} \overline{(D_h^+ f)}(x) s(x) h^3 = \langle D_h^+ f, s\rangle = 0 \\
\Rightarrow \quad &(D_h^- f)(x) = 0 \qquad \forall\, x \in \operatorname{int} G_h \qquad (\text{try} \quad s = \delta_h(x-y)). \qquad \#
\end{aligned}$$

This decomposition defines the two orthoprojections:

$$\begin{aligned}
\mathbf{P}_h^+ &: L_{2,h}(G_h) \overset{\text{onto}}{\longrightarrow} \ker D_h^+(\operatorname{int} G_h), \\
\mathbf{Q}_h^+ &: L_{2,h}(G_h) \overset{\text{onto}}{\longrightarrow} D_h^-(\overset{\circ}{W}{}^1_{2,h}(G_h)).
\end{aligned}$$

Obviously, $D_h^+ \mathbf{Q}_h^+ f = f \quad \forall\, f$. We remark that it is not necessary here to use $D_h^-(\overset{\circ}{W}{}^1_{2,h}(G_h))$ for the description of the orthogonal complement of discrete monogenic functions. If we take $\{f : \mathrm{tr} f = 0\}$ the theorem holds true, because for fixed $h > 0$ this set belongs to all discrete SOBOLEV spaces. But, finally, we have to study the behaviour of all operators and representations for $h \to 0$, and then we have to prove the approximation of $D_h^-(\overset{\circ}{W}{}^1_{2,h}(G_h))$. As in the continuous case we look for an explicit description of $\mathbf{P}_h^+$ and $\mathbf{Q}_h^+$, respectively. In preparation for this representation we study some simple boundary value problems.

Theorem 5.5. *The boundary value problem*

$$\begin{aligned} -\Delta_h u &= f \quad \textit{in} \quad \mathrm{int}\, G_h, \\ u &= 0 \quad \textit{on} \quad \partial G_h, \end{aligned}$$

has a solution.

Proof. $u = T_h^- \mathbf{Q}_h^- T_h^+ f$ is a solution. #

Theorem 5.6.

$$\begin{aligned} -\Delta_h u &= 0 \quad \textit{in} \quad \mathrm{int}\, G_h, \\ u &= g \quad \textit{on} \quad \partial G_h, \end{aligned}$$

has a solution.

Proof. $u = w + T_h^- \mathbf{Q}_h^+ T_h^- \Delta_h w$, where w is an arbitrary extension of g into G_h, is a solution. #

Consequently, we are now able to handle the inhomogeneous DIRICHLET problem.

Theorem 5.7.

$$\begin{aligned} -\Delta_h u &= f \quad \textit{in} \quad \mathrm{int}\, G_h, \\ u &= g \quad \textit{on} \quad \partial G_h, \end{aligned}$$

has a unique solution.

Proof. Clearly, the solution exists. As usual we now assume the existence of two different solutions u_1, u_2 and consider $v = u_1 - u_2$. Then,

$$\begin{aligned} -\Delta_h v &= 0 \quad \text{in} \quad \mathrm{int}\, G_h, \\ v &= 0 \quad \text{on} \quad \partial G_h. \end{aligned}$$

$$\Rightarrow \quad D_h^+ D_h^- v = 0 \quad \Rightarrow \quad \left\{ \begin{array}{l} D_h^- v \ker D_h^+ \quad = \mathrm{im}\, \mathbf{P}_h^+ \\ D_h^- v \in D_h^-(\overset{\circ}{W}{}^1_{2,h}) = \mathrm{im}\, \mathbf{Q}_h^+ \end{array} \right. \quad \Rightarrow \quad D_h^- v = 0$$

$$\Rightarrow \quad v = F_h^- v + T_h^- D_h^- v = 0. \qquad \#$$

Let us mention here that the representations as in Theorem 5.7. are much more practical than the corresponding formulas in the continuous case. The reason is that it is not hard to find an extension of given boundary data into the discrete domain.

Theorem 5.8.

$$u \in \ker \Delta_h(\operatorname{int} G_h) \quad \Rightarrow \quad u = f_1 + T_h^- f_2, \qquad \begin{array}{l} f_1 \in \ker D_h^-(\operatorname{int} G_h), \\ f_2 \in \ker D_h^+(\operatorname{int} G_h). \end{array}$$

Proof.

$$\begin{aligned} u &= w + T_h^- \mathbf{Q}_h^+ T_h^+ \Delta_h w = w - T_h^- \mathbf{Q}_h^+ T_h^+ D_h^+ D_h^- w = \\ &= w - T_h^- \mathbf{Q}_h^+ [D_h^- w - F_h^+ D_h^- w] = w - T_h^- \mathbf{Q}_h^+ D_h^- w = \\ &= w - T_h^- D_h^- w + T_h^- \mathbf{P}_h^+ D_h^- w = F_h^- w + T_h^- (\mathbf{P}_h^+ D_h^- w). \qquad \# \end{aligned}$$

Using this decomposition and taking the traces we obtain a decomposition of the boundary values.

Theorem 5.9.

$$L_{2,h}(\partial G_h) = \operatorname{im} \operatorname{tr} T_h^- + \operatorname{im} \operatorname{tr} F_h^- . \qquad \#$$

The following theorem is the basis for a well-adapted approximation of $(\operatorname{tr} TF_\Gamma)$ and $(\operatorname{tr} TF_\Gamma)^{-1}$ as well as for the explicit description of discrete orthoprojections.

Theorem 5.10.

$\operatorname{tr} T_h^- F_h^+ : \operatorname{im} \operatorname{tr} F_h^+ \longrightarrow \operatorname{im} \operatorname{tr} T_h^-$ *is an isomorphism.*

Proof.

$$\begin{aligned} &\operatorname{tr} T_h^- F_h^+ v = 0, v \in \operatorname{im} \operatorname{tr} F_h^+ \quad \Rightarrow \quad T_h^- F_h^+ v = 0 \quad \Rightarrow \quad \operatorname{tr} F_h^+ v = 0 \\ &\qquad \Rightarrow \quad v = 0 \quad \Rightarrow \quad \text{injectivity of the mapping.} \end{aligned}$$

Let $w \in \operatorname{Im} \operatorname{tr} T_h^-$. Then, $w = \operatorname{tr} T_h^- v$. We solve the boundary value problem

$$\begin{aligned} -\Delta_h u &= f \quad \text{in} \quad \operatorname{int} G_h, \\ u &= w \quad \text{on} \quad \partial G_h. \end{aligned}$$

$$\begin{aligned} &\Rightarrow \quad u = f_1 + T_h^- f_2, \quad f_1 \in \ker D_h^-, \quad f_2 \in \ker D_h^+ \\ &\Rightarrow \quad \operatorname{tr} u = w = \operatorname{tr} F_h^- f_1 + \operatorname{tr} T_h^- F_h^+ f_2 = \operatorname{tr} T_h^- v \\ &\Rightarrow \quad \operatorname{tr} F_h^- f_1 = \operatorname{tr} T_h^- (v - F_h^+ f_2) \Rightarrow f_1 = 0, \quad v = F_h^+ f_2 \\ &\Rightarrow \quad w = (\operatorname{tr} T_h^- F_h^+) f_2 \qquad \Rightarrow \qquad \text{surjectivity of the mapping.} \qquad \# \end{aligned}$$

$(\operatorname{tr} T_h^- F_h^+)^{-1}$ makes sense now and we are able to prove the representation formula

$$\mathbf{P}_h^+ = F_h^+(\operatorname{tr} T_h^- F_h^+)^{-1} \operatorname{tr} T_h^-, \quad \mathbf{Q}_h^+ = I - \mathbf{P}_h^+.$$

Arriving at this level it is now necessary to study the approximation properties of the operators $T_h^\pm$, $F_h^\pm$, $\mathbf{Q}_h^\pm$, $\mathbf{P}_h^\pm$ defined above. Let us state that this investigation is closely connected to the approximation of e by $e_h^\pm$. We refer to the next section for details. Instead of this extensive proof we want to study first the formal application of the discrete theory to the solution of a boundary value problem of the STOKES equation.

5.1.5 A discrete Stokes system

We consider

$$\begin{aligned} -\Delta u + \frac{1}{\eta}\operatorname{grad} p &= \frac{\rho}{u} f \quad \text{in} \quad G, & -\Delta_h u + \frac{1}{\eta}\operatorname{grad}_h^+ p &= \frac{\rho}{h} f \quad \text{in} \quad G_h, \\ \operatorname{div} u &= 0 \quad \text{in} \quad G, \quad \Longleftrightarrow & \operatorname{div}_h^- u &= 0 \quad \text{in} \quad G_h, \\ u &= 0 \quad \text{on} \quad \Gamma, & u &= 0 \quad \text{on} \quad \partial G_h, \end{aligned}$$

where

$$\begin{aligned} \operatorname{grad}_h^+ f &:= (D_{1,h}^+ f, D_{2,h}^+ f, D_{3,h}^+ f)^T, \qquad f(\text{ real valued!}), \\ \operatorname{div}_h^- v &:= \sum_{i=1}^{3} D_{i,h}^- v, \qquad v = (v_1, v_2, v_3)^T. \end{aligned}$$

In these equations u stands for the velocity of a fluid, p for the hydrostatical pressure, and f denotes the outer forces; ρ is the density and η the viscosity number. We then replace the vector $u = (u_1, u_2, u_3)^T$ by $(0, u)^T$ and the scalar p by $(p, 0, 0, 0)^T$. In this way it becomes possible to apply our operators defined above. The operator div_n^- has to be understood in the sense of $-\operatorname{Sc} D_h^-$ and grad_h^+ also stands for $D_h^+ \operatorname{Sc}$.

Lemma 5.11.

$$\begin{aligned} f, g \in \overset{\circ}{W}{}_{2,h}^1 (G_h) &\implies \langle D_h^+ f, g\rangle = \langle f, D_h^- g\rangle, \\ f, g \in \overset{\circ}{W}{}_{2,h}^1 (G_h) &\implies \operatorname{Sc}\langle D_h^+ \operatorname{Sc} f, g\rangle = \operatorname{Sc}\langle f, \operatorname{Sc} D_h^- g\rangle. \end{aligned}$$

Idea of the proof: For the proof we try in a first step to prove it for the one-dimensional case and real valued functions $f, g \in \overset{\circ}{W}{}_{2,h}^1 [a, b]$.

$$\begin{aligned} \langle D_{1,h}^+ f, g\rangle &= \sum_{i=0}^{N} (D_{1,h}^+ f)(x) v(x) h = \sum_{i=0}^{N-1} (D_{1,h}^+ f)(x) v(x) h \\ &= -\sum_{i=1}^{N} u(x)(D_{1,h}^- v)(x) h = -\sum_{i=0}^{N} u(x)(D_{1,h}^- v)(x) h = -\langle f, D_{1,h}^- g\rangle, \end{aligned}$$

where $x = a + ih, i = 0, ..., N, h = (b-a)/N$. The generalization to 3D should be an exercise. #
We define a new scalar product by $[f, g] = \mathrm{Sc}\,\langle f, g\rangle$. Then, $[\mathrm{grad}_h^+ u, v] = [u, -\mathrm{div}_h^- v] = [u, \mathrm{Sc}\, D_h^- v]$.
Consequently, for $u, v \in \overset{\circ}{W}{}_{2,h}^1\,(G_h)$ $(\mathrm{grad}_h^+)^* = -\mathrm{div}_h^-$ follows.
Therefore, $\ker \mathrm{div}_h^- = (\mathrm{im}\,\mathrm{grad}_h^+)^\perp$.

Lemma 5.12.

$$u \in \overset{\circ}{W}{}_{2,h}^1\,(G_h) \cap \ker \mathrm{div}_h^-, p \in L_{2,h}(G_h),\ \ \mathrm{im}\; p = 0 \Longrightarrow [D_h^- u, \mathbf{Q}_h^- p] = 0.$$

Proof.

$$[D_h^- u, \mathbf{Q}_h^+ p] = [\mathbf{Q}_h^+ D_h^- u, p] = [D_h^- u, p] = 0. \ \#$$

With the following theorem we prepare an existence result for the solution of the discrete STOKES problem.

Theorem 5.13.

$$f \in L_{2,h}(G_h) \Longrightarrow \exists\, u \in \overset{\circ}{W}{}_{2,h}^1\,(G_h) \cap \ker \mathrm{div}_h^-, \quad \exists\, p \in L_{2,h,\mathbf{R}}(G_h)$$

with

$$\frac{\rho}{\eta} \mathbf{Q}_h^+ T_h^+ f = D_h^- u + \frac{1}{\eta} \mathbf{Q}_h^+ p$$

Proof. We prove an orthogonal decomposition of $\mathbf{Q}_h^+ T_h^+ f$. Obviously,

$$D_h^- u \in \mathrm{im}\mathbf{Q}_h^+ \quad \forall u \in \overset{\circ}{W}{}_{2,h}^1\,(G_h) \cap \mathrm{kerdiv},\ \mathbf{Q}_h^+ p \in \mathrm{im}\mathbf{Q}_h^+ \quad \forall p \in L_{2,h,\mathbf{R}}(G_h).$$

Assume that

$$[D_h^- u, \mathbf{Q}_h^+ T_h^+ f] = 0 \qquad \forall\, u \in \overset{\circ}{W}{}_{2,h}^1 \cap \ker \mathrm{div}_h^-$$

and

$$[\mathbf{Q}_h^+ p, \mathbf{Q}_h^+ T_h^+ f] = 0 \qquad \forall\, p \in L_{2,h,\mathbf{R}}(G_h)$$

$$[D_h^- u, \mathbf{Q}_h^+ T_h^+ f] = [D_h^- u, T_h^+ f] = [u, f] = 0 \Rightarrow\ f \perp \ker \mathrm{div}_h^-$$

$$\Longrightarrow \quad f \in \mathrm{im}\,\mathrm{grad}_h^+ \quad \Longrightarrow \quad f = \mathrm{grad}_h^+ q \quad , \quad \mathrm{im}\, q = 0.$$

$$\Longrightarrow \quad [\mathbf{Q}_h^+ p, \mathbf{Q}_h^+ T_h^+ f] = [\mathbf{Q}_h^+ p, \mathbf{Q}_h^+ T_h^+ \mathrm{grad}_h^+ q] = [\mathbf{Q}_h^+ p, \mathbf{Q}_h^+ T_h^+ D_h^+ q]$$

$$= [\mathbf{Q}_h^+ p, \mathbf{Q}_h^+ q - \underbrace{\mathbf{Q}_h^+ F_h^+ q]}_{=0} = 0 \quad \forall\, p \in L_{2,h,\mathbf{R}}(G_h)$$

$$\Longrightarrow \|\mathbf{Q}_h^+ q\| = 0 \Longrightarrow \quad \mathbf{Q}_h^+ q = 0 \Longrightarrow\ f = \mathrm{grad}_h^+ q = D_h^+ q = D_h^+ \mathbf{Q}_h^+ q = 0. \ \#$$

Theorem 5.14.

$$
\begin{aligned}
-\Delta_h u + \frac{1}{\eta}\mathrm{grad}_h^+ p &= \frac{\rho}{\eta} f \quad \text{in} \quad \mathrm{int}\, G_h, \\
\mathrm{div}_h^- u &= 0 \quad \text{in} \quad \mathrm{int}\, G_h, \\
u &= 0 \quad \text{on} \quad \partial G_h,
\end{aligned}
$$

has a unique solution for arbitrarily given $f \in L_{2,h}(G_h)$.

Proof. Take

$$
f \in L_{2,h}(G_h) \quad \overset{(Th.5.13.)}{\Longrightarrow} \quad \exists\, u, p : D_h^- u + \frac{1}{\eta}\mathbf{Q}_h^+ p = \frac{\rho}{\eta}\mathbf{Q}_h^+ T_h^+ f,
$$

$$
u \in \overset{\circ}{W}{}_{2,h}^{1}(G_h) \cap \ker \mathrm{div}_h^-, \quad p \in L_{2,h,\mathbb{R}}(G_h).
$$

Applying T_h^- ($T_h^- D_h^- u = u$ because $\mathrm{tr}\, u = 0$!!) we obtain

$$
u + \frac{1}{\eta} T_h^- \mathbf{Q}_h^- p = \frac{\rho}{\eta} T_h^- \mathbf{Q}_h^+ T_h^+ f \tag{5.1}
$$

and this solves our problem. For the proof of uniqueness we assume again two solutions u_1, u_2 (consider $v = u_1 - u_2$) and p_1, p_2 (consider $q = p_1 - p_2$). Then,

$$
v + \frac{1}{\eta} T_h^- \mathbf{Q}_h^+ q = 0 \quad \Longrightarrow \quad D_h^- v + \frac{1}{\eta}\mathbf{Q}_h^+ q = 0
$$

by D_h^--application. Hence,

$$
D_h^- v = 0 \quad \text{and} \quad \mathbf{Q}_h^- q = 0 \quad \text{because} \quad [D_h^- u, \mathbf{Q}_h^+ q] = 0.
$$

$$
\begin{aligned}
D_h^- v = 0 \quad &\text{implies} \quad v = T_h^- D_h^- v = 0 \\
&\text{and} \quad \mathbf{Q}_h^+ q = 0 \quad \text{produces}
\end{aligned}
$$

$$
q \in \ker D_h^+, \quad \mathrm{im}\, q = 0 \quad \Longrightarrow \quad q = c \in \mathbb{R}. \qquad \#
$$

Let us mention that we have proved not only existence and uniqueness of the solution. Using our representation formula (5.1) we have also been able to investigate the stability of the solution. All our investigations are based on the same functional analytical background as the corresponding investigations in the continuous case. In particular, it was not necessary to study the entries or the special structure of the matrices resulting from the finite difference approximation. This is an essential advantage in the investigation of systems of partial differential equations (and their approximations) and also for the treatment of non-linear equations (and their nested approximations).

Of course, we have to look for a convergence result. This requires some additional investigations. We have to prove now the approximation of the fundamental solutions by the discrete analogues. After this consideration we will prove approximation properties of the discrete operators. For the final convergence result we refer to Section 5.4.

5.2 Discrete fundamental solutions

We have seen above that for constructive analytical considerations, and also in the case of numerical applications, it is necessary to have an explicit expression of the fundamental solution or of the GREEN function. In the previous section we have assumed for simplicity that fundamental solutions of finite difference operators are available.
Then, there were many connections between the differential equation and the finite difference equation and several analogies between both theories.
Of course, even in the discrete case, a function theoretic approach makes sense only if it is possible to obtain explicit expressions for fundamental solutions of the partial difference operators. First, answers have been given in [54], and some special elliptic operators were discussed in [60] and [69]. In the fundamental paper [309] the case of more general elliptic difference operators is treated and a large number of tools for further investigations and for the representation of fundamental solutions in standard situations were given. A detailed discussion of a discrete potential theory based on the concept of a discrete GREEN function is given in [250]. Further results in this direction are contained in the papers in [34] and [315]. Other approaches to the topic use algebraic tools [338] or the authors work on networks and graphs [184]. In these cases there are some difficulties in proving convergence results.
In this section we will prove the existence and the uniqueness of fundamental solutions. For simplicity we demonstrate the ideas for finite difference approximations of the Laplacian. Finally we shall prove the convergence of the discrete fundamental solutions to the known fundamental solution of the differential operator with respect to several norms.
Further results for other elliptic operators, higher order operators, systems and for parabolic equations we collected in Appendix B.
Questions connected with the relations between discrete fundamental solutions, the discrete GREEN function and the BERGMAN kernel function will be considered in the next section.

5.2.1 Definition of the discrete Fourier transform

The discrete FOURIER transform will be defined for functions $u \in l_2(\mathbb{R}_h^n)$ by the following expression (see [309]):

$$(F_h u)(\xi) = \begin{cases} \frac{h^n}{\sqrt{2\pi}^n} \sum\limits_{x \in \mathbf{R}_h^n} u(x) e^{i(x,\xi)}, & \xi \in Q_h \\ 0, & \xi \in \mathbb{R}^n \setminus Q_h, \end{cases}$$

where

$$Q_h = \left\{ \xi \in \mathbb{R}_h^n : -\frac{\pi}{h} < \xi_i < \frac{\pi}{h}, \quad i = 1, \ldots, n \right\}.$$

If we introduce, in addition, the space

$$L_2^0(Q_h) = \{u \in L_2(\mathbb{R}^n) : u = 0 \text{ in } \mathbb{R}^n \setminus Q_h\},$$

then it can be proved that the map $F_h : l_2(\mathbb{R}_h^n) \to L_2^o(Q_h)$ is an isometric isomorphism. The inverse transformation can be described by

$$F_h^{-1} = R_h F : L_2^o(Q_h) \to l_2(\mathbb{R}_h^n).$$

Here R_h stands for the operator of restriction from $\mathbb{R}^n$ to the lattice $\mathbb{R}_h^n$. With the help of the continuous FOURIER transform

$$(F F_h u)(x) = \frac{1}{\sqrt{2\pi}^n} \int\limits_{\mathbf{R}^n} F_h u(\xi) e^{-i(x,\xi)} d\xi, \quad x \in \mathbb{R}^n$$

we obtain the PARSEVAL equation

$$\langle u, v \rangle_{l_2(\mathbf{R}_h^n)} = (F_h u, F_h v)_{L_2^o(Q_h)} = (F F_h u, F F_h v)_{L_2(\mathbf{R}^n)}.$$

5.2.2 Properties of the discrete Fourier transform

From the analytical point of view the discrete FOURIER transform is now well defined. Of course, this definition is very natural. This can be seen by looking for the approximation properties of F_h for $h \mapsto 0$ (for sufficiently "good" functions). For the following we need more properties of the discrete FOURIER transform. Most of them have been well known for a long time (see again [309]) and are easy to prove. Therefore, we will list only these properties. Again, the close connection between continuous and discrete theory can be seen.
We denote by I the identity operator, by e_j as above the j-th normalized basis vector of $\mathbb{R}^n$ and $R_Q u$ stands for the restriction of u to the cube Q_h. Then we have:

(1) $R_h F F_h u = u, \quad u \in l_2(\mathbb{R}_h^n)$.

(2) $F_h R_h F u = u, \quad$ for $u \in L_2^o(Q_h)$.

(3) $F F_h R_h u = u, \quad$ for $u \in Im F\{L_2^o(Q_h)\} \cap L_2$.

(4) $F(-\Delta u) = |\xi|^2 F u$ with $|\xi|^2 = \xi_1^2 + \xi_2^2 + \ldots + \xi_n^2$.

(5) $F_h(-\Delta_h u) = d^2 F_h u$ with $d^2 = \frac{4}{h^2}\left(\sin^2 \frac{h\xi_1}{2} + \sin^2 \frac{h\xi_2}{2} + \ldots + \sin^2 \frac{h\xi_n}{2}\right)$.

(6) $-\Delta(Fu) = F|\xi|^2 u$.

(7) $-\Delta(F_h u) = F|\xi|^2 F^{-1} F_h u$.

(8) $-\Delta_h(R_h F u) = F_h^{-1} d^2 F_h R_h F u$.

(9) $-\Delta_h(R_h F_h u) = F_h^{-1} d^2 F_h R_h F_h u$.

(10) $F_h(D_{j,h}^+ u) = -\eta_{-j} F_h u, \quad j = 1, \ldots, n, \quad \eta_{-j} = h^{-1}\left(1 - e^{-ih\xi_j}\right)$.

(11) $F_h(D_{j,h}^- u) = \eta_j F_h u, \quad j = 1, \ldots, n, \quad \eta_j = h^{-1}\left(1 - e^{ih\xi_j}\right)$.

(12) $D_{j,h}^+(R_h F u) = R_h F(-u \cdot \eta_{-j}), \quad j = 1, \ldots, n$.

(13) $D_{j,h}^-(R_h F u) = R_h F(u \cdot \eta_j), \quad j = 1, \ldots, n$.

We will remark here only that functionals in spaces of lattice functions can be also considered. Then, most of the ideas considered here can be carried over to spaces of distributions.

5.2.3 The fundamental solution of the Laplacian

We consider the difference equation

$$-\Delta_h E_h(x) = -\sum_{j=1}^{n} D_{j,h}^- D_{j,h}^+ E_h(x) = \delta_h(x) = \begin{cases} h^{-n}, & x = 0 \\ 0, & x \in \mathbb{R}_h^n \setminus \{0\}. \end{cases} \tag{5.2}$$

In case $n = 2$ we look for solutions with $|E_h(x)| \leq C \ln |x|$. For $n > 2$ we require the property (see [34]) $|E_h(x)| \leq C|x|^m$ with a certain $m > 0$ depending on the properties of the difference operator.
From the properties of the discrete FOURIER transform we obtain

$$F_h(-\Delta_h E_h(x)) = d^2 F_h E_h(x) \text{ and } F_h \delta_h(x) = \frac{1}{\sqrt{2\pi}^n}.$$

The transformed equation looks like

$$F_h E_h(x) = \frac{1}{\sqrt{2\pi}^n} \frac{1}{d^2}.$$

Applying the inverse discrete FOURIER transform $F_h^{-1} = R_h F$ we get a solution of (5.2) by

$$E_h(x) = \frac{1}{\sqrt{2\pi}^n} R_h F\left(\frac{1}{d^2}\right).$$

In particular, for $n = 3$ ($\mathbb{R}_h^3$)

$$E_h(kh, jh, lh) = \frac{1}{(2\pi)^3} \int\limits_{-\frac{\pi}{h}}^{\frac{\pi}{h}} \int\limits_{-\frac{\pi}{h}}^{\frac{\pi}{h}} \int\limits_{-\frac{\pi}{h}}^{\frac{\pi}{h}} \frac{e^{-i(kh\xi_1 + jh\xi_2 + lh\xi_3)}}{d^2} d\xi_1 d\xi_2 d\xi_3, \quad k, j, l \in \mathbf{Z},$$

holds. For $n \geq 3$ this is a weakly singular integral. In case $n = 2$ the integral does not exist. Then, following [315] we can regularize the integral, subtracting the first part of the TAYLOR expansion for $\exp(ix)$ in the nominator. We obtain

$$E_h(kh, jh) = \frac{1}{(2\pi)^2} \int\limits_{-\frac{\pi}{h}}^{\frac{\pi}{h}} \int\limits_{-\frac{\pi}{h}}^{\frac{\pi}{h}} \frac{e^{-i(kh\xi_1 + jh\xi_2)} - 1}{d^2} d\xi_1 d\xi_2. \tag{5.3}$$

These fundamental solutions are well investigated. We will do the necessary calculation in the case $n = 2$. We are restricted to $n = 2$ because on the one hand the theoretical part is more complicated as for $n > 2$ (due to the logarithmic behaviour at infinity), and on the other hand the numerical approach is easier to explain.

5.2.4 Numerical computation of E_h

We investigate (5.3) for $h = 1$. Because of symmetry properties it is sufficient to evaluate $E_1(k, j) = E_1(-k, j) = E_1(k, -j) = E_1(j, k)$ only at the points (k, j) with $k, j \in \mathbf{Z}$ and $0 \leq j \leq k$. In 1952 SOBOLEV [277] proved the following formula at the main diagonal of the lattice:

$$E_1(n, n) = -\frac{1}{\pi}\left(1 + \frac{1}{3} + \frac{1}{5} + \ldots + \frac{1}{2n-1}\right), \quad n \geq 1, n \in \mathbf{N}.$$

Furthermore, $E_1(0, 0) = 0$ and $E_1(1, 0) = 1/4$ hold. Starting from these known values (evaluated by a finite number of operations at each point) the values at all the other points of $\mathbb{R}_h^2$ can be computed using the difference equation. In addition, we have

$$E_h(kh, jh) = E_1(k, j) \quad (\forall\ h > 0).$$

Therefore, it is sufficient to make all computations for only one mesh width.

5.2.5 Some properties of the fundamental solution

Theorem 5.15. *The fundamental solution (5.3) is unique up to a constant.*

Proof. SOBOLEV proved that $E_1(k, j)$ behaves at infinity like $\ln\sqrt{k^2 + j^2}$. We assume the existence of two different fundamental solutions $E_1^1(k, j)$ and $E_1^2(k, j)$. Then, $E_1^*(k, j) = E_1^1(k, j) - E_1^2(k, j)$ solves the discrete LAPLACE equation in R_h^2. From [276] we have that all solutions of the discrete LAPLACE equation which can be majorized by $\sqrt{k^2+j^2}$ have to be a constant. #. It should be remarked here that the fundamental solution (5.3) is the most investigated fundamental solution in the research papers cited above in this section. But, in almost all considerations the behaviour for $h \mapsto 0$ is not of interest because E_h is not used for discretization methods. In [309] E_h is needed for the proof of solvability of certain equations from discrete variational methods. RYABENKIJ did not realize the possibility of simple computations for E_h and used discrete GREEN functions. Later in this section and in Section 6.1 we discuss some connections between these different approaches.

But now, looking for convergence results, we have to mention that in the continuous theory there are also some other regularizations (we have the same problem with the integral $\int \frac{\cdots}{r^2} dx$) of the fundamental solution (distribution). One usual possibility is to regularize by (see e.g. [321])

$$E(x) = \frac{1}{(2\pi)^2}\left(\int\limits_{|\xi|<1} \frac{e^{-ix\xi}-1}{|\xi|^2} d\xi + \int\limits_{|\xi|>1} \frac{e^{-ix\xi}}{|\xi|^2} d\xi\right) = -\frac{1}{2\pi}(\mathbf{C} - \ln 2 + \ln |x|).$$

Here, $x\xi = x_1\xi_1 + x_2\xi_2$. and $\mathbf{C}$ stands for the EULER constant. Additionally, we will investigate the following regularization:

$$E_h^1(x) = \frac{1}{(2\pi)^2}\left(\int\limits_{|\xi|<1} \frac{e^{-ix\xi}-1}{d^2} d\xi + \int\limits_{|\xi|>1,\xi\in Q_h} \frac{e^{-ix\xi}}{d^2} d\xi\right). \tag{5.4}$$

This fundamental solution differs from (5.3) by the constant

$$K_1 = \frac{1}{(2\pi)^2} \int\limits_{|\xi|>1,\xi\in Q_h} \frac{1}{d^2}\, d\xi$$

depending on h.

Lemma 5.16. *We have* $|E_h^1(x) - E(x)| \leq C_1 h^2 + C_2 \frac{h}{|x|}$.

Proof. From the triangle inequality it follows that

$$\begin{aligned} |E_h^1(x) - E(x)| \;\leq\; & \frac{1}{(2\pi)^2}\left|\int\limits_{|\xi|<1} \left(\frac{1}{d^2} - \frac{1}{|\xi|^2}\right)\left(e^{-ix\xi}-1\right) d\xi\right| \\ & + \frac{1}{(2\pi)^2}\left|\int\limits_{|\xi|>1,\xi\in Q_h} \left(\frac{1}{d^2} - \frac{1}{|\xi|^2}\right) e^{-ix\xi} d\xi\right| \\ & + \frac{1}{(2\pi)^2}\left|\int\limits_{|\xi|\in \mathbf{R}^2\setminus Q_h} \frac{1}{|\xi|^2} e^{-ix\xi} d\xi\right|. \end{aligned} \tag{5.5}$$

We now estimate the three parts from the right-hand side of (5.5). Setting

$$I_1 = \frac{1}{(2\pi)^2}\left| \int\limits_{|\xi|<1} \left(\frac{1}{d^2} - \frac{1}{|\xi|^2}\right)\left(e^{-ix\xi} - 1\right) d\xi \right|$$

we get by the help of the TAYLOR expansion and $d^2 \geq \frac{4}{\pi^2}|\xi|^2$ the inequality $0 < 1/d^2 - 1/|\xi|^2 \leq Ch^2$. Hence,

$$I_1 \leq C_{11}\, h^2 \int\limits_{|\xi|<1} d\xi = C_{12}\, h^2 .$$

We investigate now

$$I_2 = \frac{1}{(2\pi)^2}\left| \int\limits_{|\xi|>1,\xi\in Q_h} \left(\frac{1}{d^2} - \frac{1}{|\xi|^2}\right) e^{-ix\xi} d\xi \right| .$$

Integrating by parts with respect to ξ_1 we obtain

$$\begin{aligned}
I_2 \;\leq\; & \frac{1}{(2\pi)^2}\left| \int\limits_{|\xi|=1} -\frac{1}{ix_1}\left(\frac{1}{d^2} - \frac{1}{|\xi|^2}\right) e^{-ix\xi}\cos(\vec{n},\xi_1) d\xi \right| \\
& + \frac{1}{(2\pi)^2}\left| -\frac{1}{ix_1} \int\limits_{\xi_2=-\frac{\pi}{h}}^{\frac{\pi}{h}} \left(\frac{1}{d^2} - \frac{1}{|\xi|^2}\right)_{|\xi_1=\frac{\pi}{h}} e^{-ix_2\xi_2}\left(e^{-ix_1\frac{\pi}{h}} - e^{ix_1\frac{\pi}{h}}\right) d\xi_2 \right| \\
& + \frac{1}{(2\pi)^2}\left| \frac{1}{ix_1} \int\limits_{|\xi|>1,\xi\in Q_h} \left(\frac{2\xi_1}{|\xi|^4} - \frac{2\sin(h\xi_1)}{h\, d^4}\right) e^{-ix\xi}\, d\xi \right| \\
\leq\; & C_{21}\, h^2 \frac{1}{|x_1|}\left(\int\limits_{|\xi|=1} d\xi + \int\limits_{\xi_2=-\frac{\pi}{h}}^{\frac{\pi}{h}} d\xi_2 \right) \\
& + C_{22}\frac{1}{|x_1|} \int\limits_{|\xi|>1,\xi\in Q_h} \left| \frac{2\xi_1}{|\xi|^4} - \frac{2\sin(h\xi_1)}{h\, d^4} \right| d\xi \\
\leq\; & \frac{1}{|x_1|} C_{23}\, h^2 + \frac{1}{|x_1|} C_{24}\, h .
\end{aligned}$$

Again we used the TAYLOR expansion and $d^2 \geq \frac{4}{\pi^2}|\xi|^2$. Here $\vec{n}$ stands for the outward-pointed unit normal vector of Q_h.

$$I_3 = \frac{1}{(2\pi)^2}\left| \int\limits_{|\xi|\in \mathbf{R}^2\setminus Q_h} \frac{1}{|\xi|^2} e^{-ix\xi} d\xi \right|$$

Again we integrate by parts with respect to ξ_1 and get

$$\begin{aligned}
I_3 &\leq \frac{1}{(2\pi)^2}\left|\frac{1}{ix_1}\int\limits_{\xi_2=-\frac{\pi}{h}}^{\frac{\pi}{h}} \frac{e^{-ix_2\xi_2}}{\pi^2h^{-2}+\xi_2^2}\left(e^{-ix_1\frac{\pi}{h}}-e^{ix_1\frac{\pi}{h}}\right)d\xi_2\right| \\
&+\frac{1}{(2\pi)^2}\left|\lim_{b\to\infty}\int\limits_{|\xi|=b} -\frac{1}{ix_1}\frac{1}{\xi_1^2+\xi_2^2}e^{-ix\xi}\cos(\vec{n},\xi_1)\,d\xi\right| \\
&+\frac{1}{(2\pi)^2}\left|-\frac{1}{ix_1}\int\limits_{|\xi|\in\mathbf{R}^2\setminus Q_h}\frac{2\xi_1}{|\xi|^4}e^{-ix\xi}d\xi\right| \\
&\leq C_{31}\frac{1}{|x_1|}\int\limits_{\xi_2=-\frac{\pi}{h}}^{\frac{\pi}{h}}\frac{1}{\pi^2h^{-2}+\xi_2^2}d\xi_2 + C_{32}\frac{1}{|x_1|}\lim_{b\to\infty}\int\limits_{|\xi|=b}\frac{1}{\xi_1^2+\xi_2^2}d\xi \\
&+C_{33}\frac{1}{|x_1|}\int\limits_{|\xi|\in\mathbf{R}^2\setminus Q_h}\frac{1}{|\xi|^3}d\xi \\
&\leq \frac{1}{|x_1|}C_{34}\,h.
\end{aligned}$$

If we repeat this procedure with respect to ξ_2 we obtain the upper bound $Ch|x_2|^{-1}$ for I_2 and I_3. For $k=2,3$ we get from $(|x_1|I_k)^2+(|x_2|I_k)^2\leq C^2h^2$ the estimate $|x|^2\,I_k^2\leq C^2h^2$ and hence, $I_k\leq Ch|x|^{-1}$. Lemma 5.16., is proved. #

Using Lemma 5.16. we can prove a first convergence result. Let $G\subset\mathbb{R}^2$ be a bounded domain and $G_h=(G\cap\mathbb{R}_h^2)\subset\mathbb{R}_h^2$ the corresponding discrete domain. We define $A(G_h)=\sum_{x\in G_h}h^2$. By Q_h we denote a square with $G_h\subset Q_h$, side length $L=2lh,\ l\in\mathbf{N}$, centred in (m_1h,m_2h). For simplicity we assume for the proof of the following theorem that $m_1=m_2=0$.

Theorem 5.17. *We have*

$$\|E_h^1-E\|_{l_p(G_h\setminus\{0\})}\leq\begin{cases} C(L)h & 1\leq p<2,\\ Ch\sqrt{|\ln h|} & p=2,\\ Ch^{2/p} & p>2.\end{cases}$$

Proof. From Lemma 5.16. we have

$$\begin{aligned}
\|E_h^1-E\|_{l_p(G_h)} &= \left(\sum_{(k_1h,k_2h)\subset G_h\setminus(0,0)}|E_h^1(k_1h,k_2h)-E(k_1h,k_2h)|^ph^2\right)^{1/p} \\
&\leq C_1\,h^2(A(G_h))^{1/p}+C_2\left\|\frac{h}{|x|}\right\|_{l_p(G_h)}
\end{aligned}$$

$$\leq C_1 h^2 (A(G_h))^{1/p} + 4C_2 \left(\sum_{k_1,k_2=1}^{l} (k_1^2 + k_2^2)^{-p/2} h^2 + \sum_{k=1}^{l} k^{-p} h^2 \right)^{1/p} . \quad (5.6)$$

In particular, we use

$$\sum_{k_1=1}^{l} \sum_{k_2=1}^{l} (k_1^2 + k_2^2)^{-p/2} h^2 \leq$$

$$\leq 2^{-p/2} h^2 + 2 \int_h^{lh} \left(\frac{h^2}{x^2 + h^2} \right)^{p/2} h \, dx + \int_h^{lh} \int_h^{lh} \left(\frac{h^2}{x_1^2 + x_2^2} \right)^{p/2} dx_1 dx_2$$

$$\leq \begin{cases} C(L)h & p = 1 \\ C(L)h^p & 1 < p < 2 \\ Ch^2 |\ln h| & p = 2 \\ Ch^2 & p > 2 \end{cases} \quad (5.7)$$

and

$$\sum_{k=1}^{l} k^{-p} h^2 \leq h^2 + \int_h^{lh} \left(\frac{h}{x} \right)^p h \, dx \leq \begin{cases} Ch^2 |\ln h| & p = 1 \\ Ch^2 & p > 1. \end{cases} \quad (5.8)$$

By (5.6) – (5.8) the theorem is completely proved. #

Clearly, the fundamental solution investigated above is not suitable for applications in unbounded domains because our constant C investigated above depends on $A(G_h)$. For this reason we look at

$$E_h^2(x) = \frac{1}{(2\pi)^2} \left(\int_{|\xi|<1} \frac{e^{-ix\xi} - 1}{d^2} d\xi + \int_{|\xi|>1, \xi \in Q_h} \frac{e^{-ix\xi}}{d^2} d\xi + \int_{|\xi|<1} \left(\frac{1}{d^2} - \frac{1}{|\xi|^2} \right) d\xi \right).$$

This fundamental solution differs from (5.4) by

$$K_2 = \frac{1}{(2\pi)^2} \int_{|\xi|<1} \left(\frac{1}{d^2} - \frac{1}{|\xi|^2} \right) d\xi.$$

Lemma 5.18. $|E_h^2(x) - E(x)| \leq Ch/|x|$.

Proof. We omit the details here. One has to proceed the same lines as in the proof of Lemma 5.16. #

Again we are looking for l_p-estimates and we get

Theorem 5.19. $\|E_h^2(x) - E(x)\|_{l_p(\mathbf{R}_h^2\setminus(0,0))} \le C\, h^{2/p},\ 2 < p < \infty\,.$

Proof. Using Lemma 5.18. and Theorem 5.17. we obtain

$$\|E_h^2(x) - E(x)\|_{l_p(\mathbf{R}_h^2\setminus(0,0))} \tag{5.9}$$
$$\le C_1 \left\| \frac{h}{|x|} \right\|_{l_p(\mathbf{R}_h^2\setminus(0,0))}$$
$$\le C_2 \left(h^2 + 3 \int\limits_{x=h}^{\infty} \left(\frac{h}{x}\right)^p h\, dx + \int\limits_{r=h}^{\infty} \int\limits_{\varphi=0}^{\pi/2} \left(\frac{h}{r}\right)^p r\, d\varphi dr \right)^{1/p} . \tag{5.10}$$

The condition $2 < p < \infty$ ensures the existence of $\int_{r=h}^{\infty} \left(\frac{h}{r}\right)^p r\, dr$. Computing then the integrals from (5.9) we come to our assertion. #

Looking to the difference

$$E_h^2(x) - E_h(x) = E_h^1(x) + K_2 - E_h(x) = K_1 + K_2 = \frac{1}{2\pi}\left(\frac{5}{2}\ln 2 - \ln h\right) \tag{5.11}$$

it can be seen that even the well known fundamental solution (5.3) does not converge to the fundamental solution of the (continuous) Laplacian. For practical use the values of our constants are of some interest. An estimate of K_2 can be proved (by the help of a TAYLOR expansion) $0.005\, h^2 < K_2 < 0.0164\, h^2$. A numerical value has been computed by $K_2 \approx 0.005\, h^2$. The constant K_1 can be estimated by the help of (5.11).

The ideas presented above allow immediately a generalization to fundamental solutions of more general second-order elliptic operators.
Let $a_{jk} = a_{kj}$ be the (constant) coefficients of an elliptic difference operator

$$P_h u_h = - \sum_{j,k=1}^{2} a_{jk} D_{j,h}^{+} D_{k,h}^{-} u_h ,$$

where for $C > 0$ and $\xi = (\xi_1, \xi_2)$ the inequality $\sum_{j,k=1}^{2} a_{jk}\, \xi_j\, \xi_k \ge C\,|\xi|^2$ has to be fulfilled. Using the properties 10 and 11 of the discrete FOURIER transform we obtain analogous to (5.4) the fundamental solution

$$E_h^1(x) = \frac{1}{(2\pi)^2} \left(\int\limits_{|\xi|<1} \left(\sum_{j,k=1}^{2} a_{jk}\, \eta_{-j}\, \eta_k \right)^{-1} (e^{-ix\xi} - 1) d\xi \right.$$
$$\left. + \int\limits_{|\xi|>1, \xi\in Q_h} \left(\sum_{j,k=1}^{2} a_{jk}\, \eta_{-j}\, \eta_k \right)^{-1} e^{-ix\xi} d\xi \right) .$$

Here

$$\eta_{-j}\,\eta_k = \left(2h^{-1}\sin\left(\frac{h\xi_j}{2}\right)\right)\left(2h^{-1}\sin\left(\frac{h\xi_k}{2}\right)\right)\left(\cos\left(\frac{h(\xi_j-\xi_k)}{2}\right)\right)$$

holds. This fundamental solution converges to

$$\frac{1}{(2\pi)^2}\left(\int\limits_{|\xi|<1}\left(\sum_{j,k=1}^{2} a_{jk}\xi_j\xi_k\right)^{-1}(e^{-ix\xi}-1)d\xi + \int\limits_{|\xi|>1}\left(\sum_{j,k=1}^{2} a_{jk}\xi_j\xi_k\right)^{-1} e^{-ix\xi}d\xi\right).$$

We will denote this limit function by $E(x)$. Clearly, E is a fundamental solution of the elliptic differential operator

$$Pu = -\sum_{j,k=1}^{2} a_{jk}D_jD_ku,$$

which will be approximated by our difference operator P_h. For the proof we have to use

$$d^2 \geq \frac{4}{\pi^2}|\xi|^2 \text{ and } \sum_{j,k=1}^{2} a_{jk}\,\xi_j\,\xi_k \geq C\,|\xi|^2.$$

Analogous to Lemma 5.18. and Theorem 5.19. results can be proved using the regularized fundamental solution $E_h^2(x) = E_h^1(x) + K_3(h)$ with

$$K_3(h) = \frac{1}{(2\pi)^2}\int\limits_{|\xi|<1}\left[\left(\sum_{j,k=1}^{2} a_{jk}\,\eta_{-j}\,\eta_k\right)^{-1} - \left(\sum_{j,k=1}^{2} a_{jk}\,\xi_j\,\xi_k\right)^{-1}\right]d\xi.$$

In the following we are looking for pointwise upper bounds of the fundamental solutions considered.

Lemma 5.20. *Let $G \subset \mathbb{R}^2$ be a bounded domain and $h > 0$ the fixed mesh width. Then*

$$|E_h(x)| \leq C_1(G) + C_2|\ln h|$$

and

$$|E_h^2(x)| \leq C_3(G) + C_4|\ln h|$$

hold.

Proof. Let χ be the characteristic function of G, $J_0(u)$ the Bessel function of order 0 and $\mathbf{C}$ again Euler's constant. From $d^2 \geq \frac{4}{\pi^2}|\xi|^2$ it follows that

$$|E_h(x)| \;=\; \frac{1}{\pi^2}\int\limits_{\xi_1=0}^{\frac{\pi}{h}}\int\limits_{\xi_2=0}^{\frac{\pi}{h}} \frac{1-\cos(x_1\xi_1)\cos(x_2\xi_2)}{d^2}d\xi_1 d\xi_2$$

$$
\begin{aligned}
&\leq \frac{1}{4} \int_{\xi_1=0}^{\frac{\pi}{h}} \int_{\xi_2=0}^{\frac{\pi}{h}} \frac{1-\cos(x_1\xi_1)\cos(x_2\xi_2)}{|\xi|^2} d\xi_1 d\xi_2 \\
&\leq \frac{1}{16} \int_{\xi_1=-\sqrt{2}\frac{\pi}{h}}^{\sqrt{2}\frac{\pi}{h}} \int_{\xi_2=-\sqrt{2}\frac{\pi}{h}}^{\sqrt{2}\frac{\pi}{h}} \frac{1-e^{ix\xi}}{|\xi|^2} \chi\left(\sqrt{2}\frac{\pi}{h} - |\xi|\right) d\xi_1 d\xi_2 \qquad (5.12) \\
&= \frac{1}{16}\left[\int_{u=0}^{\sqrt{2}\frac{\pi}{h}|x|} \frac{1-J_0(u)}{u} du - \int_{u=\sqrt{2}\frac{\pi}{h}|x|}^{\infty} \frac{J_0(u)}{u} du \right] + \frac{1}{16} \int_{u=\sqrt{2}\frac{\pi}{h}|x|}^{\infty} \frac{J_0(u)}{u} du \\
&\leq \left[\mathbf{C} - \ln 2 + \ln(\sqrt{2}\frac{\pi}{h}|x|)\right] + \frac{1}{16}\left| \int_{u=\sqrt{2}\frac{\pi}{h}|x|}^{\infty} \frac{J_0(u)}{u} du \right| .
\end{aligned}
$$

It remains the estimation of

$$
\left| \int_{u=\sqrt{2}\frac{\pi}{h}|x|}^{\infty} \frac{J_0(u)}{u} du \right|.
$$

Let C_G be a constant depending on G and G_h with

$$
\sqrt{2}\pi h^{-1}|x| < C_G < \infty \quad (\forall\, x \in G \cup G_h)
$$

. From $|J_0(u)| \leq 1$ we get

$$
\begin{aligned}
\left| \int_{u=\sqrt{2}\frac{\pi}{h}|x|}^{\infty} \frac{J_0(u)}{u} du \right| &\leq \int_{u=\sqrt{2}\frac{\pi}{h}|x|}^{h^{-1}C_G} \frac{|J_0(u)|}{u} du + \left| \int_{u=h^{-1}C_G}^{\infty} \frac{J_0(u)}{u} du \right| \\
&\leq \ln\left(\frac{C_G}{\sqrt{2}\pi|x|}\right) + \left| \int_{u=h^{-1}C_G}^{\infty} \frac{J_0(u)}{u} du \right|. \qquad (5.13)
\end{aligned}
$$

The asymptotic formula

$$
J_0(u) = \sqrt{\frac{2}{\pi u}} \cos(u - \frac{\pi}{4}) + \frac{\varphi_0(u)}{u^{3/2}}
$$

is used for proving the convergence of

$$
\int_{u=h^{-1}C_G}^{\infty} \frac{J_0(u)}{u} du.
$$

We remember that $\varphi_0(u)$ is bounded for $u \to \infty$. From (5.12) and (5.13) and

$$\Big| \int\limits_{u=h^{-1}C_G}^{\infty} \frac{J_0(u)}{u} du \Big| \leq C_1$$

we obtain for $h < 1$ the upper bound

$$\begin{aligned} |E_h(x)| &\leq \frac{1}{16}\left(\mathbf{C} - \ln 2 + \ln(\sqrt{2}\,\frac{\pi}{h}\,|x|) + \ln\left(\frac{C_G}{\sqrt{2}\pi|x|}\right) + C_1\right) \\ &= C_2(G) + C_3|\ln h|. \end{aligned}$$

The second part of the assertion follows from (5.11). #

Remark 5.21. If we look to the formulas defining our fundamental solutions, then it can be observed that all integrals (sometimes after regularization) exist in the classical sense. It is possible to omit the restriction operator R_h and after this operation we have the C^∞-function defined in $\mathbb{R}^n$. That means we have C^∞-extensions of the fundamental solutions in a very natural way. This natural extension is useful at the end of numerical procedures where in many cases an interpolation from finite dimensional spaces to the original function space is necessary. Recalling to mind the properties of discrete and continuous Fourier transform it seems possible to carry over most of the norm estimates from the l_p-spaces to corresponding L_p-spaces. This impression will be confirmed. If we look, for example, to Lemma 5.20. it is not necessary for the proof that x should belong to the lattice. The estimates hold true in the whole domain G if we cancel the action of R_h in $F_h^{-1} = R_h F$.

More details are required for the behaviour of the partial derivatives and difference quotients, respectively, of the fundamental solutions. Because the asymptotic behaviour of the finite differences is much better than the behaviour of the fundamental solution itself, this problem has been studied a long time ago. We refer to [315] for second-order elliptic operators. To avoid the again very expensive computations we will only give the results. For higher order operators, parabolic operators and several generalizations of CAUCHY–RIEMANN operators we refer the reader to [103], [100], [102], [110], and for a nearly complete collection to [135].

Lemma 5.22. *The extensions to* $\mathbb{R}^2$ *of the fundamental solutions* $E_h(x)$, $E_h^1(x)$, *and* $E_h^2(x)$ *belong to* $C^{1,\mathrm{loc}}(\mathbb{R}^2)$ *[135].*

Preparing the estimation of the approximation error for the first derivatives we have to look to the inequality

$$\begin{aligned} \left|\frac{\partial}{\partial x_j}(E_h^* - E)(x)\right| &\leq \left|\int\limits_{\xi\in Q_h}\left(\frac{1}{d^2} - \frac{1}{|\xi|^2}\right)(-i\xi_j)e^{-ix\xi}d\xi\right| \\ &\quad + \left|\int\limits_{\xi\in\mathbf{R}^2\setminus Q_h}\frac{(-i\xi_j)e^{-ix\xi}}{|\xi|^2}d\xi\right|, \end{aligned}$$

where $j = 1, 2$. $E_h^*(x)$ is one of $E_h(x)$, $E_h^1(x)$, and $E_h^2(x)$, respectively.

Lemma 5.23. $\left| \frac{\partial}{\partial x_j}(E_h^*(x) - E(x)) \right| \leq \frac{Ch}{|x|^2}, \quad j = 1, 2.$

Proof. Integrate by parts twice (and pay attention to the singularity). #

Remark 5.24. With the results proved above we are able to consider approximations and the approximation error of NEWTONian-like potentials. The method described in this book requires the boundary reduction of the problem. Therefore, it is necessary to repeat the whole procedure on the boundary of G and G_h, respectively. We will skip this step here. It should be remarked that in principle one has to use the same ideas. But, due to the lower dimension of the boundary (and the same singularity of the kernel function) one has to be careful in handling the fundamental solution near the singularity. The rate of convergence will be reduced.
We refer to [110] for the treatment of the DIRAC operator and to [135] for second-order problems with constant coefficients.

For a better understanding we will repeat here without proof some very natural results concerning properties of the discrete fundamental solution of the DIRAC difference operator. We will not distinguish between the fundamental solution e_h^+ of D_h^+ and e_h^- of D_h^-, respectively. The proofs of the following properties are contained in [110] and are carried out analogously to the properties of E_h proved above.

Theorem 5.25. *Let G_h be a bounded discrete domain. Then we have for $p < \frac{3}{2}$*

$$||e_h^{\pm}||_{p,h,G_h} \leq K_{p,G_h}, \quad \forall\, h \leq h_0. \tag{5.14}$$

This estimate corresponds to the continuous case where analogously for $p < 3/2$ the inequality $||e||_p < \infty$ holds. In the same way as in Theorems 5.17. and 5.19. we can prove a convergence result in discrete l_p-spaces.

Theorem 5.26. *The estimate*

$$||e_h^{\pm}||_{p,h,G_h} \leq Ch^{-2+\frac{3}{p}} \tag{5.15}$$

holds for all $p < \frac{3}{2}$ with a constant C uniformly with respect to h.

5.3 Discrete reprokernels

In the previous section we have looked into parts of a well-developed theory of fundamental solutions for several classes of partial difference operators. In particular, for $n = 2$ efficient numerical algorithms are included. From a theoretical point of view the same ideas will again be useful in the higher dimensional case. But for $n > 2$ we have either to evaluate the fundamental solution on $(n-1)$-dimensional hyperplanes and then to extend it or to evaluate the fundamental solution using quadrature formulas for the integral representations. Both possibilities are very expensive. At least two ideas for avoiding these problems should be mentioned here. One way is the replacement of the fundamental solution by the discrete Green function. Then, all assumptions from Theorem 5.1. can be also fulfilled, and we can repeat all the procedure from Section 5.1. This leads to a discrete potential theory (see e.g. [250], [103]). The disadvantage of this approach is that, of course, the Green function depends on the domain. That means that in preparing the numerical solution of a boundary value problem one has to evaluate the GREEN function for each domain we want to deal with (this problem has been excluded in the fundamental book [250] on discrete function theory). The second idea is the approximation of the fundamental solution by a sequence of GREEN functions for cubes with increasing side length (this discrete GREEN function is explicitly known!). This approach has been used already in [110]. Because of its practical importance we will repeat it here. Meanwhile, new questions arise. Now, we look for the relationships between discrete fundamental solutions and the discrete GREEN function. Furthermore, we have to seek for a discrete BERGMAN kernel function and its connections to the fundamental solution and the GREEN function, respectively. If we want to have a full analogy between continuous theory and its discretization these useful kernel functions are needed. For comparison we refer to Section 4.16.

5.3.1 Discrete GREEN function

The discrete GREEN function of a domain G_h is defined to be the solution of the boundary value problem

$$\begin{aligned} -\Delta_h u &= \delta_h \quad \text{in} \quad \operatorname{int} G_h, \\ u &= 0 \quad \text{on} \quad \partial G_h. \end{aligned}$$

To solve this boundary value problem is very easy because we have from Theorem 5.7. the representation formula

$$G_h(x,y) = (T_h^- Q_h^+ T_h^+ \delta_x)(y) \tag{5.16}$$

Writing again $e_h^{\pm}$ for the fundamental solutions of the discrete CAUCHY–RIEMANN operators $D_h^{\pm}$ and E_h for the fundamental solution of the discrete Laplacian Δ_h we obtain

$$\begin{aligned} T_h^- Q_h^+ T_h^+ \delta_{h,x} &= T_h^- Q_h^+ T_h^+ D_h^+ D_h^- E_h = T_h^- Q_h^+ (I - F_h^+) D_h^- E_h \\ &= T_h^- Q_h^+ D_h^- Eh = T_h^- Q_h^+ e_h^+ . \end{aligned} \tag{5.17}$$

Therefore, we have proved a first representation of the GREEN function by the help of the fundamental solution.

Theorem 5.27. *The discrete* GREEN *function can be represented by*

$$G_h(x,y) = T_h^- Q_h^+ e_h^+ . \tag{5.18}$$

This theorem is very helpful in theoretical investigation of the discrete GREEN function, because the properties of e_h^+, Q_h^+, and T_h^- are well known. Using this knowledge we are enabled to answer the questions of norm estimates and convergence of the GREEN function. For practical purposes the best idea probably is to solve the boundary value problem by the help of a fast solver for the system of linear algebraic equations. Unfortunately, equation (5.18) is a one-way street because it is impossible to obtain the fundamental solution from the knowledge of G_h. Here it helps to repeat the considerations from [110] (written down therein only from the numerical point of view). The main idea is that not one GREEN function but an infinite sequence of GREEN functions determines the fundamental solution. This sequence can be used also for the numerical approximation of fundamental solutions.

Let us look for the solutions of the boundary value problems

$$\begin{aligned} -\Delta_h u &= \delta_h \quad \text{in } \operatorname{int} G_h, \\ u &= 0 \quad \text{on } \partial G_h . \end{aligned}$$

If we choose $G_h = Q_l = [-l,l]^3 \cap \mathbb{R}_h^3$ then the solution (GREEN's function for Q_l) can be described as follows:

$$G_{h,l}(x) = \sum_{i,j,k=0}^{n-1} l^{-3/2} {\lambda_{i,j,k}^{(h,l)}}^{-1} v_{i,j,k}^{(h,l)}(x), \tag{5.19}$$

where

$$\lambda_{i,j,k}^{(h,l)} = 4h^{-2}\Big(\sin^2\Big(\frac{2i+1}{4l}\pi h\Big) + \sin^2\Big(\frac{2j+1}{4l}\pi h\Big) + \sin^2\Big(\frac{2k+1}{4l}\pi h\Big)\Big),$$

$$v_{i,j,k}^{(h,l)} = l^{-3/2}\cos\Big(\frac{2i+1}{2l}\pi x_1\Big)\cos\Big(\frac{2j+1}{2l}\pi x_2\Big)\cos\Big(\frac{2k+1}{2l}\pi x_3\Big)$$

and $i,j,k = 0,\dots n-1$; $(x_1,x_2,x_3)^T \in Q_l$.
Based on this representation formula it was proved in [110] that

$$G_{h,l}(x) \geq 0, \quad \forall\, x \in Q_l,$$

$$G_{h,l}(x) \leq G_{h,L}(x), \quad \forall\, x \in \mathbb{R}_h^3 \;\forall L > l,$$

$$G_{h,l}(0) \geq G_{h,l}(x), \quad \forall\, x \in \mathbb{R}_h^3$$

and, finally,

$$G_{h,l}(x) \mapsto E_h(x) \quad \text{for } l \to \infty \quad \forall\, x \in \mathbb{R}_h^3. \tag{5.20}$$

Furthermore, this convergence is true also in the uniform sense. This is the desired relationship between a sequence of GREEN functions and the fundamental solution. Let us remark that these ideas include, for numerical purposes, essential properties. We do not only have the uniform convergence but we also have the motonicity of the convergence. Due to the simple structure of Q_l we can use efficient algorithms for the computation of $G_{h,l}$ from standard program packages as well as the explicit formula (5.19).

5.3.2 Discrete BERGMAN kernel

Now we require a discrete BERGMAN kernel. There is no problem in proving the existence of such a kernel because in our discrete spaces the functionals $f_x \mapsto f(x)$ are continuous in each case. We will start with the GREEN function already investigated

$$G_h(x,y) = (T_h^- Q_h^+ T_h^+ \delta_{h,x})(y)$$

where $\delta_{h,x}$ stands for the δ_h function supported at the point x. We "differentiate" from the right and obtain

$$\omega_x = T_h^- Q_h^+ T_h^+ D_{h,y}^+ \delta_{h,x} = -T_h^- Q_h^+ T_h^+ \delta_{h,x} D_{h,x}^- = -G_h(x,y) D_{h,x}^-.$$

Here, we have used $D_{h,y}^+ \delta_x(y) = -D_{h,x}^- \delta_x(y) = -(\delta_x(y)) D_{h,x}^-$. Then,

$$D_{h,y}\omega_x(y) = Q_h^+ \delta_x(y) = -D_{h,y}^- G_h(x,y) D_{h,x}^-.$$

We define a function $P_h^+(x,y)$ by

$$P_h^+(x,y) = \delta_x \delta_y + D_{h,y}^- G_h(x,y) D_{h,x}^-.$$

Theorem 5.28.

$$P_h^+(x,y) = \delta_x \delta_y + D_{h,y}^- G_h(x,y) D_{h,x}^-$$

is the reproducing kernel of $l_{2,h}(G_h) \cap \ker D_h^+(\operatorname{int} G_h)$, *i.e.,*

$$\langle P_h^+(x,y), f(y)\rangle = f(x) \quad \forall\, f \in \ker D_h^+(G_h)$$

Proof. We denote by $Q_h^+(x,y)$ the function $\langle Q_h^+\delta_x, \delta_y\rangle$. Then, $P_h^+(x,y) = \delta_x\delta_y - Q_h^+(x,y)$.

$$\begin{aligned}
\langle P_h^+(x,y), f(y)\rangle &= \langle \delta_x\delta_y, f(y)\rangle - \langle Q_h^+(x,y), f(y)\rangle \\
&= f(x) - \langle Q_h^+(x,y), f(y)\rangle \\
&= f(x) - \langle\langle Q_h^+\delta_x, \delta_y\rangle, f(y)\rangle \\
&= f(x) - \sum_y \sum_\xi Q_h^+\delta_x(\xi)\delta_y(\xi)h^3 f(y)h^3 \\
&= f(x) - \langle Q_h^+\delta_x, f(y)\rangle = f(x) - \langle \delta_x, Q_h^+ f\rangle \\
&= f(x) - (Q_h^+ f)(x) = (P_h^+ f)(x) \\
&= f(x), \ \forall\, f \in \ker D_h^+. \qquad \#
\end{aligned}$$

It should be remarked here that the representation of the discrete BERGMAN kernel corresponds with the representations proved in [236] and [101] for the continuous case.

Furthermore, the procedure proved above seems to be useful in practical computations, too. This is because the function ω from above is the solution of the boundary value problem

$$\begin{aligned}
-\Delta_h\omega_x(y) &= D_{h,y}^+\delta_x \quad \text{in } G_h, \\
\omega_x(y) &= 0 \qquad\quad \text{on } \partial G_h.
\end{aligned}$$

Again this solution (and hence, the function $P_h^+(x,y)$) can be evaluated by well-developed standard software.

5.4 Convergence results

In the foregoing sections we defined discrete analogues of the operators T_G, D and F_Γ and we studied elements of a theory for finite difference boundary value problems. We were able to transfer the advantages from the continuous theory to the discretized model. In particular, it was possible to find representations of the solutions, results on existence, uniqueness, and stability without a detailed investigation of the system matrices of our linear systems of algebraic equations. Now, the last problem is to find the convergence if $h \to 0$. Preparing these investigations we have already studied the behaviour of the kernel functions (fundamental solutions) for $h \to 0$. First in this section we will study all of our discrete operators alone and prove some convergence properties. Using these properties we will be able to prove the approximation and convergence properties for the solutions of discrete boundary value problems in a very simple and formal way. This last-mentioned step will be demonstrated by the help of a non-trivial example (NAVIER–STOKES equations).

5.4.1 Approximation of integral operators

Theorem 5.29. *The operator*

$T_h^+ : L_{p,h,H}(G_h) \mapsto L_{q,h,H}(G_h)$

is continuous for $1 < p < 3$ *and* $q < \dfrac{3p}{3-p}$.

Proof. We assume that $p < q$. The application of HÖLDER's inequality (for sums) to a product consisting of three factors yields

$$\begin{aligned}
|T_h^+ f(x)| &\leq \Big(\sum_{y \in G_h} \{ |f(y)|^{\frac{p}{q}} |x-y|^{\frac{-3}{q} + \frac{3}{2q} + \frac{3}{2p'} - 1} \}^q h^3 \Big)^{\frac{1}{q}} \\
&\times \Big(\sum_{y \in G_h} \{ |f(y)|^{p(\frac{1}{p} - \frac{1}{q})} \}^{\frac{pq}{q-p}} h^3 \Big)^{\frac{1}{p} - \frac{1}{q}} \\
&\times \Big(\sum_{y \in G_h} |x-y|^{\frac{-3}{p'} + \frac{3}{2q} + \frac{3}{2p'} - 1} \}^{p'} h^3 \Big)^{\frac{1}{p'}} \\
&\leq \|f\|_{p,h,G}^{1-\frac{p}{q}} (C (\operatorname{diam} G)^{\frac{3p}{2q} + \frac{3}{2} - p'} \Big(\frac{3p'}{2q} - \frac{3}{2} - p' \Big)^{-1})^{\frac{1}{p'}} \\
&\times \Big(\sum_{y \in G_h} \{ |f(y)|^p |x-y|^{-\frac{3}{2} + \frac{3q}{2p'} - q} h^3 \Big)^{\frac{1}{q}} \\
&= \|f\|_{p,h,G}^{1-\frac{p}{q}} K(p', q, G)^{\frac{1}{p'}} \times \Big(\sum_{y \in G_h} |f(y)|^p |x-y|^{-\frac{3}{2} + \frac{3q}{2p'} - q} h^3 \Big)^{\frac{1}{q}} .
\end{aligned}$$

Hence,

$$\|T_h^+ f\|_{q,h,G}^q \leq \|f\|_{p,h,G}^{q-p} K(p',q,G)^{\frac{q}{p'}} \sum_{y\in G_h} |f(y)|^p h^3 K(q,p',G)$$

and

$$\|T_h^+ f\|_{q,h,G} \leq K(p',q,G)^{\frac{1}{p'}} K(q,p',G)^{\frac{1}{q}} \|f\|_{p,h,G}. \tag{5.21}$$

In the proof we have used the necessary conditions

$$\frac{3}{2} + \frac{3q}{2p'} - q > 0 \quad \text{and} \quad \frac{3p'}{2q} + \frac{3}{2} - p' > 0.$$

Both inequalities are fulfilled for $q < 3p/(3-p)$. Here p' is defined by $1/p + 1/p' = 1$. The function $K(r,s,G)$ is given by

$$K(r,s,G) = C(\operatorname{diam} G)^{\frac{3r}{2s}+\frac{3}{2-r}} \left(\frac{3r}{2s} + \frac{3}{2} - r\right)^{-1}.$$

The idea of the proof is similar to the continuous case [273]. #

Making special choices we immediately get from the previous theorem:

Corollary 5.30.

$T_h^+ \in \mathcal{L}(L_{p,h,G}; L_{2,h,G})$ *for* $p > \frac{6}{5}$

$T_h^+ \in \mathcal{L}(L_{2,h,G}; L_{q,h,G})$ *for* $q < 6$.

Theorem 5.31.

$T_h^+ \in \mathcal{L}(L_{p,h,G}; C_{h,G})$ *for* $p > 3$

Proof.

$$\begin{aligned} |T_h^+ f(x)| &= |\sum_{y\in G_h} e_h^+(x-y) f(y) h^3| \\ &\leq \Big(\sum_{y\in G_h} |e_h^+(x-y)|^q h^3\Big)^{\frac{1}{q}} \Big(\sum_{y\in G_h} |f(y)|^p h^3\Big)^{\frac{1}{p}} \\ &\leq \|f\|_{p,h,G} \|e_h^+\|_{q,h,G} \end{aligned}$$

Here we have used the norm estimate for the fundamental solution e_h^+ and again HÖLDER's inequality. #

Now, we restrict the domain of T_h^+ to $D_h^-(W_{2,h}^{1,-}(G_h))$ and look for analogues of the well-known embedding theorems.

Theorem 5.32. *Let $u \in W^{1,-}_{2,h,H(G_h)}$ and $q < 6$. Then,*

$$\|u\|_{q,h,G} \leq C\|u\|_{W^{1,-}_{2,h,H(G_h)}}$$

Proof. Clearly,$D_h^- u \in L_{2,h,G}$. Hence,

$$\|u\|_{q,h} \leq \|T_h^- D_h^- u\|_{q,h} \leq K(2,q,G)^{\frac{1}{2}} K(q,2,G)^{\frac{1}{q}} \|D_h^- u\|_{2,h} \leq C\|u\|_{W^{1,-}_{2,h}}$$

We used the discrete BOREL–POMPEIU formula and Theorem 5.29. #

In the discrete case it is also possible to deduce a relationship between the smallest eigenvalue $\lambda_{1,h(G_h)}$ of $-\Delta_h$ and the norm of T_h^+. A simple calculation yields
$\langle -\Delta_h u, v \rangle = \langle u, -\Delta_h v \rangle$ for $u, v \in W^{1,-}_{2,h}(G_h)$.
In addition we obtain

$$\langle -\Delta_h u, u \rangle = \langle D_h^+ D_h^- u, u \rangle = \langle D_h^- u, D_h^- u \rangle \geq 0 \text{ for } u \in W^{1,-}_{2,h}(G_h).$$

Because we have for $u \in W^{1,-}_{2,h}(G_h) \cap \ker D_h^-(intG_h)$ that $u = 0$ (by the help of the discrete BOREL–POMPEIU formula) the positive definiteness of $-\Delta_h$ has been shown. Furthermore, we get

$$\sup\{\|T_h^- f\|_{2,h,G}/\|f\|_{2,h,G}; f \in \operatorname{im} \mathbf{Q}_h^+\}$$
$$= \sup\{\|g\|_{2,h,G}/\|D_h^- g\|_{2,h,G}; g \in W^{1,-}_{2,h}(G)\} = (\lambda_{1,h(G_h)})^{-\frac{1}{2}}$$

Theorem 5.33.

$$\|T_h^- f\|_{2,h,G} \leq (\lambda_{1,h}(G_h))^{-\frac{1}{2}} \|f\|_{2,h,G} \quad f \in \operatorname{im} \mathbf{Q}_h \tag{5.22}$$
$$\|T_h^- f\|_{2,1,h,G} \leq H \quad (1 + 1/\lambda_{1,h}(G_h))^{\frac{1}{2}} \|f\|_{2,h,G} \quad f \in \operatorname{im} \mathbf{Q}_h^+. \tag{5.23}$$

Proof. The first inequality has already been verified. For the proof of the second estimate we refer to the corresponding idea from the continuous case. #

It can be verified that estimates in Theorem 5.33. cannot be improved. By the aid of (5.21) it is possible to find out lower bounds for $\lambda_{1,h}(G_h)$ (independently of h). These problems may be solved with methods which are applied in the continuous case, too. One can prove that $\lambda_{1,h(G_h)} \to \lambda_1(G)$ for $h \to 0$. Obviously, $\|\mathbf{Q}_h^+\|_{[L_{2,h,G};L_{2,h,G}]} = 1$ holds. Therefore, we can obtain all the estimates which are needed for investigating elliptic boundary value problems even in the discrete case (in corresponding spaces) for fixed h, and we are now enabled to formulate our continuous results in a discrete language. Note that the norms of T_h^+ and T_h^- as well as the embedding constants of Theorem 5.32. may be uniformly estimated with respect to h. By the aid of a lower estimate of $\lambda_{1,h}$ the assertion of Theorem 5.33. becomes independent of h. To generate a relation between the discrete and the continuous boundary value problems, we have to consider the behaviour of the operators $T_h^+, T_h^-, \mathbf{Q}_h^+, F_h^+, F_h^-, D_h^-$, and D_h^+ for $h \to 0$.

Theorem 5.34. *Let f be a* RIEMANN*-integrable function which belongs to $L_\infty(G)$. Then*

$$\|T_h^+ f - T_G f\|_{C_h(G)} \to 0.$$

If $f \in C_H^{0,\beta}(\overline{G}), 0 < \beta \leq 1$, then we have

$$\begin{aligned}\|T_h^+ f - T_G f\|_{C_h(G)} &\leq C(G, \|f\|_{q,h,H}, \|f\|_{q,H}) h^{-2+\frac{3}{p}} |\ln h| \\ &\quad + K_{p,G} |h|^\beta \|f\|_{C_h(G)}\end{aligned}$$

for $p < 3/2$, $1/p + 1/q = 1$.

Proof. Let $W(y)$ be a cube with edge length h and midpoint y. We obtain the following estimates:

$$\begin{aligned}
|(T_h^+ f)(x) - (Tf)(x)| &\leq |\int_{W(x)} e(x-y) f(y)\, dG_y| \\
&\quad + |\sum_{y \neq x} e_h^+(x-y) f(y) h^3 - \sum_{x \in W(y)} \int_{W(y)} e(x-z) f(z)\, dG_z| \\
&\leq \sum_{y \neq x} |e_h^+(x-y) - e(x-y)| |f(y)| h^3 + \sum_{y \neq x} |e(x-y) f(y) h^3 \\
&\quad - \int_{W(y)} e(x-z) f(z) dG_z| + \int_{W(x)} |x-y|^{-2} |f(y)|\, dG_y \\
&\leq (\sum_{y \neq x} |e_h^+(x-y) - e(x-y)|^p h^3)^{\frac{1}{p}} \|f\|_{q,h,G} + h^{\frac{3}{p}-2} \|f\|_{q,W(x)} \\
&\quad + \sum_{y \neq x} |e(x-y) f(y) h^3 - \int_{W(y)} e(x-z) f(z) dG_z|.
\end{aligned}$$

Furthermore, we decompose the last sum

$$\sum_{y \neq x} |e(x-y) f(y) h^3 - \int_{W(y)} e(x-z) f(z)\, dG_z|$$

$$\leq \underbrace{\sum_{y \neq x} |\int_{W(y)} e(x-y)[f(y) - f(z)] dG_z|}_{S_1} + \underbrace{\sum_{y \neq x} \int_{W(y)} |e(x-y) - e(x-z)| |f(z)| dG_z}_{S_2}.$$

By the help of a TAYLOR expansion of the function $e(x-z)$ we obtain the following estimate of S_2:

$$\sum_{y \neq x} \int_{W(y)} |e(x-y) - e(x-z)| |f(z)| dG_z$$

$$\leq \sum_{y\neq x} \int_{W(y)} \sum_{i=1}^{3} |D_{i,z} e(x-z)|_{z=\theta(y)} |z_i - y_i| |f(z)| dG_z$$

$$\leq 3C^* \sum_{y\neq x} \Big(\int_{W(y)} h^p 2^{-p} (|x-y| - \sqrt{3}0.5h)^{-3p}\, dG_z \Big)^{\frac{1}{p}} ||f||_{q,W(y)}$$

$$\leq h^{-2+\frac{3}{p}} 1.5C' \sum_{y\neq x} h^3 (|x-y| - 3^{\frac{1}{2}} 0.5h)^{-3} ||f||_{q,W(y)}$$

$$\leq h^{-2+\frac{3}{p}} 1.5C' \Big[C \sum_{y\neq x} h^3 |x-y|^{-3} ||f||_{q,W(y)} \Big]$$

$$\leq h^{-2+\frac{3}{p}} C^* ||f||_{q,G} [C_1 + C_2(\operatorname{diam} G) - C_3 |\ln(h)|] \qquad (5.24)$$

For the following we need the inequality

$$\sum_{i,j,k-0}^{n-1} ((2i+1)^2 + (2j+1)^2 + (2k+1)^2)^{-\frac{p}{2}} \leq \frac{C_1 h^{p-3} l^{3-p}}{3-p} + C_2.$$

Comparing this estimate with estimates for the integral $\int_G 1/|x|^p\, dx$ the inequality seems to be very natural. With the help of an adapted quadrature formula this has been proved in [110] and we will omit the proof here. Deducing (5.24) we use this inequality and

$$(|x-y| - \frac{1}{2}\sqrt{3}h)^{-1} \leq C|x-y|^{-1} \text{ for } x, y \in \mathbb{R}_h^3, x \neq y$$

with $C > (1 - \frac{1}{2}\sqrt{3})^{-1}$.

Now the estimation of the term S_1 follows .

$$S_1(x) \leq \sum_{y\neq x} \Big(\int_{W(y)} |e(x-y)|^p\, dG_z \Big)^{\frac{1}{p}} \Big(\int_{W(y)} |f(y) - f(z)|^q\, dG_z \Big)^{\frac{1}{q}}$$

$$\leq \sum_{y\neq x} \Big(|e(x-y)|^p h^3 \Big)^{\frac{1}{p}} \Big(\int_{W(y)} |f(y) - f(z)|^q\, dG_z \Big)^{\frac{1}{q}}$$

$$\leq ||e^*||_{p,h,G} \Big(\sum_{y\neq x} \int_{W(y)} |f(y) - f(z)|^q\, dG_z \Big)^{\frac{1}{q}}$$

$$\leq ||e^*||_{p,h,G} \Big(\sum_{y\neq x} | \sup_{z\in W(y)} |f(z)| - \inf_{z\in W(y)} |f(z)| |h^3 \Big)^{\frac{1}{q}} \times (2V)^{1-\frac{1}{q}} \to 0$$

for $h \to 0$ if f is RIEMANN-integrable, $|f| \leq V$, and $p\langle \frac{3}{2}$. Summarizing all estimates and using inequality (5.14) and the convergence result (5.15) we get the inequality

$$\begin{aligned}|(T_h^+ f - Tf)(x)| &\leq C(\|f\|_{q,h}, \|f\|_q) h^{-2+\frac{3}{p}} |\ln h| + \\ &+ K_{p,G} \Big(\sum_{y \in G_h} |f(y) - f(z)|^q dG_z \Big)^{\frac{1}{q}},\end{aligned}$$

where $x \in G_h$. Since the right-hand side of the last inequality does not depend on x, we may find uniform estimates in G_h. #

Now, learning from this proof, it can be shown similarly that corresponding results also hold true in L_p-spaces. We will present these results without explanation of the proof (see [110]).

Theorem 5.35. *Let $f \in \mathcal{R}_H \cap L_{\infty,H}(G)$. For $p \in (\frac{6}{5}, \frac{3}{2})$ we have that*

$$\|T_h^+ f - Tf\|_{p,G} \to 0$$

and under the assumptions that $f \in C_H^{0,\beta}(\overline{G}), p \in (\frac{6}{5}, \frac{3}{2}), 0 \langle \beta \leq 1$

$$\begin{aligned}\|T_h^+ f - Tf\|_{p,G} &\leq C\|f\|_{\infty,G} h^{-2+\frac{3}{p}} + C_1(\|f\|_{q',h,G}, \|f\|_{q',G}) h^{-2+\frac{3}{p'}} |\ln h| \\ &\quad + K_{p',G} \|f\|_{C_H^{0,\beta}}(G) h^\beta.\end{aligned}$$

As we have seen above the approximation properties of $T_h^\pm$ are as expected. The proof of these properties requires some work, but there are only technical difficulties. This is very natural in approximation of a weakly singular integral by an adapted qudrature formula. Much more interesting is the investigation of the projection operator $\mathbf{Q}_h^+$. We argued above in Section 5.1 that the orthoprojections remove all our problems and put them into an operator with known complicated structure, but with norm equals 1. In fact $\mathbf{Q}_h^+$ approximates (we hope) a composition of two integral operators and the inverse of a boundary integral operator. Here it is not obvious what happens if $h \to 0$. To prepare a statement of convergence for $\mathbf{Q}_h^+$, we consider the operator F_h^+. For simplicity we deal only with the case $F_h^+ f$ for $f \in C_H^{1,\beta}(\overline{G})$ in order to prove convergence with a certain rate.

Theorem 5.36. *Let $f \in C_H^{1,\beta}(\overline{G}) \cap \ker D(G)$. Then, we have*

$$\|f - F_h^+ f\|_{2,h,G} \leq Ch^\beta$$

where the constant C does not depend on h.

Proof. BOREL–POMPEIU's formula yields

$$f(x) - (F_h^+ f)(x) = (T_h^+ D_h^+ f)(x) = (T_h^+ (D_h^+ f - Df))(x).$$

In addition we have

$$|(D_h^+ f)(x) - (Df)(x)| \leq C\|f\|_{C_H^{1,\beta}} h^\beta$$

and, therefore,

$$||f - F_h^+ f||_{2,h,G} \leq ||T_h^+||_{[C,L_{2,h}]}||D_h^+ f - Df||_{2,h,G} \leq C||f||_{C_H^{1,\beta}} h^\beta. \qquad \#$$

Theorem 5.37. *Let* $f \in C_H^{1,\beta}(\overline{G})$. *Then,*

$$||\mathbf{Q}_h^+ f - \mathbf{Q}f||_{2,h,G} \leq Ch^\beta 0\langle \beta \leq 1.$$

Proof. We have

$$\mathbf{Q}_h^+ f - \mathbf{Q}f = \mathbf{Q}_h^+ Pf + \mathbf{Q}_h^+ \mathbf{Q}f - \mathbf{Q}\mathbf{Q}f.$$

On the other hand,

$$\mathbf{Q}_h^+ Pf = \mathbf{Q}_h^+ FPf = \mathbf{Q}_h^+ FPf - \mathbf{Q}_h^+ F_h^+ Pf = \mathbf{Q}_h^+ (F - F_h^+)Pf.$$

It follows that

$$||\mathbf{Q}_h^+ Pf|| \leq Ch^\beta.$$

Using $\mathbf{Q}f = Dg$ with $g \in W_2^1(G) \cap C_H^{2,\beta}(\overline{G})$ we obtain

$$\begin{aligned} \mathbf{Q}_h^+ Dg - Dg &= \mathbf{Q}_h^+ (Dg - D_h^- g) + \mathbf{Q}_h^+ D_h^- g - Dg = \\ &= \mathbf{Q}_h^+ (Dg - D_h^- g) + (D_h^- g - Dg) \end{aligned}$$

because $D_h^- g$ im $\mathbf{Q}_h^+$. Hence,

$$||\mathbf{Q}_h^+ \mathbf{Q}f - \mathbf{Q}\mathbf{Q}f||_{2,h,G} \leq C_2 h^\beta. \qquad \#$$

In the case $f \in C_H^{0,\beta}(\overline{G})$ we can prove a weaker result.

Theorem 5.38.

$$||\mathbf{Q}_h^+ f - \mathbf{Q}f||_{2,h,G} \to 0 \quad \textit{for } h \to 0 \quad \textit{if } f \in C_H^{0,\beta}(\overline{G}).$$

Proof. Let $f \in C_H^{0,\beta}(\overline{G})$ and $\varepsilon > 0$. Then

$$||\mathbf{Q}_h^+ f - \mathbf{Q}f||_{2,h} \leq ||\mathbf{Q}_h^+ f - \mathbf{Q}_h^+ f^*||_{2,h} + ||\mathbf{Q}_h^+ f^* - \mathbf{Q}f^*||_{2,h} + ||\mathbf{Q}f^* - \mathbf{Q}f||_{2,h}$$

holds, where

$$f^* \in C_H^{1,\beta}(\overline{G}) \quad \text{with} \quad ||f^* - f||_{C_H^{0,\beta}} < \varepsilon/4.$$

Furthermore, we obtain

$$||\mathbf{Q}_h^+ (f - f^*)||_{2,h,G} \leq ||f - f^*|_{2,h,G} < \varepsilon/4.$$

Because of Theorem 5.37. we have for all $h > 0$ that

$$||\mathbf{Q}_h^+ f^* - \mathbf{Q}f^*||_{2,h,G} \leq C(f^*)h^\beta$$

and

$$\|\mathbf{Q}(f^* - f)\|_{2,h,G} \leq 2\|\mathbf{Q}(f^* - f)\|_{2,G} \leq \varepsilon/2.$$

The last estimate holds for $h < h_1$ because of the RIEMANN-integrability of the function $\mathbf{Q}(f^* - f)$. With $C(f^*)h^\beta < \varepsilon/4$ for $h < h_2$ we obtain

$$\|\mathbf{Q}_h^+ f - \mathbf{Q}f\|_{2,h,G} < \varepsilon \quad \text{for } h < \min\{h_1, h_2\}. \qquad \#$$

Consideration of the operator $\mathbf{Q}_h^+$ emphasizes the efficiency of this approach. In particular, the part $(\text{tr } T_h^- F_h^+)^{-1}$ of the projection $\mathbf{Q}_h^+$ cannot be mastered using other inadequate approximations of T_G and F_Γ, because the conditions of the invertibility in most of the cases are not fulfilled or not fulfilled uniformly.

5.4.2 Convergence of approximative solutions

The discrete boundary value problem from Theorem 5.14. may be interpreted by a scheme of finite differences. That means that in Section 5.3 the discrete function theory presented can be seen as a new approach to the construction and analytical investigation of finite difference methods. In a simple way a priori estimates, for instance implied by (5.1), allow the investigation of stability problems. It has been known in the long term that stability and approximation of a finite difference scheme imply the convergence of the scheme. Of course, this is true only for linear problems. That means for linear boundary value problems our considerations are complete. The approximation of the finite difference operators can be proved by standard methods (TAYLOR-expansions, etc.) and the stability will be obtained by the help of our representation formulas. As we have seen above, all the operator norms are bounded independently of h and therefore, there are no further problems. But, even more has been proved above. We have got not only the boundedness of the operator sequences, but also the convergence in corresponding spaces. These convergence results will now be used to demonstrate that the approach to finite difference schemes presented here works for non-linear problems, too, where the "approximation + stability=convergence"-principle fails. For example, we shall deal with DIRICHLET's problem for the NAVIER–STOKES equations. Considering the results for the STOKES problem obtained in Section 5.3, the non-linear term $M^*(u) = \rho/\eta(u, \text{grad})u$ remains to be discretized in a proper way. Define

$$M_h^{*,-}(u) := \frac{\rho}{\eta}(u, \text{grad}_h^-)u \quad \text{and} \quad M_h^-(u) := M_h^{*,-}(u) - \frac{\rho}{\eta}f$$

Then, we have

$$\|\frac{\rho}{\eta}M_h^{*,-}(u)\|_{p,h}^p \leq \|\sum_{i,j=1}^{3} u_i D_{i,h}^- u_j e_j\|_{p,h}^p \leq \sum_{i,j=1}^{3} \|u_i D_{i,h}^- u_j\|_{p,h}^p$$

$$= \sum_{i,j=1}^{3} \sum_{y \in G_h} |u_i(y) D_{i,h}^{-} u_j(y)|^p h^3 \leq \sum_{i,j=1}^{3} ||u_i||_{q,h}^p ||D_{i,h}^{-} u_j||_{2,h}^p$$

$$\leq \sum_{i,j=1}^{3} C^p ||u_i||_{2,1,h}^p ||u_j||_{2,1,h}^p \leq 9C^p ||u||_{2,1,h}^{2p}$$

with $q < 6$, $p = 2q/(2+q)$. To obtain this estimate, we used Theorem 5.32. and HÖLDER's inequality for sums. Consider the non-linear boundary value problem

$$\begin{aligned} -\Delta_h u + \frac{1}{\eta}\mathrm{grad}_h^{+} p + \frac{\rho}{\eta}(u, \mathrm{grad}_h^{-})u &= \frac{\rho}{\eta} f \quad \text{in int } G_h \\ \mathrm{div}_h^{-} u &= 0 \quad \text{in int } G_h \\ u &= 0 \quad \text{on } \partial G_h \end{aligned} \tag{5.25}$$

Using the discrete operator calculus from Section 5.3 we obtain the following problem:

$$u = -T_h^{-} \mathbf{Q}_h^{+} T_h^{+} M_h^{-}(u) - \frac{1}{\eta} T_h^{-} \mathbf{Q}_h^{+} p \quad \text{in } G_h \tag{5.26}$$

$$\mathrm{Sc}\, \mathbf{Q}_h^{+} T_h^{+} M_h^{-}(u) = \frac{1}{\eta} \mathrm{Sc}\, \mathbf{Q}_h^{+} p \quad \text{in } G_h. \tag{5.27}$$

Lemma 5.39. *Problems (5.25) and (5.26,5.27) are equivalent.*

Proof. Let (u,p) be a solution of (5.26,5.27), then applying $-\Delta_h$ and $-\mathrm{Sc}\, D_h^{-}$ the equations (5.25) immediately follow. Because of the definition of $\mathbf{Q}_h^{+}$ the boundary condition is satisfied. Conversely, let (u,p) be a solution of the discrete boundary value problem (5.25). We have to apply T_h^{+}, the discrete BOREL–POMPEIU formula, $\mathbf{Q}_h^{+}$, $D_h^{-} u \in \operatorname{im} \mathbf{Q}_h^{+}, T_h^{-}$, and finally, BOREL–POMPEIU's formula again to (5.25). #

Remark 5.40. It should be mentioned that this proof is much easier than the corresponding proof in the continuous case. No weak formulation of the problem is necessary to obtain the result. Furthermore, there is no connection to regularity properties. That means we can split the theoretical problems and look for them step by step.

Consider the following iteration procedure (it is the same procedure as in the continuous case):

$$u_n = -T_h^{-} \mathbf{Q}_h^{+} T_h^{+} M_h^{-}(u_{n-1}) - \frac{1}{\eta} T_h^{-} \mathbf{Q}_h^{+} p_n \tag{5.28}$$

$$\mathrm{Sc}\, \mathbf{Q}_h^{+} T_h^{+} M_h^{-}(u_{n-1}) = \frac{1}{\eta} \mathrm{Sc}\, \mathbf{Q}_h^{+} p_n \tag{5.29}$$

$u_0 \in \overset{\circ}{W}{}^{1,-}_{2,h}(G_h) \cap \ker \operatorname{div}_h^-, n = 1, 2, 3, ...$

This defines a sequence of STOKES problems. We know from Theorem 5.14. that all these problems are uniquely solvable. Therefore, our sequence is well defined and the procedure goes through. Since all the norms of $T_h^-, T_h^+, \mathbf{Q}_h^+$, and M_h^- can be estimated in a similar manner as in the continuous case, the whole proof of convergence of iteration can be repeated analogously. For this reason we will omit the proof of the following theorem.

Theorem 5.41. *The non-linear system (5.25) has a unique solution* (u, p), *where*

$$u \in \overset{\circ}{W}{}^{1,-}_{2,h}(G_h) \cap \ker \operatorname{div}_h^-, p \in L_{2,h,G}$$

(p is uniquely defined up to a real constant) if

$$\frac{\rho}{\eta}||f||_{p,h,G} \leq (16K_h^2 C_{1,h})^{-1}. \tag{5.30}$$

For every function

$$u_0 \in \overset{\circ}{W}{}^{1,-}_{2,h}(G_h) \cap \ker \operatorname{div}_h^- \tag{5.31}$$

with $||u_0||_{2,1,h,G} \leq R_h$ *the procedure (5.28,5.29) converges in* $W^{1,-}_{2,h}(G_h) \times L_{2,h,G}$ *to the solution of the problem (5.26,5.27).*

The constants arising in the theorem are defined by

$$K_h = ||T_h^-||_{[L_{2,h,G} \cap \text{ im } \mathbf{Q}_h^+, \overset{\circ}{W}{}^{1,-}_{2,h}]} ||T_h^+||_{[L_{p,h,G}, L_{2,h,G}]},$$

$$C_{1,h} = 9^{\frac{1}{p}} C_h \frac{\rho}{\eta} \text{ (embedding constant)},$$

$$W_h = ((4K_h C_{1,h})^{-2} - \rho ||f||_{p,h,G} (\eta C_{1,h})^{-1})^{\frac{1}{2}},$$

$$R_h = (4K_h C_{1,h})^{-1}.$$

Looking at the proofs of Section 4.8 one can believe that the constants listed above arise if we repeat these considerations for lattice functions. We have already discussed that the norms of T_h^- and T_h^+ can be uniformly estimated with respect to h. To understand that the embedding constant $C_{1,h}$ is also bounded uniformly we can use the monotonicity of the eigenvalues $\lambda_{1,h}(G_h)$ in the case of the embedding in a larger domain (for instance, in a described cube). If f is RIEMANN-integrable, then $||f||_{L_{p,h}}(G)$ converges and W_h is uniformly bounded, too. This enables us to reformulate Theorem 5.41. in such a way that the right-hand sides contain only terms which do not depend on h.

Theorem 5.42. *Suppose that f is* RIEMANN-*integrable,*

$$\frac{\rho}{\eta}||f||_{p,G} < (16K^2C_1)^{-1},$$

$$||u_0||_{2,1,G} < (4KC_1)^{-1}. \tag{5.32}$$

Then, the system (5.26,5.27) has for $h < h_0$ a unique solution

$$u \in \overset{\circ}{W}{}^{1,-}_{2,h}(G_h) \cap \ker \operatorname{div}_h^-, \quad p \in L_{2,h,G}$$

(p is unique up to a real constant).
For every function $u_0 \in \overset{\circ}{W}{}^{1,-}_{2,h}(G_h) \cap \ker \operatorname{div}_h^-$ with (5.32) the iteration procedure (5.28,5.29) converges in $W^{1,-}_{2,h}(G_h) \times L_{2,h,G}$ to the solution of (5.26,5.27).

Proof. We have only to prove that $\lambda_{1,h}(G_h)$ converges to $\lambda_1(G)$ if h tends to zero. Then, using the integrability of $|x|^p, p > -3$ in bounded domains of $\mathbb{R}^3$ it follows that $K_h \to K, C_{1,h} \to C_1, W_h \to W, R_h \to R$ and our assertion follows. #

Remark 5.43. Comparing the assumptions of Theorem 4.94. and Theorem 5.42. we recognize that we have almost the same supposed conditions. Only some "$<$" are now replaced by "$\leq$". This very small lack of information during the discretization underlines again the advantages of the discrete operator calculus described. It should be mentioned again that we do not need detailed knowledge about the entries of the matrix representations which are related to our operators.

The following corollary is an immediate consequence of the fixed-point theorem.

Corollary 5.44. *Let u be the solution of (5.25). Then*

$$||u||_{2,1,h,G} \leq (4K_hC_{1,h})^{-1} - W_h.$$

Let $L_h = (4K_hC_{1,h})^{-1} - 4K_hC_{1,h}W$. Then we have

$$||u_n - u||_{2,1,h,G} \leq L_h^n||u_0 - u||_{2,1,h,G}.$$

If $u_0 = 0$, then

$$||u_n - u||_{2,1,h,G} \leq L_h^n\{(4K_hC_{1,h})^{-1} - W_h\}.$$

At this point we are able to solve linear STOKES problems. We can approximate for each fixed $h > 0$ the solution of the non-linear NAVIER–STOKES problem. Up to now we can prove all steps we carry out and we are able to describe the rate of convergence by the help of the fixed-point theorem. But, we are also interested in the behaviour of our solutions u_h if $h \to 0$. For this reason we need the following two theorems. We will see that for this part of the convergence results, knowledge of the approximation properties of $T_h^\pm, D_h^\pm, \mathbf{Q}_h^\pm, F_h^\pm$ is absolutely necessary.

Theorem 5.45. *Set* $v_h = Th - \mathbf{Q}_h^+ T_h^+ M_h^-(u) + u + \frac{1}{\eta} Th - \mathbf{Q}_h^+ p$.
If

$f \in L_{p,H}(G)$ *then* $v_h \to 0$ *for* $h \to 0$.

If

$f \in C_H^{0,\beta}(\overline{G}), 0 < \beta <, 1$ *then* $\|v_h\|_{W_{2,h}^{1,-}(G)} \leq C_1 h^\beta + C_2 h |\ln h|$.

Proof. Denote the solution of (5.28,5.29) for fixed h by u_h. We have to consider the behaviour of u_h for $h \to 0$. In the first step we estimate the approximation error by applying the discrete operators to the solution (u, p) of problem (5.25).

$$\|\mathbf{Q}_h^+ T_h^+ M_h^-(u) - \mathbf{Q} T M(u) + D_h^- u - Du + \frac{1}{\eta}\mathbf{Q}_h^+ p - \frac{1}{\eta}\mathbf{Q} p\|_{2,h,G}$$

$$\leq \|\mathbf{Q}_h^+ T_h^+ (M_h^{*,-}(u) - M^*(u))\|_{2,h,G} + \|\mathbf{Q}_h^+ (T_h^+ - T) M^*(u)\|_{2,h,G}$$

$$+ \|(\mathbf{Q}_h^+ - \mathbf{Q}) T M^*(u)\|_{2,h,G} + \|D_h^- u - Du\|_{2,h,G} + \frac{1}{\eta}\|\mathbf{Q}_h^+ p - \mathbf{Q} p\|_{2,h,G}$$

$$+ \frac{\rho}{\eta}\|\mathbf{Q}_h^+ T_h^+ f - \mathbf{Q}_h^+ T f\|_{2,h,G} + \frac{\rho}{\eta}\|(\mathbf{Q}_h^+ - \mathbf{Q}) T f\|_{2,h,G}$$

Assuming $f \in L_p, p > 3$ we obtain by the help of Theorem 34.5 that $u \in W_p^2$. From embedding theorems $M^*(u) \in C_H^{0,\beta}(\overline{G})$, $\beta < 1 - 3/p$ follows. In more detail, we find the following estimates for the above-mentioned norms. Because $u \in C_H^{1,\beta}(\overline{G})$ we have

$$\|D_h^- u - Du\|_{2,h,G} \leq C h^\beta, \beta \langle 1 - \frac{3}{p}$$

and

$$\|\mathbf{Q}_h^+ T_h^+ (M_h^{*,-}(u) - M^*(u))\|_{2,h} \leq C_1 \|M_h^{*,-}(u) - M^*(u)\|_{2,h} \leq C_2 h^\beta$$

for $\beta < 1 - 3/p$. Furthermore, we get from Theorems 5.35. and 5.37.

$$\|\mathbf{Q}_h^+ (T_h^+ - T) M^*(u)\|_{2,h,G} \leq C_3 h^\beta + C_4 h^{2 - \frac{3}{p}} |\ln h||, \beta < 1 - \frac{3}{p}$$

From $TM^*(u) \in C_H^{1,\beta}(G)$ and Theorem 5.37., we find

$$\|(\mathbf{Q}_h^+ - \mathbf{Q}) T M^*(u)\|_{2,h,G} \leq C_5 h^\beta, \beta < 1 - \frac{3}{p}$$

If $f \in L_{p,H}, p > 3$, then $Tf \in C_H^{0,\beta}(\overline{G})$ follows. In this case Theorem 5.38. yields

$$\|(\mathbf{Q}_h^+ - \mathbf{Q}) T f\|_{2,h,G} \to 0 \quad \text{for } h \to 0.$$

In connection with Theorem 5.35. and $||\mathbf{Q}_h^+|| = 1$ we get

$$||\mathbf{Q}_h^+ T_h^+ f - \mathbf{Q}_h^+ T f||_{2,h,G} \to 0 \quad \text{for } h \to 0.$$

Assuming that f belongs to $C_H^{\beta}(\overline{G})$ we obtain the estimates

$$||(\mathbf{Q}_h^+ - \mathbf{Q}) T f||_{2,h,G} \leq C h^{\beta}$$

$$||\mathbf{Q}_h^+ T_h^+ f - \mathbf{Q}_h^+ T f||_{2,h,G} \leq C_1 h |\ln h| + C_2 h^{\beta}.$$

Finally, from Theorem 5.38. and the corresponding regularity theorem we get

$$||\mathbf{Q}_h^+ p - \mathbf{Q} p||_{2,h,G} \to 0 \quad \text{for } h \to 0.$$

Assume $p \in C_H^{1,\beta}(\overline{G})$. Then it follows by using Theorem 5.37. that

$$||\mathbf{Q}_h^+ p - \mathbf{Q} p||_{2,h,G} \leq C h^{\beta}.$$

Summarizing all these estimates we arrive at

$$\mathbf{Q}_h^+ T_h^+ M_h^-(u) + D_h^- u + \frac{1}{\eta} \mathbf{Q}_h^+ p = \mathbf{Q} T M(u) + D u + \frac{1}{\eta} \mathbf{Q} p + w_h,$$

where

$$||w_h||_{2,h,G} \to 0 \quad \text{for } h \to 0 \text{ and } f \in L_{p,H}, p > 3$$

and

$$||w_h||_{2,h,G} \leq C_1 h^{\beta} + C_2 h |\ln h|| \quad \text{for } f \in C_H^{0,\beta}(\overline{G}), 0 \langle \beta \langle 1.$$

Recalling to mind that (u, p) is a solution of the problem (4.84,4.85) we find

$$\mathbf{Q}_h^+ T_h^+ M_h^-(u) + D_h^- u + \frac{1}{\eta} \mathbf{Q}_h^+ p = w_h.$$

Obviously, $w_h \in \operatorname{im} \mathbf{Q}_h^+$. Therefore, we can apply T_h^- to w_h and obtain an estimate of the approximation error. #

Our next aim is to find an error estimate in the space $W_{2,h,H}^{1,-}(G_h)$. We prove the following result.

Theorem 5.46. *Let $f \in L_\infty \cap \mathcal{R}_H$. Then*

$$||u - u_h||_{W_{2,h}^1(G)} \to 0 \quad \textit{for } h \to 0.$$

If $f \in C_H^{0,\beta}(\overline{G}), 0 \langle \beta \langle 1$ we even have

$$||u - u_h||_{W_{2,h}^1(G)} \leq C(h^{\beta} + h|\ln h|), h \langle h_0.$$

Proof. Theorem 5.45. and (5.32) yield

$$D_h^-(u-u_h)+\frac{1}{\eta}\mathbf{Q}_h^+(p-p_h)=\mathbf{Q}_h^+T_h^+(M_h^-(u)-M_h^-(u_h))+w_h. \tag{5.33}$$

Applying D_h^+ to (5.32) it then follows that Sc $D_h^+ w = 0$. Furthermore, we have $w_h = \mathbf{Q}_h^+T_h^+D_h^+w_h = \mathbf{Q}_h^+T_h^+g_h$.

Denote $M_h^-(u)-M_h^-(u_h)+g_h$ by f_h, then (5.33) can be written as

$$D_h^-(u-u_h)+\frac{1}{\eta}\mathbf{Q}_h^+(p-p_h)=\mathbf{Q}_h^+T_h^+f_h$$

where $(u-u_h, p-p_h)$ is the solution of a discrete STOKES problem. Lemma 5.12. and Theorem 5.13. yield the orthogonality of $D_h^-(u-u_h)$ and $\frac{1}{\eta}\mathbf{Q}_h^+(p-p_h)$. Scalar multiplication of this equation by $\mathbf{Q}_h^+(p-p_h)$ leads to

$$\frac{1}{\eta}||\mathbf{Q}_h^+(p-p_h)||_{2,h,G}^2=-[D_h^-(u-u_h),\mathbf{Q}_h^+(p-p_h)]$$

$$+[\mathbf{Q}_h^+T_h^+(M_h^-(u)-M_h^-(u_h)),\mathbf{Q}_h^+(p-p_h)]+[w_h,\mathbf{Q}_h^+(p-p_h)]\le$$

$$\le ||\mathbf{Q}_h^+T_h^+(M_h^-(u)-M_h^-(u_h))||_{2,h,G}+||w_h||_{2,h,G}||\mathbf{Q}_h^+(p-p_h)||_{2,h,G}$$

and, therefore,

$$\frac{1}{\eta}||\mathbf{Q}_h^+(p-p_h)||_{2,h,G}\le ||\mathbf{Q}_h^+T_h^+(M_h^-(u)-M_h^-(u_h))||_{2,h,G}+||w_h||_{2,h,G}.$$

The identity

$$u-u_h=-T\mathbf{Q}TM(u)-\frac{1}{\eta}T\mathbf{Q}p+T_h^-\mathbf{Q}_h^+T_h^+M_h^-(u_h)+\frac{1}{\eta}T_h^-\mathbf{Q}_h^+p_h$$

yields

$$||u-u_h||_{2,1,h}\le$$

$$\le ||T\mathbf{Q}TM(u)-T_h^-\mathbf{Q}_h^+T_h^+M_h^-(u)+\frac{1}{\eta}(T\mathbf{Q}p-T_h^-\mathbf{Q}_h^+p)||_{2,1,h}$$

$$+||T_h^-\mathbf{Q}_h^+T_h^+(M_h^-(u)-M_h^-(u_h))||_{2,1,h}+\frac{1}{\eta}||T_h^-\mathbf{Q}_h^+(p-p_h)||_{2,1,h}$$

$$\le ||v_h||_{2,1,h}+||T_h^-\mathbf{Q}_h^+T_h^+(M_h^-(u)-M_h^-(u_h))||_{2,1,h}$$

$$+\frac{1}{\eta}(1+\lambda_{1,h}^{-1})^{\frac{1}{2}}||\mathbf{Q}_h^+(p-p_h)||_{2,h}$$

$$\le ||v_h||_{2,1,h}+2(1+\lambda_{1,h}^{-1})^{\frac{1}{2}}||\mathbf{Q}_h^+T_h^+(M_h^-(u)-M_h^-(u_h))||_{2,h}$$

$$+(1+\lambda_{1,h}^{-1})^{\frac{1}{2}}||w_h||_{2,h}$$

$$\leq 2K_h||w_h||_{2,h} + +2K_hC_{1,h}||u-u_h||_{2,1,h}(||u||_{2,1,h} + ||u_h||_{2,1,h}).$$

With the condition $6/5 < p < 3/2$ it immediately follows that
$||u-u_h||_{2,1,h,G} \leq$

$$\leq 2K_h||w_h||_{2,h,G}\{1-2K_hC_{1,h}(||u||_{2,1,h,G} + ||u_h||_{2,1,h,G})\}^{-1}.$$

Supposing (5.30) and (5.31) we have

$$||u_h||_{2,1,h,G} \leq (4K_hC_{1,h})^{-1} - W_h$$

and for the function u Corollary 4.99. yields

$$||u||_{2,1,G} \leq (4KC_1)^{-1} - W.$$

These two inequalities, the possibility of a uniform estimate of K_h and $C_{1,h}$ and the RIEMANN-integrability of u and D_iu $(i = 1,2,3)$ ensure that for sufficiently small h

$$1-2K_hC_{1,h}(||u||_{2,1,h,G} + ||u_h||_{2,1,h,G}) > c > 0$$

uniformly holds with respect to h. Now we can describe the convergence $u_h \to u$ in dependence on the properties of w_h which we have already considered. #

5.5 Functional analytic theorems

In Section 2.1 spaces of $C\ell_{0,n}$-valued functions were introduced. With respect to our aim to show that essential parts of the $C\ell_{0,n}$-theory together with analytical and numerical applications can be developed analogously to the one-dimensional complex function theory we have to look for generalizations of the main theorems from functional analysis. So, we are mainly interested in generalizations of HAHN–BANACH's theorem and in a RIESZ representation theorem for functionals. Two remarks at least are necessary. Firstly, we mention that the method of investigation and the structure of the results seems to be clear. Our theorems will confirm this impression. But, besides all the advantages, we have to be careful because our function spaces are only right- or left-linear spaces, respectively. Secondly, the results strongly depend on the definition of the inner product in $L_2(G)$. We will use here only the CLIFFORD-valued inner product

$$(u,v)_{L_2(G)} = (u,v)_2 = \int_G \overline{u}(x)\, v(x)\, dx. \tag{5.34}$$

Another often investigated inner product is defined by Sc $(\cdot,\cdot)$. We refer to [36] for corresponding results. To show the principles of the consideration it is sufficient to handle only the case $n = 2$ $(C\ell_{0,2})$. Even in the complex case considered also some restriction seems to be necessary for the sake of a finite length of this section.

Theorem 5.47. (HAHN–BANACH's theorem). *Let X be a normed right-vector space over $\mathbb{H}$, $X_0 \subset X$ a closed subspace and f a bounded right-linear $\mathbb{H}$-valued functional. Then, there exists a right-linear extension F of f bounded on X.*

Proof. Let $X_{0,R}$ be a subset considered as a vector space over the field of real numbers. Denote by f_0 an R-valued bounded linear functional which is defined on X_0. This may be extended onto X_R and will be denoted by F_0. We define an $\mathbb{H}$-valued functional F by

$$F(x) = F_0(x)e_0 - F_0(xe_1)e_1 - F_0(xe_2)e_2 - F_0(xe_3)e_3. \tag{5.35}$$

Because of $|F(x)| \leq 4||F_0||||x||$ the functional F is bounded. Obviously, F is additive. A simple consideration of the terms $F_0(x)$, $F_0(xe_1)$, $F_0(xe_2)$, and $F_0(xe_3)$ leads to the verification of the homogeneity of F. Thus, F is a bounded linear functional. Similar to the proof in [145] for complex-valued functionals, we can prove that $||F|| = ||f||$. We only have to replace $sign f(z_0)$ by $\overline{F}(x_0)|F(x_0)|^{-1}$ and $z_1 = \text{sign}\,\overline{\text{f}}(\text{z}_0)\text{z}_0$ by $x_1 = x_0\overline{F}(x_0)|F(x_0)|^{-1}$. To prove that every bounded linear $\mathbb{H}$-valued functional can be written in the form (5.35), we start from the representation

$$f(x) = f_0(x)e_0 + f_1(x)e_1 + f_2(x)e_2 + f_3(x)e_3.$$

Using the relations $f(xe_i) = f(x)e_i$ $(i = 1, 2, 3)$, we find

$$\begin{aligned} f_0(x) &= f_1(xe_1) = f_2(xe_2) = f_3(xe_3) \\ f_1(x) &= -f_0(xe_1) = f_3(xe_2) = -f_2(xe_3) \\ f_2(x) &= -f_3(xe_1) = -f_0(xe_2) = f_1(xe_3) \\ f_3(x) &= f_2(xe_1) = -f_1(xe_2) = -f_0(xe_3). \end{aligned}$$

Hence, it follows that $f_i(x) = -f_0(xe_i)$, $i = 1, 2, 3$. The functional F constructed above is an extension of f. #

Remark 5.48. A corresponding theorem for locally convex spaces of $C\ell_{0,n}$-valued functions is contained in [36]. G.A. SUCHOMLINOV also proved a HAHN–BANACH-type theorem in quaternionic spaces [312].

Theorem 5.49. *(*RIESZ *theorem) Every bounded right-linear* $\mathbb{H}$*-valued functional* F *over* $L_p(G, \mathbb{H})$ *admits the representation*

$$F(u) - \int_G \overline{f}\, u\, dG, \ u \in L_p(G, \mathbb{H}) \tag{5.36}$$

with $f \in L_q(G, \mathbb{H})$, $1/p + 1/q = 1$.

For the proof, see [36]. #

In the case of a complex CLIFFORD algebra $\mathbb{C} \otimes C\ell_{0,n}$ F. KIPPIG proved in [149] the corresponding result. For convenience of the representation we will consider only to the theorem and the basic idea of the proof. We consider a bounded domain $U \subset \mathbb{C}^n$ and functions defined on U with values in $\mathbb{C} \otimes C\ell_{0,n}$.

Theorem 5.50. *Let* X *be a normed right-vector space over* $\mathbb{C} \otimes C\ell_{0,n}$, X_0 *a closed subspace of* X, *and* f *a bounded right-linear* $C\ell_{0,n}$*-valued functional. Then, there exists a right-linear bounded extension* F *of* f *with* $||F|| = ||f||$.

Proof. (Idea) We denote again by X_R, and $X_{0,R}$ real vector spaces. Let $X_{0,R}$ be a closed subspace of X_R. Each real-valued functional f_0^0 defined on $X_{0,R}$ can be extended to X_R by the help of the real version of HAHN–BANACH's theorem. This extension will be denoted by F_0^0. Then, we define a $C\ell_{0,n}$-valued functional by

$$\begin{aligned} F(\mathbf{z}) &= F_0^0(\mathbf{z})e_0 - F_0^0(\mathbf{z}e_1) - \dots - F_0^0(\mathbf{z}e_n) + \dots \\ &+(-1)^{\frac{(r+1)r}{2}} F_0^0(\mathbf{z}e_{1\dots r})e_{1\dots r} + \dots + (-1)^{\frac{(n+1)n}{2}} F_0^0(\mathbf{z}e_{1\dots n}) - \mathbf{i}F_0^0(\mathbf{iz}e_1)e_1 + \dots \\ &+\mathbf{i}F_0^0(\mathbf{iz}e_n)e_n - \dots - (-1)^{\frac{r+1)r}{2}} \mathbf{i}F_0^0(\mathbf{iz}e_{1\dots r})e_{1\dots r} - \dots \\ &-(-1)^{\frac{(n+1)n}{2}} \mathbf{i}F_0^0(\mathbf{iz}e_{1\dots n})e_{1\dots n}. \end{aligned} \tag{5.37}$$

Now, we have to prove that each bounded right-linear $C\ell_{0,n}$-valued functional can be represented in the form (5.37). Ideas of the real case can be used. But these computations are very time-consuming, and for further explanations it is sufficient to know the kind of representation. Therefore, the end of the proof will be omitted here. #

5.6 Special systems of $Cl_{0,n}$-regular functions

We will study in this section very special systems of regular functions. Looking at the aim of the book, we need some tools for theoretical investigations of our function spaces. One of the most essential properties is the separability of function spaces. With respect to the other tools, mainly the integral and differential operators of the calculus described above, we do not only need the answer "Yes, the space is separable". It is desired to know explicitly some countable sets of functions for the approximation of arbitrary elements from the function spaces. Finally, we will also study numerical procedures for the solution of boundary value problems. In this context we have to look for countable sets of special functions with a simple structure to keep the numerical computation small. The polynomial systems and generalizations of other elementary functions have been will investigated. We refer to Sections 2.1 and 2.3 and to [36]. But these functions are not adapted to requirements of numerical processes. If we take, for instance, polynomials then some difficulties occur. With increasing degree the costs for the evaluation of function values increase very fast. Polynomials are not adapted for approximation of regular functions in unbounded domains. Similar difficulties arise if we use trigonometric functions. It should be remarked again explicitly that we speak here about difficulties in practical handling of such systems.
We will consider here systems of functions defined by shifting the fundamental solution of the DIRAC operator. Independently from the number of one element of the system (they differ only in the location of the singularity) the evaluation requires the same computation. These functions can be used in bounded as well as in unbounded domains (particularly in the complement of a bounded domain). Last but not least, shifted fundamental solutions of the DIRAC operator (their images under certain mappings) allow physical interpretations.

We begin with the definition of $\mathbb{H}$-completeness and $\mathbb{H}$-closedness. These notions generalize known concepts from the classical real analysis.

Definition 5.51. *Let X be a normed right-vector space over $\mathbb{H}$. A system of points $\{x^{(i)}\}_{i\in\mathbb{N}} \subset X$ is called $\mathbb{H}$-complete if and only if every element $x \in X$ may be approximated arbitrarily closely by finite right-linear combinations of the elements $\{x^{(i)}\}_{i\in\mathbb{N}}$. $\{x^{(i)}\}_{i\in\mathbb{N}}$ is called closed in X if for every bounded right-linear functional F over X with values in $\mathbb{H}$ from $F(x^{(i)}) = 0,\ \forall\, i \in \mathbb{N}$ it follows that $F = 0$.*

Paying attention to the right linearity of X, the following corollary can be proved analogously to the real case.

Corollary 5.52. *Let X be a normed right-vector space over $\mathbb{H}$. The system $\{x^{(i)}\}_{i\in\mathbb{N}} \subset X$ is closed if and only if it is $\mathbb{H}$-complete in X.*

Now, we can prove a first result about $\mathbb{H}$-complete systems in $L_2(G,\mathbb{H})$.

Theorem 5.53. *Let G and G_e be domains with sufficiently smooth boundaries $\Gamma = \partial G$ and $\Gamma_e = \partial G_e$, respectively. Suppose that $\overline{G} \subset G_e$. Let $\{x^{(i)}\}_{i\in\mathbb{N}}$ be a dense subset of Γ_e,*

$$\phi_i(x) = \frac{x - x^{(i)}}{|x - x^{(i)}|^3}.$$

Then the system $\{\phi_i\}_{i\in\mathbb{N}} \subset L_2(G,\mathbb{H}) \cap \ker D$ is $\mathbb{H}$-complete in $L_2(G,\mathbb{H}) \cap \ker D$.

Proof. We assume that there exists a non-trivial function $f \in L_2 \cap \ker D(G) = \mathrm{IM}\,\mathbf{P}$ with the property that $(\phi_i, f)_2 = 0$ for all $i \in \mathbb{N}$. Comparing the definitions of the inner product and of the operator T_G, respectively, we recognize that $(T_G f)(x^{(i)}) = 0$ for all $i \in \mathbb{N}$. With respect to the smoothness of $T_G f$ in the complement of G we obtain $T_G f = 0$ on Γ_e. Now, using in sequel that $T_G f$ belongs to $\ker D(co\,\overline{G})$,

$$\lim_{|x|\to\infty} T_G f(x) = 0,$$

and the uniqueness of monogenic extensions we obtain $\mathrm{tr}_\Gamma T_G f = 0$. Taking into account the orthogonal decomposition of f we have $\mathrm{tr}_\Gamma T_G \mathbf{P} f + \mathrm{tr}_\Gamma T_G \mathbf{Q} f = 0$. Because $\mathrm{tr}_\Gamma T_G \mathbf{Q} f = 0$ for all $f \in L_2$ it remains that $\mathrm{tr}_\Gamma T_G \mathbf{P} f = 0$. $T_G \mathbf{P} f$ belongs to $\ker \Delta \cap W_2^1(G) \cap \ker \mathrm{tr}_\Gamma$. That means that $T_G \mathbf{P} f$ and then $\mathbf{P} f$ have to be zero. From our assumption we have that $\mathbf{P} f = f = 0$ and this contradiction proves our theorem. #

The idea used in the proof can be extended to the statement

$$\mathrm{tr}_\Gamma T_G f = 0 \Longleftrightarrow f \in \mathrm{im}\,\mathbf{Q} \cap L_2(G).$$

In this way we have "*en passant*" got a criterion to prove whether or not functions belong to $\mathrm{im}\mathbf{Q}$. Of course we are interested in completeness results in Sobolev spaces, too. Preparing such a theorem we prove the following lemma.

Lemma 5.54. *Let $G \subset \mathbb{R}^3$ be a bounded domain with sufficiently smooth boundary, which is star-shaped with respect to zero. Furthermore, we assume that $G_e \subset \mathbb{R}^3$ is a bounded domain with $\overline{G} \subset G_e$. If a function system*

$$\{h_i\}_{i\in\mathbb{N}} \subset W_2^k(G) \cap L_2(G_e) \cap \ker D(G_e)$$

is $\mathbb{H}$-complete in $L_2(G_e) \cap \ker D(G_e)$, then $\{h_i\}_{i\in\mathbb{N}}$ is $\mathbb{H}$-complete in $W_2^k(G) \cap \ker D(G)$.

Proof. Let

$$G_t = \{x \in \mathbb{R}^3 : tx \in G, t \in (t_0, 1)\}, g_t(x) = f(tx),$$

where $t_0 = \inf\{t : G_t \subset G_e\}$ and $f \in W_2^k(G) \cap \ker D(G)$. We verify that

$$W_2^k(G) - \operatorname{clos}\Big(\bigcup_{t\in(t_0,1)} W_2^k(G_t) \cap \ker D(G_t)\Big) = W_2^k(G) \cap \ker D(G) \tag{5.38}$$

First, let $f \in W_2^k(G) \cap \ker D(G)$, then it follows that $g_t \in W_2^k(G_t) \cap \ker D(G_t)$. In addition

$$\lim_{t\to 1} ||g_t||_{W_2^k(G)} = ||f||_{W_2^k(G)}.$$

Therefore, it is possible to chose a subsequence $\{g_{t'}\}$ of $\{g_t\}$ such that $\{g_{t'}\}$ weakly converges to a certain $\mathbb{H}$-valued function g if t' tends to 1. Consequently, $\lim_{t'\to 1} g_{t'}(x) = g(x)$ for any $x \in G$. Obviously,

$$\lim_{t\to 1} g_t(x) = \lim_{t\to 1} f(tx) = f(x)$$

holds. Hence, $g(x) = f(x)$ for all $x \in G$. Now we obtain

$$W_2^k(G) - \lim_{t'\to 1} g_{t'} = f \tag{5.39}$$

and the identity (5.38) is shown. Using similar ideas as in the proof of Theorem 5.53. we can prove that the system $\{h_i\}_{i\in\mathbb{N}}$ is also $\mathbb{H}$-complete in $L_2(G_t) \cap \ker D(G_t)$ for all $t \in (t_0, 1)$. Monogenic functions are harmonic and therefore the L_2-convergence implies the uniform convergence of all derivatives in compact subsets of G_t. Hence, we have

$$\operatorname{tr}_G g_t \in C^k(\overline{G}) - \operatorname{clos}(\operatorname{span}\{h_i\}_{i\in\mathbb{N}}),$$

and

$$\operatorname{tr}_G g_t \in W_2^k(\overline{G}) - \operatorname{clos}(\operatorname{span}\{h_i\}_{i\in\mathbb{N}}).$$

Here, tr_G stands for the restriction operator to the domain G. From (5.39) it now follows that $f \in W_2^k(G) - \operatorname{clos}[\operatorname{span}\{h_i\}_{i\in\mathbb{N}}]$. #
Using the lemma the following theorem immediately follows.

Theorem 5.55. *Let $G, G_e, \Gamma, \Gamma_e, \{x^{(i)}\}_{i\in\mathbb{N}}, \{\phi_i\}_{i\in\mathbb{N}}$ be defined as in Theorem 5.53. Then the system $\{\phi_i\}_{i\in\mathbb{N}} \subset W_2^k(G) \cap \ker D(G)$ is $\mathbb{H}$-complete in $W_2^k(G) \cap \ker D(G)$ for arbitrary $k \in \mathbb{N}$.*

Theorem 5.56. *Let $G, G_e, \Gamma, \Gamma_e, \{x^{(i)}\}_{i\in\mathbb{N}}, \{\phi_i\}_{i\in\mathbb{N}}$ be defined as in Theorem 5.53. Then the system $\{\operatorname{tr}_\Gamma \phi_i\}_{i\in\mathbb{N}}$ is $\mathbb{H}$-complete in $W_2^{k-\frac{1}{2}}(\Gamma) \cap \operatorname{im} P_\Gamma$ for arbitrary $k \in \mathbb{N}$.*

Proof. The theorem is a consequence of the well-known theorems about traces (see e.g. [330]) and Theorem 5.55.. #

Remark 5.57. The importance of Theorem 5.55. and Theorem 5.56. consists in the possibility of applying the same basis of approximation or interpolation independently of the smoothness of the given function (compare approximations with splines!).

Using similar ideas the whole thing can be repeated for basic functions with singularities in G. Then we obtain the following result for the traces of such functions.

Theorem 5.58. *Let G, G_i be bounded star-shaped domains with the boundaries Γ and γ_i, respectively, and let $\overline{G}_i \subset G$. Let $\{y_i\}_{i\in\mathbb{N}}$ be a dense subset of Γ_i and ψ_i defined by*

$$\psi_i(x) = \frac{x - y^{(i)}}{|x - y^{(i)}|^3}.$$

Then the system $\mathrm{tr}_\Gamma\{\psi_i\}_{i\in\mathbb{N}}$ is $\mathbb{H}$-complete in $W_2^{k-\frac{1}{2}}(\Gamma) \cap \mathrm{im}\, Q_\Gamma$ for every $k \in \mathbb{N}$.

Both theorems about approximations on the boundary ensure that arbitrary functions on the boundary can be approximated. That is because $\mathrm{im}\, P_\Gamma \cap W_2^s(\Gamma) \oplus \mathrm{im}\, Q_\Gamma \cap W_2^s(\Gamma) = W_2^s(\Gamma)$.

5.7 A Gram-Schmidt procedure

The results of Sections 5.6 and 5.7 enable us to approximate functions from $\ker D \cap L_2(G, C\ell_{0,n})$ by the help of certain complete function systems. If we look to the set of polynomials then we have to remark that polynomials are well adapted to use in series expansions (TAYLOR series, FOURIER series). Unfortunately, in the higher-dimensional case, polynomials are not convenient for use in interpolation formulas. The reason is the more complicated structure of the set of zeros. On the other hand, if we use polynomials, then the computations of basic functions with higher powers will be very time-consuming. A way out of this is to consider other systems of basic functions.

In this way we will investigate systems produced by shifting the fundamental solutions of the CAUCHY–FUETER and DIRAC operator, respectively. The idea to approximate functions from the kernel of a partial differential operator by the help of shifted fundamental solutions is not new. In 1960 F.E. BROWDER proved that this approximation is possible for a large class of (scalar) elliptic operators. In [36] systems of shifted fundamental solutions or of derivatives of fundamental solutions were investigated in the sense of convergence on compact subsets. We have seen in Section 5.6 that the completeness can be also proved with respect to the usual $L_2(G, C\ell_{0,n})$-norm.

Therefore, the necessary condition for the interpolation problem is satisfied. An advantage for the practical use of

$$\left\{ \frac{x - x_i}{|x - x_i|^3} \right\}_{i \in N}$$

as interpolation basis is that the length (and accuracy) of computations do not depend on the index of the basis function.

5.7.1 Minimal systems of Clifford regular functions

Looking to the results of some numerical experiments one can see that there remain some difficulties. One problem consists in the fact that the system matrix of the interpolation problem becomes almost singular if the number of used basic functions increases (the rows of the matrix are then almost parallel vectors). The idea is now to find a way between "doing nothing" and "orthogonalization", because the very bad condition number in the first case cannot be accepted and the numerical computation for the orthogonalization procedure are too complex (integrals over complicated domains in $\mathbb{R}^n$). We refer here to MICHLIN's [200] idea of so-called minimal systems.

Definition 5.59. *Let $\{h_i\}_{i\in N}$ be a system in a normed right-vector space X. If for each $j \in N$ $h_j \notin$ clos (span $\{h_i\}_{i\in N, i\neq j}$) then the system $\{h_i\}_{i\in N}$ is called minimal in X.*

Remark 5.60. The function system

$$\{\varphi_i\}_{i\in N} = \left\{ \frac{x - x^{(i)}}{|x - x^{(i)}|^{n+1}} \right\}_{i\in N},$$

where the set of points $\{x^{(i)}\}_{i\in \mathbf{N}}$ is dense in $\Gamma_\alpha \subset \mathbb{R}^{n+1} \setminus \overline{G}$, is complete in $L_2 \cap \ker D(G)$ but not minimal. If we omit a finite number of points $\{x^{(i_k)}\}_{k=1,\ldots,n} \subset \{x^{(i)}\}_{i\in N}$ then the density of $\{x^{(i)}\}_{i\in N\setminus\{i_1,\ldots,i_n\}}$ remains true and, therefore, the closures of $\{\varphi_i\}_{i\in N}$ and of $\{\varphi_i\}_{i\in N\setminus\{i_1,\ldots,i_n\}}$ coincide.

It should be mentioned here without proof (see, e.g. [200]) that the system matrices of our interpolation problems possess the property that the smallest eigenvalue $\lambda_1^{(n)}$ does not converge to zero (but the greatest eigenvalue $\lambda_n^{(n)}$ can tend to infinity).
Now, the problem arises how a given non-minimal system can be transformed to a minimal system. Of course, each orthogonal system is minimal, too. But we will look for algorithms which allow us to avoid the evaluation of inner products. Preparing this consideration, we have to introduce the notion of unisolvence.

Definition 5.61. *Let X be a normed right-vector space of $C\ell_{0,n}$-valued functions, $\{y^{(i)}\}_{i=1}^n \subset G$, $G \subset \mathbb{R}^{n+1}$, and $\{h_i\}_{i=1}^n \subset X$. The system $\{h_i\}_{i=1}^n$ is called unisolvent with respect to $\{y^{(i)}\}_{i=1}^n$ if from $\sum_{i=1}^n h_i(y)^{(k)} c_i = 0$ for $k = 1, ..., n$ it follows that $c_1 = c_2 = ... = c_n = 0$.*

This notion generalizes well-known properties of real or complex-valued polynomials in the one-dimensional case, and the notion of so-called CHEBYSHEV systems. In practical applications it is not easy to prove this property (in advance) for general sets $\{y^{(i)}\}_{i=1}^n$. But, assuming this property, we will describe an algorithm for minimalizing function systems.

This algorithm is useful for the proof of unisolvence, too, because it stops with a division by zero in the case of a non-unisolvent system (replacing zero by some small ε the criterion can be adapted to the, "real-life" situation).

5.7.2 A special Gram–Schmidt procedure

As mentioned above, our aim is to transfer complete systems to minimal systems avoiding inner products. Let $\{h_i\}_{i\in \mathbf{N}} \subset X$ and $\{y^{(i)}\}_{i\in N} \subset G$. We consider the following attachment (motivated by the known GRAM–SCHMIDT algorithm)

$$g_1 = h_1$$

$$g_2 = h_2 - a_{12}$$

$$\vdots$$

$$g_n = h_n - g_1 a_{1n} - \ldots - g_{n-1} a_{n-1,n},$$

where $a_{ik} \in C\ell_{0,n}$. To compute the coefficients a_{ik} we look for the values of our functions at the points $\{y^{(i)}\}$ and solve n systems of algebraic equations.

$$g_1(y^{(1)})a_{12} = h_2(y^{(1)}) \quad \text{for} \quad g_2$$

$$\ldots$$

$$\begin{bmatrix} g_1(y^{(1)}) & 0 & \ldots & 0 \\ g_1(y^{(2)}) & g_2(y^{(2)}) & \ldots & 0 \\ \ldots & \ldots & \ldots & \ldots \\ g_1(y^{(n)}) & g_2(y^{(n)}) & \ldots & g_n(y^{(n)}) \end{bmatrix} \begin{bmatrix} a_{1,n+1} \\ a_{2,n+1} \\ \ldots \\ a_{n,n+1} \end{bmatrix} = \begin{bmatrix} h_{n+1}(y^{(1)}) \\ \ldots \\ \ldots \\ h_{n+1}(y^{(n)}) \end{bmatrix} \quad \text{for } g_{n+1}.$$

We have to prove that the system $\{g_i\}_{i \in \mathbb{N}}$ is a minimal system and we have to ensure that $g_i(y^{(i)}) \neq 0 \ \forall\, i \in \mathbb{N}$ (the necessary and sufficient condition for the solvability of the triangular systems).

If we assume the existence of $i_0 \in \mathbb{N}$ with $g_{i_0}(y^{(i_0)}) = 0$ then we have

$$g_{i_0}(y^{(i_0)}) = h_{i_0}(y^{(i_0)}) - \sum_{i=1}^{i_0-1} g_i(y^{(i_0)})a_{in} = 0. \tag{5.40}$$

When the algorithm runs over all functions h_i it implies the property

$$g_i(y^{(j)}) = 0 \quad (\forall\, j < i).$$

Together with equation (5.40) the latter relation contradicts the unisolvence of the system $\{h_i\}_{i \in \mathbb{N}}$. Here, we understand that the prescribed algorithm can also be used during the transformation for proving the unisolvence.

Theorem 5.62. *Let $X = L_2(G, C\ell_{0,n}) \cap \ker D$. We assume $\{h_i\}_{i \in N} \subset X$ to be complete in X, $\{y^{(i)}\}_{i \in N} \subset G$, and let $\{h_i\}_{i \in N}$ be a unisolvent system with respect to $\{y^{(i)}\}_{i=1}^n (\forall\, n \in N)$. Then, the system $\{g_i\}_{i \in N}$ is $C\ell_{0,n}$-complete and $C\ell_{0,n}$-minimal in X.*

Proof. We assume that $\{g_i\}_{i \in N}$ is non-minimal. Then, we have at least one $i_0 \in \mathbb{N}$ with

$$g_{i_0} \in \operatorname{clos}\{\operatorname{span}\{g_i\}_{i \in \mathbb{N} \setminus \{i_0\}}\}.$$

Hence, there exist coefficients $c_i^{(n)}$ with

$$g_i = \lim_{n\to\infty} \sum_{\substack{i=1 \\ i\neq i_0}}^{n} g_i c_i^{(n)}$$

From Proposition 3.73. we have the pointwise convergence

$$g_{i_0}(x) = \lim_{n\to\infty} \sum_{\substack{i=1 \\ i\neq i_0}}^{n} g_i(x) c_i^{(n)} \quad (\forall\, x \in G). \tag{5.41}$$

Looking to the definition of the functions g_i we have $g_i(y^{(1)}) = 0 \quad \forall\, i > 1$ und using (5.41) we obtain $\lim_{n\to\infty} c_1^{(n)} = 0$. Analogously, we prove $\lim_{n\to\infty} c_j^{(n)} = 0 \quad (\forall\, j < i_0)$. Finally,

$$g_{i_0}(y^{(i_0)}) = \lim_{n\to\infty} \sum_{\substack{i=1 \\ i\neq i_0}}^{n} g_i(y^{(i_0)}) c_i^{(n)} = \lim_{n\to\infty} \sum_{i=1}^{i_0-1} g_i(y^{(i_0)}) c_i^{(n)} = 0.$$

This contradicts the assumption of unisolvence. The proof of the $C\ell_{0,n}$-completeness is not difficult because from $(g_i, f)_X = 0 \quad (\forall\, i \in N)$ it immediately follows that $(h_i f) = 0\ (\forall\, i \in N)$. Having in mind numerical applications we have to mention that both the algorithm, and the solution of the interpolation problems based on $\{g_i\}$ and the nodes $\{y^{(i)}\}$ require only the resolution of triangular systems. The entries of the system matrices are simple function values.

5.8 Boundary collocation methods

The idea of boundary collocation methods is very old. Engineers and physicists often are interested in a special kind of (approximative) solutions to boundary value problems. They are interested in solutions which satisfy the differential equation at all points of the given domain G. Then it will be accepted that the boundary condition is not fulfilled (of course the hope is to have only a small error on the boundary). Solutions of this type have some advantages. All information from the model which is contained in the differential equation is retained in the approximative solution. For linear models the superposition principle holds and can be used for the construction of "better" solutions. Special elementary (or fundamental) problems allow a direct (physical, mechanical,...) interpretation. In this way the user is enabled to estimate the quality of numerical solutions from his knowledge of the model.
The idea of the method can be described as follows. Let A be an elliptic partial differential operator and R a boundary operator. We consider the boundary value problem

$$Au = 0 \quad \text{in } G$$

$$Ru = g \quad \text{on } \Gamma$$

We are looking for solutions u_n of $Au = 0$. For this reason we assume that a function system $\{\phi_i\}_{i\in\mathbf{N}} \subset \ker A$ is known. Then for arbitrarily chosen coefficients c_i all functions $u_n(x) = \sum_{i=1}^{n} \phi_i(x)c_i$ belong to the kernel of A. Depending on the special choice of c_i we hope to obtain acceptable errors on the boundary, i.e., we hope that $|Ru_n - g|$ is small. Then at least three problems are of some interest:

- The functions ϕ_i should be explicitly known and simple to compute at arbitrary points.
- The function system $\{\phi_i\}_{i\in\mathbf{N}}$ has to be a complete system. This is a necessary condition for convergence results (as for instance $u_n - u \to 0, Ru_n - g \to 0$ for $n \to \infty$.
- How the coefficients c_i have to be chosen for a fast convergent sequence of numerical solutions.

The first two problems are already solved if we look for the system of shifted fundamental solutions. But for practical purposes we also have to look for algorithms to obtain 'good' coefficients c_i.
There are two possibilities different in principle. The first idea is to look for the coefficients c_i of the best approximation to g in the subspace $X_n = \text{span}\{R\phi_1, ..., R\phi_n\}$ (see e.g., the well-known TREFFTZ method). If we work in adapted function spaces (L_p, W_p^k)

with strictly convex norms, then there is no problem in proving the existence and uniqueness of the best approximation, and also the completeness of $\{\phi_i\}$ is now sufficient for the convergence. To improve the numerical properties of this approach we have to repeat the investigations of 'completeness $\Rightarrow$ minimality $\Rightarrow$ strong minimality $\Rightarrow$ almost orthogonal systems $\Rightarrow$ orthogonal systems'. But one of the disadvantages here is that we cannot avoid the evaluation of complicated integrals (over higher-dimensional domains, hypersurfaces). In the most cases a second discretization level is necessary when the integrals cannot be computed exactly.
The second approach consists in finding the coefficients c_i via an interpolation problem. More exactly, we solve

$$\sum_{i=1}^{n}(R\phi_i(x_j))c_i = g(x_j) \quad (j = 1, ..., n), \quad \{x_j\}_{j=1,\ldots,n} \subset \Gamma$$

From the viewpoint of applications this idea has some advantages. The numerical method is very simple to use. We have only to distribute the interpolation nodes x_i (perhaps using our experience in dealing with the solution). Then we have to solve a linear system of algebraical equations. The entries of the system matrix $\left(R(\Phi_i)(x_j)\right)_{i,j=1}^{n}$ are simple values of our basic functions at the interpolation nodes.
Now we have to investigate three new problems:

- How do we choose the interpolation nodes?
- Does the solution of the interpolation problem exist?
- Can we prove the convergence of the sequence of interpolants? Of course the solution of the interpolation problem cannot be better than the best approximation. We have to find the relationships between best approximation and interpolation.

Unfortunately, not all of these problems have as yet been solved. Nevertheless, looking to the advantages and the practical importance we will describe this method here and what is known from the theoretical point of view.

5.8.1 Interpolation and best approximation

In the following we will look for connections between the best approximation of $f \in X$ in $X_n = \operatorname{span}\{\Phi_1, ..., \Phi_n\}$ and the solution of the generalized collocation problem

$$\sum_{i=1}^{n} \phi_i(y^{(j)})c_i = f(y^{(j)}) \quad (j = 1, ..., n). \tag{5.42}$$

Introducing the algebraic interpolation function

$$(\Pi_n f)(x) = \sum_{i=1}^{n} \phi_i(x) c_i, \tag{5.43}$$

(5.42) may be rewritten as

$$(\Pi_n f)(y^{(j)}) = f(y^{(j)}) \quad (j = 1, ..., n). \tag{5.44}$$

For the remainder of this section we will abbreviate $L_{2,H}(G) \cap \ker D(G)$ by X. Most of our considerations can be repeated analogously if X is a HILBERT space with reproducing kernel.

Theorem 5.63. *Let $\{\phi_i\}_{i\in\mathbb{N}} \subset X$ be a complete system in X with the property that*

$$\phi_i(y^{(j)}) = 0 \;\; \text{for } j < i, i \in \mathbb{N} \;\; \text{and} \;\; \phi_i(y^{(i)}) \neq 0 \;\; \text{for } i \in \mathbb{N}.$$

Let $P_n f = \sum_{i=1}^{n} \phi_i a_{i,n}$ be the best ($L_{2,H}$-) approximation of $f \in X$ in X_n. Then we have

$$\lim_{n\to\infty} a_{i,n} = c_i, \quad \forall\, i \in \mathbb{N},$$

where $(c_1, ..., c_n) \in H^n$ fulfils the collocation conditions (5.44).

Proof. The completeness of the system $\{\phi_i\}_{i\in\mathbb{N}} \subset X$ yields

$$f(x) = \lim_{n\to\infty} \sum_{i=1}^{n} \phi_i(x) a_{i,n}.$$

Replacing x by $y_1, y_2, ..., y_n$ we obtain

$$f(y) = \lim_{n\to\infty} g_1(y^{(1)}) a_{1,n} = g_1(y^{(1)}) c_1,$$

$$f(y) = \lim_{n\to\infty} (g_1(y^{(2)}) a_{1,n} + g_2(y^{(2)}) a_{2,n}) = g_1(y^{(2)}) c_1 + g_2(y^{(2)}) c_2$$

. . . .

$$f(y) = \lim_{n\to\infty} \sum_{k=1}^{n} g_k(y^{(n)}) a_{k,n} = g_1(y^{(n)}) c_1 + ... + g_n(y^{(n)}) c_n$$

Remark 5.64. Theorem 5.63. remains true if $P_n f$ is replaced by another sequence of approximations of f in X_n, which converges to f.

Theorem 5.65. *Let $\{h_i\}_{i\in\mathbb{N}}$ be an $\mathbb{H}$-complete orthonormal system in X and $\{y_i\}_{i\in\mathbb{N}} \subset G$ a total set of collocation points. Furthermore, let $\{h_i\}_{i=1}^{n}$ be $\mathbb{H}$-unisolvent with respect to $\{y_i\}_{i=1}^{n}, n \in \mathbb{N}$. Then we have*

$$\Pi_n f = P_n f + \sum_{k=n+1}^{\infty} \Pi_n h_k (h_k, f)_X, \tag{5.45}$$

where $\Pi_n f = \sum_{k=1}^{\infty} h_k c_{k,n}$, $c_{k,n} \in \mathbb{H}$.

Proof. f admits the generalized FOURIER expansion $f = \sum_{k=1}^{\infty} h_k (h_k, f)_X$ for every $f \in X$. Using the fact that for monogenic functions the L_2-convergence implies the uniform convergence on compact subsets of G we get

$$f(x) = \sum_{k=1}^{\infty} h_k(x)(h_k, f)_X \quad \text{for each} \quad f \in X,\ x \in G.$$

This implies

$$\Pi_n f = \sum_{k=1}^{\infty} \Pi_n h_k (h_k, f)_X$$

because $\Pi_n \in L(L_2(G, \mathbb{H}), L_2(G, \mathbb{H})$. Using the unisolvence of $\{h_i\}_{i=1}^{n}$ we obtain

$$\begin{aligned} \Pi_n f &= \sum_{k=1}^{n} \Pi_n h_k (h_k, f)_X + \sum_{k=n+1}^{\infty} \Pi_n h_k (h_k, f)_X \\ &= \sum_{k=1}^{n} h_k (h_k, f)_X + \sum_{k=n+1}^{\infty} \Pi_n h_k (h_k, f)_X . \end{aligned}$$

Hence the assertion follows. #

Corollary 5.66. *The conditions* $\Pi_n h_k = 0$ *for* $k > n,\ n \in \mathbb{N}$ *are necessary and sufficient to have*

$$\Pi_n f = P_n f \quad \forall\, n \in \mathbb{N}. \tag{5.46}$$

Remark 5.67. Obviously, $\Pi_n h_k = 0$ for $k > n, n \in \mathbb{N}$ means that $h_k(y^{(j)}) = 0$ for $j < k$ and $k \in \mathbb{N}$. Moreover, we mention that that the conditions for the equivalence of the described interpolation $\Pi_n f$ and the best $L_2 G, \mathbb{H}$-approximation $P_n f$ for every n coincide with the conditions for the location of zeros in our minimizing procedure from Section 5.7.

These conditions seem to be very natural at several positions in our study of special function systems. But it is not obvious that there exist orthonormal systems with the described set of zeros. Clearly, it is not difficult to obtain basic systems with the necessary distribution of zeros (e.g., by the minimalizing procedure). But the usual method to orthogonalize these systems will destroy the set of zeros (because it is an L_2-process). On the other hand, if we start with an orthonormal system and look for linear combinations with the desired distribution of zeros then the orthogonality will be destroyed. That means it is a serious problem to ensure at first the existence of such orthonormal systems and then to look for algorithms leading to such systems. Connected with these considerations is the notion of *optimal interpolation.*

Definition 5.68. *Let $\{\phi_i\}_{i\in\mathbb{N}}$ be an $\mathbb{H}$-complete system in X. Let $\{y^{(i)}\}_{i\in\mathbb{N}} \subset G$ and $X_n = \operatorname{span}\{g_1, ..., g_n\}$. If for all $f \in X$ and all $n \in \mathbb{N}$ the interpolant*

$$\Pi_n f = \sum_{i=1}^{n} \phi_i a_{i,n}, \quad a_{i,n} \in \mathbb{H}$$

with $(\Pi_n f)(y^{(i)}) = f(y^{(i)})$, $i = 1, ..., n$, and the best $L_2G, \mathbb{H}$-approximation, $P_n f$, of f in X_n coincide, then $\{\phi_i\}_{i\in\mathbb{N}}$ is called an optimal basis of interpolation with respect to $\{y^{(i)}\}_{i\in\mathbb{N}}$.

The existence problem for optimal bases is solved by the following theorem.

Theorem 5.69. *For all sets $\{y^{(j)}\}_{j\in\mathbb{N}} \subset G$ there exists an orthonormal system $\{\phi_i\}_{i\in\mathbb{N}} \subset L_2(G, \mathbb{H})(G) \cap \ker D(G)$ with $\phi_i(y^{(j)}) = 0$ for all $j < i$, $i = 2, 3, \ldots$, and $\phi_i(y^{(i)}) \neq 0$ for $i = 1, 2, 3, \ldots$. If $\{y^{(j)}\}_{j\in\mathbb{N}}$ is a total set in G, then the system $\{\phi_i\}_{i\in\mathbb{N}}$ is complete in $L_2(G, \mathbb{H}) \cap \ker D(G)$.*

Proof. For the proof we refer to [95]. #

We will mention here only that the proof is constructive. The main idea is to define the i-th basic function to be the solution of

$$\inf\{||u||, u(y^{(1)}) = 0, ..., u(y^{(i-1)}) = 0, u(y^{(i)}) = 1\}.$$

In some cases we can avoid the solution of (infinitely many) variational problems.

Theorem 5.70. *Let $\{y^{(j)}\}_{j\in\mathbb{N}} \subset co\overline{B}_1(0)$, $z^{(j)} = -y^{(j)}/|y^{(j)}|^2$ and*

$$(\Pi_n f)(x) = \sum_{i=1}^{n} e(x - y^{(i)}) a_i,$$

with $a_i \in \mathbb{H}$. Then,

$$(f - \Pi_n f, \overline{e}(x - y^{(j)}))_{L_2(S_1, \mathbf{H})} = 0, \quad \forall j \leq n$$

iff

$$f(y^{(j)}) = (\Pi_n f)(y^{(j)}), \quad \forall j \leq n.$$

Proof. Suppose that $(f - \Pi_n f, \overline{e}(x - y^{(j)}))_{L_2(S_1, \mathbf{H})} = 0, \quad j \leq n$. Then it follows that

$$\left(e(x - y^{(j)}), f - \Pi_n f\right)_{L_2(S_1, \mathbf{H})} = 0, \quad j \leq n$$

and

$$\left(\frac{1 - y^{(j)}\overline{x}}{|1 - y^{(j)}\overline{x}|^3} e(x), f - \Pi_n f\right)_{L_2(S_1)} = 0, \quad j \leq n.$$

Since $x \in B_1(0)$, $|x| = 1$, we have that

$$\left(\frac{z^{(j)^{-1}} - \overline{x}}{|z^{(j)^{-1}} - \overline{x}|^3} \frac{x}{|x|}, f - \Pi_n f \right)_{L_2(S_1)} = 0, \quad j \leq n.$$

Hence,

$$F_\Gamma(f - \Pi_n f)(z(j)^{-1}) = 0, \quad j \leq n.$$

Using $F_\Gamma u = u$ for $u \in \ker D(G)$ we finally get

$$f(z^{(j)^{-1}}) - (\Pi_n f)(z^{(j)^{-1}}) = 0, \quad j \leq n.$$

The consideration in the reverse order makes the proof complete. #

The statement of Theorem 5.70. remains true in the case of a ball $B_r(0)$, if we use the singularities $y^{(i)}$, $|y^{(i)}| > r$, and the collocation points $y^{(i)^{-1}} r^2$. We mention that Theorem 5.70. generalizes results of the complex function theory concerning the approximation and interpolation with rational functions [83], [326]. The theorem seems to be very restrictive because the domain has to be a ball. From the theoretical point of view the importance of this result is the possibility of proving the solvability of the interpolation problem in the chosen subspace if the interpolation nodes can be arranged as in the theorem required. On the other hand, for applications we should mention at least the papers [258] and [260] of J. Ryan who has studied the problem of conformal mappings intensively. The results are not as general as in the complex case but sufficient for many cases. In particular, for the whole space $\mathbb{R}^{n+1}$ we have no problems with the assumptions of the theorem.
We mention that in [95] it has been suggested that optimal bases can also be obtained by the help of power series.

Looking for convergence results we recognize that in case, of complete optimal bases there is no problem. If we have only complete orthonormal systems the following result can be obtained.

Theorem 5.71. *Let $\{\phi_i\}_{i \in \mathbb{N}}$ be a complete orthonormal system in $L_2(G, \mathbb{H}) \cap \ker D(G)$, $\{y^{(j)}\}_{j \in \mathbb{N}} \subset G$ be a total set and let n be defined by (5.43). Then for all $f \in L_2(G) \cap \ker D(G)$*

$$\lim_{n \to \infty} ||\Pi_n f - f||_{L_2(G)} = 0$$

holds.

We omit the proof (see e.g. [95]) because interpolation with orthonormal basic functions is not so interesting in practice. Having an orthonormal basis the best way is to use the best approximation.
In the following we return to the elementary system of shifted fundamental solutions and look for their properties.

Proposition 5.72. *Let $G, G_e, \Gamma, \Gamma_e, \{x^{(i)}\}_{i\in\mathbf{N}}, \{\phi_i\}_{i\in\mathbf{N}}$ be defined as in Theorem 5.53.. Then the function system $\{\phi_i\}_{i\in\mathbf{N}}$ is linear independent in G, and the system $\{tr_\Gamma\phi_i\}_{i\in\mathbf{N}}$ is linear independent on Γ.*

Proof. Clearly, $\{\phi_i\}_{i\in\mathbf{N}}$ is right-linear independent in G. We suppose that

$$\sum_{i=1}^{n} \mathrm{tr}\phi_i(x)c_i = 0, x \in \Gamma, c_i \in H, \sum_{i=1}^{n} |c_i|^2 \neq 0.$$

Using the maximum modulus theorem we get

$$\sum_{i=1}^{n} \phi_i(x)c_i = 0, \quad \forall\, x \in G,$$

and $c_i = 0$, $i = 1, ..., n$, immediately follows. #

Using the previous proposition we get

Proposition 5.73. *Let $\{x^{(i)}\}_{i\in\mathbf{N}} \subset \Gamma_a$, $\{y^{(i)}\}_{i\in\mathbf{N}} \subset G$ $(\{y^{(i)}\}_{i\in\mathbf{N}} \subset \Gamma)$ and $\{\phi_i\}_{i\in\mathbf{N}}$ $(\{\mathrm{tr}\phi_i\}_{i\in\mathbf{N}})$ be $\mathbb{H}$-unisolvent with respect to $\{y^{(i)}\}_{i=1}^{n}$. Then there exists a point $y^{(n+1)} \in G$ $(y^{(n+1)} \in \Gamma)$ such that $\{\phi_i\}_{i=1}^{n+1}$ $(\{\mathrm{tr}\ \phi_i\}_{i=1}^{n+1})$ is $\mathbb{H}$-unisolvent with respect to $\{y^{(i)}\}_{i=1}^{n+1}$.*

Proposition 5.73. is essential for the interpolation on Γ (the heart of the method). Similar results are obtained by ALEKZIDSE [6] for other functions. We remember that for polynomials of several variables surfaces exist that for an arbitrary choice of n points on such a surface these polynomials never form a unisolvent system. This case is excluded by the help of Proposition 5.73.. One case where the unisolvence of a given system is ensured describes the next proposition.

Proposition 5.74. *Let $G \subset G_e$, $\{x^{(i)}\}_{i=1}^{n} \subset \Gamma_e$, $\{y^{(i)}\}_{i=1}^{n} \subset \Gamma$, and $\alpha^{(i)}$ the outward-pointing normal unit vector to Γ at the point $y^{(i)}$, $x^{(i)} = y^{(i)} + t\alpha^{(i)}$, $i = 1, ..., n$, $t > 0$. Then a real number $\delta > 0$ exists with the property that the system $\{\phi_i\}_{i=1}^{n}$ with*

$$\phi_i = \frac{x - x^{(i)}}{|x - x^{(i)}|^3}$$

is $\mathbb{H}$-unisolvent with respect to $\{y^{(i)}\}_{i=1}^{n}$ for all $t \in (0, \delta)$.

We omit the proof here because, clearly, the unisolvence is caused by the diagonal dominance of the corresponding matrix of the interpolation problem. These functions are "local". That means we have to pay for the unisolvence with a lack of approximation properties. Recall to mind that theorem 5.70. also includes a result on the unisolvence.

5.8.2 Collocation Method for Several Elliptic BVP's

Let L be an elliptic differential operator of second order with constant coefficients. The following boundary value problem has to be solved numerically.

$$\begin{aligned} Lu &= 0 \quad \text{in } G \\ Ru &= g \quad \text{on } \Gamma \end{aligned} \tag{5.47}$$

Of course we could try to repeat all the above considerations. But having in mind the representation formulas for the solutions of several elliptic boundary value problems we will now describe, how the collocation method in $\ker D$ and collocation methods in $\ker L$ are connected.
Suppose that a continuous mapping M_L with

$$M_L \; : \; \ker D(G) \times \ker D(G) \to \ker L$$

is known in suitably chosen spaces. We will use our representation formulas for the following. In all cases we have a combination of two functions belonging to $\ker D$ (and some continuous operators) as description of the solution. These representations may be understood to be mappings of the kind described above. On account of the continuity then it follows that the system $\{M_L(h_i, h_j)\}_{i,j\in\mathbb{N}} \subset \ker L$ is complete in $\ker L$ if the system $\{h_i\}_{i\in\mathbb{N}} \subset \ker D(G)$ is complete in $\ker D(G)$. If we succeed in describing functions $M_L\{h_i, h_j\}$ by simple analytic terms, then all constructions made above may be renewed in $\ker L, R(\ker L)$, respectively. This problem has been stated in [110]. We will repeat here only the considerations concerning boundary value problems of the Laplace equation. Looking to the necessary steps we will see that one of the main problems is the explicit description of systems $\{T_G h_i\}_{i\in\mathbb{N}}$ for special systems $\{h_i\}_{i\in\mathbb{N}}$. Therefore, in [110] it was necessary to study all the problems and their representation formulas again. Nowadays, tabulars of T-correspondences are available and the theoretical investigations may be stretched. For a large class of functions such correspondences are included in Appendix A.

Theorem 5.75. *Let $G \subset \mathbb{R}^3$ be a sufficiently smooth bounded domain, $\{x^{(i)}\}_{i\in\mathbb{N}} \subset \Gamma_e$, $\Gamma_e \subset co\overline{G}$, $x \in G$. Then,*

$$T_G\left(\sum_{k=1}^{3} \frac{x_k - x_k^{(i)}}{|x - x^{(i)}|^3} e_k\right) = -\frac{1}{|x - x^{(i)}|} + \Phi_{G,i}, \quad \Phi_{G,i} \in \ker D(G)$$

$$T_G\left(\frac{x_k - x_k^{(i)}}{|x - x^{(i)}|^3} e_k\right) = \frac{1}{2}\sum_{k=1}^{3} \frac{(x_k - x_k^{(i)})(x_k^{(i)} - x_k)}{|x - x^{(i)}|^3} e_k - \frac{1}{2|x - x^{(i)}|} + \Psi_{G,k},$$

with $\Psi_{G,k} \in \ker D(G)$

Proof. Application of D proves the theorem by a straightforward computation.
#

Theorem 5.76. *Let $G \subset \mathbb{R}^3$ be a bounded star-shaped domain with sufficiently smooth boundary Γ, $G_e \subset \mathbb{R}^3$ with $G \subset G_e$, $\Gamma_e = \partial G_e$, and $\{x^{(i)}\} \subset \Gamma_e$ be a dense subset on Γ_e. Then, the system*

$$\left\{\sum_{k=1}^{3} \frac{x_k^{(i)} - x_k}{|x^{(i)} - x|^3} e_k\right\}_{i\in\mathbb{N}} \cup \left\{\frac{1}{|x^{(i)} - x|} e_0\right\}_{i\in\mathbb{N}}$$

is complete in $L_2(G, \mathbb{H}) \cap \ker \Delta(G)$.

Proof. Let $u \in L_2(G,\mathbb{H}) \cap \ker \Delta(G)$, $\varepsilon > 0$ arbitrarily chosen, and $k > 2$. Then there exist $\mathbb{H}$-valued functions $u_n \in W_2^k(G,\mathbb{H}) \cap \ker \Delta(G)$ with the property $u_n \to u$ in $L_2(G,\mathbb{H})$ for $n \to \infty$ (see Lemma 5.54.). Therefore, an index $n_0 \in \mathbb{N}$ with $\|u_n - u\|_{L_2} \leq \varepsilon/2$ exists. Now, we can use the representation formula

$$u_n = \Phi_1 + T_G \Phi_2, \quad \Phi_1, \Phi_2 \in \ker D(G).$$

Applying the completeness result in ker D, the continuity of T_G and the known images of the basic functions under the transformation T_G, we obtain that there exist coefficients with the property that

$$\left\| u_{n_0} - \sum_{i=1}^{n_1} \left(- \frac{1}{|x^{(i)} - x|} e_0 a_{i,n_1} \right) - \sum_{i=1}^{n_2} \sum_{k=1}^{3} \frac{x_k - x_k^{(i)}}{|x - x^{(i)}|^3} e_k b_{i,n_2} \right\| \leq \varepsilon/2$$

and this proves the statement. #

Let us mention that we need the approximating sequence $\{u_n\}$ for technical reasons because our representation formulas require that we have a harmonic W_2^1-function. The result from Theorem 5.76. can be generalized to SOBOLEV spaces. If we assume that $\operatorname{dist}(\Gamma, \Gamma_e) > 0$ then the function system described above is complete in the L_2-sense in a larger domain G_ε ($G \subset G_\varepsilon \subset G_e$), too. Following BROWDER [37] we obtain the following result.

Theorem 5.77. *Under the suppositions of Theorem 5.76. the system*

$$\left\{\sum_{k=1}^{3} \frac{x_k^{(i)} - x_k}{|x^{(i)} - x|^3} e_k\right\}_{i\in\mathbb{N}} \cup \left\{\frac{1}{|x^{(i)} - x|} e_0\right\}_{i\in\mathbb{N}}$$

is complete in $W_2^k(G) \cap \ker \Delta(G)$ for arbitrary $k \in \mathbb{N}$.

Now, using Theorem 5.77. and the theorems on traces, we can prove corresponding completeness results on the boundary.

Theorem 5.78. *Under the suppositions of Theorem 5.76. the system*

$$\left\{\operatorname{tr} \sum_{k=1}^{3} \frac{x_k^{(i)} - x_k}{|x^{(i)} - x|^3} e_k\right\}_{i\in\mathbb{N}} \cup \left\{\operatorname{tr} \frac{1}{|x^{(i)} - x|} e_0\right\}_{i\in\mathbb{N}}$$

is complete in $W_2^k(\Gamma)$, $k > 3/2$, *and the system*

$$\left\{ \mathrm{tr} \sum_{k=1}^{3} \frac{\partial}{\partial \alpha} \frac{x_k^{(i)} - x_k}{|x^{(i)} - x|^3} e_k \right\}_{i \in \mathbb{N}} \cup \left\{ \mathrm{tr} \frac{\partial}{\partial \alpha} \frac{1}{|x^{(i)} - x|} e_0 \right\}_{i \in \mathbb{N}}$$

is complete in $W_2^k(\Gamma, \mathbb{H})/span\{1\}$ *for* $k > 3/2$.

The second part of the theorem ensures that with collocation methods based on the function systems described, Neumann boundary conditions can also be treated. This is true from the numerical point of view. Analytical considerations (of NEUMANN problems) using hypercomplex function theoretic methods are restricted to some special cases (see Section 4.4 and [333], [339]).
Analogously, all the other boundary value problems where representation formulas of the solutions are available can be treated. We will not discuss these results in detail and refer to [110] for a list of examples.

Chapter 6

Further results and research problems

6.1 Parabolic problems

In this section we will show how parabolic problems can be dealt with hypercomplex methods and for this reason we consider some typical problems. A physical process will take place within a period a. By the method of semi-discretization we will have a sequence of disturbed elliptic problems. The aim of this section is to describe our method in the case of the most simple situation.

6.1.1 A time-discrete method

We have the following standard problem:

Let $[0,a]$ be a segment on the time axis and G a bounded domain with a piecewise LIAPUNOV *boundary. Then we consider the equation:*

$$\partial_t u - \Delta u = f \quad in \quad G \times (0,a]. \tag{6.1}$$

Furthermore, we have an initial value condition:

$$u(x,0) = u_0(x) \quad on \quad G \tag{6.2}$$

and the boundary condition:

$$u(x,t) = g(x) \quad on \quad \partial G = \Gamma. \tag{6.3}$$

Now, we substitute the time-derivative $(\partial_t u)(x,t)$ at $(k-1)h$ by

$$\frac{u(x,kh) - u(x_1(k-1)h)}{h},$$

where $h = a/N$ is the mesh width ($N \in \mathbb{N}$). We abbreviate $u(x,kh) = u_k(x)$ $(k = 0,...,N)$. Then we get the discretized equation

$$\frac{u_k}{h} + DDu_k = f_k + \frac{u_{k-1}}{h}.$$

Here D describes again the DIRAC operator $D = \sum_{j=1}^{3} e_j \partial_j$. The function f_k is defined by

$$f_k(x) = \frac{1}{h} \int_k^{k+1} f(x,\tau) d\tau \ .$$

With the denotation $\mu = h^{-\frac{1}{2}}$ we obtain the YUKAWA equation

$$(D - i\mu)(D + i\mu)u_k = f_k + \mu^2 u_{k-1} \quad (k = 0,...,N). \tag{6.4}$$

Note that $ie_j = e_j i$ $(j = 1,2,3)$. In order to solve this equation we have to introduce right inverses to the generalized VEKUA-type operators $D \pm i\mu$. The corresponding fundamental solutions read as follows:

$$e_{\pm i\mu}(x) = e(\underline{x})(1 + \mu|\underline{x}|)e^{-\mu|\underline{x}|} \pm i\mu \frac{e^{-\mu|\underline{x}|}}{4\pi|\underline{x}|},$$

such that we can define

$$(T_{\pm i\mu}u)(x) := -\int\limits_G e_{\pm i\mu}(y-x)u(y)dy \tag{6.5}$$

and

$$(F_{\pm i\mu}u)(x) := +\int\limits_\Gamma e_{\pm i\mu}(y-x)\alpha(y)u(y)d\Gamma_y, \tag{6.6}$$

where α is the outward pointed unit normal vector to Γ at y.
Now it provides the representation

$$u_k(x) = T_{-i\mu}\mathbf{Q}_{i\mu}T_{i\mu}(\mu^2 u_{k-1} + f_k)(x) + F_{-i\mu}g + T_{-i\mu}\mathbf{P}_{i\mu}(D-i\mu)h. \tag{6.7}$$

The operators $\mathbf{Q}_{i\mu}, \mathbf{P}_{i\mu}$ are in accordance with Section 4.3 orthoprojections of $L_2(G,\mathbb{C}\mathbb{H})$ onto the subspaces $(D-i\mu)\overset{\circ}{W}{}_2^1(G,\mathbb{C}\mathbb{H})$; $\ker(D+i\mu)\cap L_2(G,\mathbb{C}\mathbb{H})$, respectively. For more information we refer to [327].

Proposition 6.1. *The following estimates are valid:*

(i) $\|u_k\|_{L_2(G,\mathbb{C}\mathbf{H})} \le \|T_{-i\mu}\mathbf{Q}_{i\mu}T_{i\mu}\|_{[\mathrm{im}\mathbf{Q}_{-i\mu},\mathrm{im}\mathbf{Q}_{i\mu}]}\|f_k + \mu^2 u_{k-1}\|_{L_2(G,\mathbb{C}\mathbf{H})}$,

(ii) $\|u_k\|_{L_2(G,\mathbb{C}\mathbf{H})} - \|u_{k-1}\|_{L_2(G,\mathbb{C}\mathbf{H})} \le \frac{1}{\mu}\|f_k\|_{L_2(G,\mathbb{C}\mathbf{H})}$,

(iii) $\|u_N\|_{L_2(G,\mathbb{C}\mathbf{H})} \le \int\limits_0^T \|f(\cdot,\tau)\|_{L_2(G,\mathbb{C}\mathbf{H})}d\tau + \|u_0\|_{L_2(G,\mathbb{C}\mathbf{H})}$.

It remains to verify the convergence of the iteration procedure 6.6 in $W_2^1(G,\mathbb{C}\mathbb{H})$. For this reason we need estimations of the integral operators $T_{\pm i\mu}$.

6.1.2 An estimation of the operator $\mathbf{T}_{\pm i\mu}$

We start with an inequality of L. HEDBERG.

Lemma 6.2. *[123] Let Q be a n-dimensional cube. Further, let $u \in L_1(Q)$ and $\nu > 0$ then there exists a constant C independent of x and u such that*

$$\int\limits_Q \frac{|u(y)|}{|x-y|^{n-\alpha}}dy \le C(\operatorname{diam} Q)^\nu M_{\operatorname{diam} Q}|u|(x),$$

where $M_R u(x) = \sup_{r<R} \oint_{B_r(x)} |u(y)|dy$ which describes a version of the HARDY–LITTLEWOOD *maximal function. The integral $\oint$ is understood as the* STEKLOV *mean.*

Lemma 6.3. *Let $u \in C(G)$. Then the integral*

$$I(x) = \int_G \mu^k \frac{e^{-\mu|x-y|}}{|x-y|^{n-1}} u(y)dy \qquad (k \geq 2, \mu \in \mathbb{R}^+)$$

can be estimated by

$$|I(x)| \leq \left[2 \max_{\mu} \sup_{r<d} \int_{B_r(x)} \mu^k ||x| - |y|| e^{-\mu\left||x|-|y|\right|} dy + C_2\right] ||u||_{C(G)},$$

where d denotes the diameter of the smallest cube which contains G and $C_2 > 0$.

Proof. Our aim is to get an estimate which is uniformly relative to $\mu \in \mathbb{R}^+$. We split G into two subsets, namely

$$G_1 = G \cap \{y : |x-y| < 1\} \quad \text{and} \quad G_2 = G \cap \{y : |y-x| \geq 1\}.$$

Clearly, that the integral I over G_2 which will be denoted by I_2 has a uniform estimation of the form

$$|I_2| \leq C_2 ||u||_{C(G)}.$$

The corresponding integral I_1 over G_1 we have to split into two integrals over

$$G_{11} = \{G_1 \cap \{y : |y| < |x|\} \text{ and } G_{12} = G_1 \cap \{y : |y| \geq, |x|\}$$

which will be denoted by I_{11}, and I_{12}. Since $|x-y| \geq ||x| - |y||$ follows for I_{11} the estimate

$$|I_{11}| \leq \mu^k e^{-\mu|x|} \int_{B_r(x)} \frac{1}{|x-y|^{n-1}} u(y) e^{\mu|y|} dy.$$

Our function u will now be extended to the smallest cube which contains G. This continuation will be denoted by u_1.
In application of Lemma 6.2. we have

$$I_{11}| \leq \mu^k e^{-\mu|x|} \int_{B_r(x)} u_1(y) e^{\mu|y|} dy \leq 2 \sup_{r<d} \int_{B_r(x) \cap G_{11}} \mu^k e^{-||x|-|y||} u(y) dy.$$

Using L'Hospital's rule we get with $\mu \to \infty$ for the right-hand-side zero. Hence

$$I_{11}| \leq 2 \max_{\mu} \sup_{r<d} \int_{B_r(x) \cap G_{11}} \mu^k \, ||x| - |y|| \, e^{-\mu||x|-|y||} u(y) dy.$$

The estimation of I_{12} can be obtained similarly. Both inequalities together provide the result required. #

Theorem 6.4. *The iteration procedure (6.6) converges to the unique solution of the problem (6.1–6.3).*

Proof. We have only to state that

$$||T_{\pm i\mu}u||_{L_2(G,\mathbb{C}\mathbf{H})} \leq \frac{C|G|}{\mu^{k-1}}||u||_{L_2(G,\mathbb{C}\mathbf{H})} \qquad (C > 0;\ k > 1))$$

and the operator $\mathbf{Q}_{i\mu}$ as orthoprojection is bounded. #

Remark 6.5. This method can be transferred without problems to initial boundary value problems of equations of the type

$$\partial_t u - DbDu = f \quad \text{in} \quad G,$$

where b is a multiplication operator of the type as described in Section 3.5. Furthermore, we see good chances to treat time-dependent STOKES equations and also NAVIER–STOKES equations by this method.

6.2 On field equations

From the first tensors have been used extensively in electromagnetic theory. We recall the famous book of H. WEYL [329] in 1921. Much later appear fundamental works by E.J. POST [234] and G. KRON [162], who in 1959 used skew-symmetric tensors in order to investigate network systems. In 1957 F. BOLINDER could express MAXWELL equations (cf. [32]) in a complex 4D-form. We should also note that in 1912–1914 papers appear by L. SILBERSTEIN (cf. [270]) where complex quaternions were used to represent electromagnetic fields. More and more tensors were substituted by elements of suitable CLIFFORD algebras. We mention here the pioneer work of G. JUVET [142] also in [140] in 1930 as well as the thesis of M. MERCIER at the University of Gent. M. RIESZ [242] and D. HESTENES [125] defined the further development in the treatment of general field equations by CLIFFORD algebra methods. Combined with CLIFFORD analytic methods in 1979, P. LOUNESTO (cf. [176]) discussed in his thesis qualitative properties of spinor-valued regular functions. In this section we will try to give a minimun of information on recent work in this field without explaining the topics in detail. General gauge and chiral theories (see e.g. [115]) would exceed the purpose of the book

First let us recall the definitions of the dot-product and also the vector product for k-vectors in a CLIFFORD algebra.

Definition 6.6. *Let x be a k-vector and y be a ℓ-vector from $Cl_{p,q}$ then the Clifford product decomposes as follows:*

$$xy = (xy)_{k+\ell} + (xy)_{k+\ell-2} + \ldots + (xy)_{|k-\ell|},$$

where $(u)_m$ denotes the m-vector part of $u \in Cl_{p,q}$. Then the dot-product and the vector product are defined by

$$x \cdot y := (xy)_{|k-\ell|} \quad \textit{and} \quad x \wedge y := (xy)_{k+\ell}.$$

Let us now describe field equations and some of their remarkable properties.

6.2.1 Maxwell equations

In Section 4.5 we have considered under some conditions weakly time-dependent MAXWELL equations. Representations of the electric and magnetic field within the algebra of real quaternions could be obtained. Using now the complex MAJORANA algebra $Cl_{3,1}$ which can be introduced similarly to the real DIRAC algebra $Cl_{1,3}$ we intend to consider MAXWELL equations.

Remark 6.7. Note that D. HESTENES used $Cl_{1,3}$ and could with his investigations show that it is not necessary to use complex CLIFFORD algebras.

For the basis $\{e_0, e_1, e_2, e_3\}$ in $Cl_{3,1}$ we can take

$$e_0 = \begin{pmatrix} i & 0 \\ 0 & i \end{pmatrix}, \quad e_k = \begin{pmatrix} 0 & \sigma_k \\ \sigma_k & 0 \end{pmatrix} \quad (k = 1, 2, 3),$$

where σ_k are the PAULI matrices. Let

$$D = \sum_{k=1}^{3} e_k \partial_k \quad \text{and} \quad \partial = D - e_0 \partial_0,$$

where $e_0^2 = -1$ and $e_k^2 = 1 \ (k = 1, 2, 3)$. It can easily be seen that the basis elements $e_k \ (k = 1, 2, 3)$ fulfil anticommutator relations. We now introduce the *electro-magnetic field* $\mathbf{F}$, which is given by the bivector

$$\mathbf{F} := \mathbf{E} e_0 - \mathbf{B} e_{123}$$

with $e_{123}^2 = (e_1 e_2 e_3)^2 = -1$. Hereby $\mathbf{E}$ is defined as the bivector

$$\mathbf{E} = E_1 e_1 e_0 + E_2 e_2 e_0 + E_3 e_3 e_0$$

and $\mathbf{B}$ is just the vector $\mathbf{B} = B_1 e_1 + B_2 e_2 + B_3 e_3$. $\mathbf{E}$ and $\mathbf{B}$ are functions of the position $X = x_1 e_1 + x_2 e_2 + x_3 e_3$ (GIBBS *vector*) and the time t. Thus we have the coordinates

$$(X, t) = \sum_{k=1}^{3} e_k x_k + t e_0.$$

Now MAXWELL equations can be written in one single equation namely

$$\partial \mathbf{F} = \mathbf{J},$$

where the right-hand side is the sum of the trivector $J_0 = \rho$ (*charge density*) and the vector J (*current vector*). Applying Definition 6.6. we obtain

$$\partial \cdot \mathbf{F} = \mathbf{J} \quad \text{and} \quad \partial \wedge \mathbf{F} = \rho.$$

For the electric field yields now

$$\mathbf{E} = (\mathbf{F} \cdot e_0) e_0^{-1}.$$

Often one will have the field in the form of a vector potential $\mathbf{A}$. Then the equation

$$\partial \mathbf{A} = \mathbf{F}$$

yields the relations

$$\partial \wedge \mathbf{A} = -\mathbf{F} \quad \text{and} \quad \partial \cdot \mathbf{A} = 0.$$

The latter relation is called LORENZ *condition*.

From the electro-magnetic field $\mathbf{F}$ one can also deduce other important quantities (cf. [125]) such as

$$\mathbf{T}_k = -\frac{1}{2}\mathbf{F}e_k\mathbf{F} \quad (k = 0, 1, 2, 3).$$

For $k = 0$ follows (cf. [179])

$$\mathbf{T}_0 = \frac{1}{2}(\mathbf{E}^2 + \mathbf{B}^2)e_0 - e_1e_2e_3\mathbf{E}\cdot\mathbf{B}.$$

The coefficient of the first term reads *energy density* and the second one POYNTING *vector*. By scalar multiplication of $\mathbf{T}_k$ ($k = 1, 2, 3$) with e_ℓ ($\ell = 1, 2, 3$) from the right we get the so-called MAXWELL *stress tensor*:

$$\mathcal{T} := (T_{k\ell}) \quad \text{with} \quad T_{k\ell} = \mathbf{T}_k \cdot e_\ell,$$

which had been introduced by M. RIESZ in 1947.

When we look for the flow of the energy momentum through a surface with $\alpha = \alpha$ as the unit vector of the normal at the point x, then it is given by the *energy momentum tensor*

$$\mathbf{T}(\alpha) := -\frac{1}{2}\mathbf{F}\alpha\mathbf{F}.$$

It is nothing other than as a vector valued linear function on the tangent space at each space point (cf. [128]).

In the time-harmonic case in [158] is to find a further approach by the so-called quaternionic MAXWELL operator. For MAXWELL equations in the vacuum

$$\text{rot}\,\mathbf{H} = \sigma\mathbf{E}, \quad \text{rot}\,\mathbf{E} = i\omega\mu\mathbf{H}$$

$$\text{div}\,\mathbf{E} = 0, \qquad \text{div}\,\mathbf{E} = 0$$

with the *complex electrical conductivity* $\sigma := \rho^{-1} - i\omega\varepsilon$ one obtains the HELMHOLTZ equations

$$\Delta\mathbf{E} + \lambda\mathbf{E} = 0 \quad \text{and} \quad \Delta\mathbf{H} + \lambda\mathbf{H} = 0$$

with $\lambda := i\omega\mu\rho^{-1} + \omega^2\mu\varepsilon$. Here $\sqrt{\lambda} = \alpha$ is called *medium wave number*. The MAXWELL *operator* is defined by

$$M := \begin{pmatrix} \sigma & \text{div} - \text{rot} \\ -\text{div} + \text{rot} & -\, i\omega\mu \end{pmatrix}.$$

V.V KRAVSHENKO and M.V. SHAPIRO worked out the following connection between MAXWELL equations and the DIRAC equation. It reads as follows:

$$AMB = \begin{pmatrix} \overline{D-\alpha} & 0 \\ 0 & D-\alpha \end{pmatrix}$$

with

$$A := \begin{pmatrix} \alpha & -\sigma \\ -\alpha & -\sigma \end{pmatrix}, \quad B := \frac{1}{2}\begin{pmatrix} \sigma^{-1} & -\sigma^{-1} \\ \alpha^{-1} & \alpha^{-1} \end{pmatrix}.$$

6.2.2 Causality principle

Following the LORENTZ invariance demand we have to organize that the basic element of the time-coordinate also has to anticommute with all the others. It gives the so-called *four vectors*:

$$\mathbf{x} = \sum_{k=1}^{3} x_k \gamma_k + t\gamma_0.$$

The corresponding basis can be realized by the real matrices

$$\gamma_0 = \gamma^0 = \begin{pmatrix} 1 & 0 \\ 0 & -1 \end{pmatrix}, \quad \gamma_k = -\gamma^k = \begin{pmatrix} 0 & -\sigma_k \\ \sigma_k & 0 \end{pmatrix} \quad (k = 1, 2, 3),$$

where σ_k $(k = 1, 2, 3)$ are again PAULI matrices. Setting now

$$x_k = x^k \quad (k = 1, 2, 3) \quad \text{and} \quad x_0 = -x^0 = -t,$$

then

$$|x|^2 = \sum_{k=0}^{3} x_k x^k = x_1^2 + x_2^2 + x_3^2 - t^2.$$

This characterizes only the causality principle, i.e. a disturbance at the origin caused another one at the point x. One can say that for $|x|^2 < 0$ the "event" at $x = 0$ is *space-like* and for $|x|^2 > 0$ the "event" is *time-like*. The hypersurface $|x|^2 = 0$ is called *light-cone*.

6.2.3 Equations of Klein-Gordon and Dirac type

One can show that in $Cl_{3,1}$ the operator D^2 is given by

$$D^2 = \partial_1^2 + \partial_2^2 + \partial_3^2 - \partial_t^2.$$

The equation

$$(D^2 - m^2)\Phi(x) = 0, \tag{6.8}$$

is called SCHRÖDINGER *quantum wave equation*, if Φ is a complex valued scalar function which describes a charged spinless particle. Having again the equation (6.8), but looking for a vector function $\mathbf{\Phi}$, it is called PROCA *equation*. In this case this equation represents a spin-one massive field of the so-called "vector-boson".

We are speaking of the KLEIN–GORDON *equation* (in the true sense), if we search for solutions of equation (6.8) in $Cl_{3,1}$ or $Cl_{1,3}$.

By factorization of equation (6.8) first-order differential equations of "DIRAC type" appear. In this way we get the *meta-monogenic massive* DIRAC *equation*

$$(D - m)\Psi(x) = 0, \tag{6.9}$$

where Ψ represents a paravector, which is of the structure

$$\Psi(x) = (D + m)\Phi(x).$$

Here Φ is a scalar function. Now the operator D may be defined by $D = \sum_{k=0}^{3} \gamma^k \partial_k$. Looking now for solutions of equation (6.9) with $m = 0$ in $Cl_{p,q}^{(2)}$ then we have the case of a *sourceless* MAXWELL *equation*. The solution Ψ characterizes a spin-one massless photon. If the solution of equation (6.9) with $m = 0$ is defined as a column spinor

$$\Psi = \begin{pmatrix} \psi_1 \\ \psi_2 \\ \psi_3 \\ \psi_4 \end{pmatrix} \in \mathbb{C}^4,$$

then equation (6.9) is called the MAJORANA *equation*. The DIRAC spinor Ψ can also appear as square matrix spinor

$$\Psi = \begin{pmatrix} \psi_1 & 0 & 0 & 0 \\ \psi_2 & 0 & 0 & 0 \\ \psi_3 & 0 & 0 & 0 \\ \psi_4 & 0 & 0 & 0 \end{pmatrix} \in M_4(\mathbb{C})f$$

with $f = 1/2(1 + \gamma_0)1/2(1 + i\gamma_1\gamma_2)$. It describes a massless spin-half neutrino. For more information see [179].

A DIRAC *equation* (in the true sense) is given in the following way. Let e, m be defined by charge and eigenmass of the electron, respectively. Further, let $A \in Cl_{1,3}$ be a given vector potential. Then the equation

$$D\Psi = (m\Psi + eA\Psi)i$$

characterizes the behaviour of electrons. Often this equation is written in the so-called BJØRKEN–DRELL formulation [28], which reads as follows:

$$\sum_{k=0}^{3} \gamma^k (i\partial_k - eA_k)\Psi = m\Psi.$$

Using now the so-called *operator representation of spinors*, we get for any $\Psi \in (\mathbb{C} \otimes Cl_{1,3})$

$$\Psi = \begin{pmatrix} \psi_1 & -\psi_2^* & \psi_3 & \psi_4^* \\ \psi_2 & \psi_1^* & \psi_4 & -\psi_3^* \\ \psi_3 & \psi_4^* & \psi_1 & -\psi_2^* \\ \psi_4 & -\psi_3^* & \psi_2 & \psi_1^* \end{pmatrix}.$$

Here "$*$" denotes the complex conjugation. Then Ψ fulfils a so-called *multivector form* of the DIRAC equation ([126])

$$\partial\Psi = (eA\Psi + m\Psi\gamma_0)\gamma_1\gamma_2 \quad (\Psi \in Cl_{3,1}^{+}),$$

where $Cl_{3,1}^{+}$ means that only even order elements are containing.

Remark 6.8. Replacing in the BJØRKEN–DRELL representation

$$e := e^{-}\frac{2\pi}{\hbar c}, \quad m := m_0\frac{2\pi c}{\hbar}, \quad \kappa := 2i\frac{\pi}{\hbar}$$

one obtains the classical formulation of DIRAC's equation

$$\left[\partial_t - \kappa e^{-}A_0 - \sum_{k=0}^{3}(\gamma_0\gamma_k\partial_k + \kappa e^{-}A^k\gamma_0\gamma_k) - \kappa c^2 m_0\gamma_0\right]\psi = 0,$$

where c is the *light velocity*, m_0 the *eigenmass*, e^{-} the *charge* and $\hbar$ PLANCK's *constant*. In [42] one can find, for instance, the wave function for nucleons. For central fields solutions of the HESTENES version of DIRAC's equation, these can be found in [163].

6.2.4 Qualitative properties

From the column spinor $\Psi \in \mathbb{C}^4$ we obtain the scalars

$$J_k = \Psi^{\dagger}\gamma_0\gamma_k\Psi \quad (k = 0, 1, 2, 3),$$

where $\Psi^{\dagger} := \Psi^{*T}$ is the Hermitean conjugate. Obviously,

$$J_0 = \Psi^{\dagger}\Psi \quad (\textit{probability density}).$$

The vector $\mathbf{J} = \sum_{k=0}^{3} J_k\gamma^k$ is called *current density*. There are real characteristic sizes (bilinear covariants), which can be used to recover the DIRAC spinors. Some are listed below:

$$\sigma := \Psi^{\dagger}\gamma_0\Psi, \qquad \omega := -\Psi^{\dagger}\gamma_0\gamma_{0123}\Psi,$$

$$J_k := \Psi^{\dagger}\gamma_0\gamma_k\Psi, \qquad S_{k\ell} := \Psi^{\dagger}\gamma_0 i\gamma_k\gamma_\ell\Psi,$$

$$K_\ell = \Psi^{\dagger}\gamma_0 i\gamma_{0123}\gamma_k\Psi.$$

With the denotations $\mathbf{S} = 1/2\sum_{k,\ell}\gamma^{k\ell}S_{k\ell}$ and $\mathbf{K} = \sum_{k=0}^{3} K_k\gamma^k$ we are able to formulate the so-called FIERZ *identities*. We have

$$\mathbf{J}^2 = \sigma^2 + \omega^2, \quad \mathbf{K}^2 = -\mathbf{J}^2, \quad \mathbf{J}\cdot\mathbf{K} = 0, \quad \mathbf{J}\wedge\mathbf{K} = -(\omega + \gamma_{0123}\sigma)\mathbf{S}.$$

For more information we refer to [55] and [148]. We will finish our explanations on this topic with the definition of a boomerang.

Definition 6.9. *[179] Let $\sigma, \mathbf{J}, \mathbf{S}, \mathbf{K}$, and ω be bilinear covariants for some spinor Ψ. Then*

$$Z = \sigma + \mathbf{J} + i\mathbf{S} - i\gamma_{0123}\mathbf{K} + \gamma_{0123}\omega$$

is called boomerang.

One can verify (see [179]) that by the help of such a boomerang DIRAC spinors can be generated. Further work in this field was done in [147], [245], and many other papers.

6.2.5 Concluding remarks

Another subject of investigation are the so-called BELTRAMI–MOSES *fields* (see [133]). The corresponding equation is given by

$$D \wedge u(x,p,t) = \lambda u(x,p,t).$$

In a special case of an electro-magnetic field $\mathbf{F} = \sqrt{\varepsilon}\mathbf{E} + i\sqrt{\mu}\mathbf{H}$, where ε denotes the *permittivity* and μ is the *permeability* we obtain the MAXWELL–HEAVISIDE *equation*

$$D \wedge \mathbf{F} = -i\sqrt{\varepsilon\mu}\partial_{it}\mathbf{F}.$$

Nowadays topics such as investigation of MAXWELL equations with memory terms (see [29],[132]) or with non-linear additive convection terms, as normally to find in NAVIER-STOKES theory (see [271]) play a decisive role. M. PARRA outlined [221], [222] suitable connections between the pedestrian and CLIFFORD algebraic approach to the DIRAC and more general equations and gave historical and philosphical contributions on this topic. Furthermore, CLIFFORD algebra methods in general field theories are also used by W. PEZZAGLIA [227], [228], R. CHISHOLM and R. FAREWELL [44]. We also refer to the introductory volume of B. JANCEWICZ [139].

6.3 On recent problems in Clifford analysis

In the preface we mentioned that a growing number of people are using CLIFFORD structures in their research work. Hence, we have also a growing number of very specific problems which will be considered using CLIFFORD or quaternionic analysis. It seems to be completely impossible to list all these specific problems and, additionally, all possible applications. A collection of these problems is written down in J. RYAN's book [258], which we will not repeat here. Without any claim to completeness, we intend to introduce classes of functions which are attractive enough to investigate with CLIFFORD algebraic methods, and problems the solutions of which are of current interest.

6.3.1 On quasiregular mappings

Quasiregular functions of the classical function theory are closely connected with the complex BELTRAMI equation

$$\partial_{\overline{z}} u = \mu(z) \partial_z u. \tag{6.10}$$

Here $\mu(z)$ is called *complex dilatation.* For $\mu(z) = 0$ we have exactly the CAUCHY–RIEMANN equations. Therefore, (6.10) can be seen as a generalized CAUCHY–RIEMANN system. In even dimensions the following generalization is possible:

Definition 6.10. (Cf. [138]) *Let $G \subset \mathbb{R}^n$ be a domain and $u \in W_1^{p,\mathrm{loc}}(G, \mathbb{R}^n)$. Then u is said to be weakly K-quasiregular* $(1 \leq K < \infty)$ *if*

(i) $\det J_u(x) \geq 0$ *or* $\det J_u(x) \leq 0$ *(almost everywhere, respectively),*

(ii) $\max\limits_{|h|=1} |J_u(x)h| \leq K \min |J_u(x)h|$ *(almost everywhere),*

where $J_u(x) = (\partial_i u_j) \in L_p^{\mathrm{loc}}(G, M_n(\mathbb{R}))$ *stands for the* JACOBI*an. K is called the dilatation of u. For $p = n$ we have the K-quasiregular case. The K-quasiregular function u is called to be K-quasiconformal if u is also an homeomorphism.*

Note here that quasiconformal mappings appear first in 1928 in H. GRÖTSCH [94]. A very good reference in the 2-dimensional case is the booklet of L. AHLFORS [3]. The system

$$J_u(x) J_u^T(x) = (\det J_u(x))^{\frac{n}{2}} Id,$$

where J_u^T denotes the transpose of J_u is called the *generalized* CAUCHY–RIEMANN *system.* Substituting on the right-hand side Id by a matrix function $R : G \to S_n$ with

$$K^{2-2n} \leq (R(x)\xi, \xi)^n \leq K^{2n-2}$$

we get the *n-dimensional* BELTRAMI *equation*, where $(x, \xi) \in G \times S_n$ and $\det R = 1$.

The study of both weakly K-quasiregular and K-quasiconformal mappings (also in odd dimensions) can produce new ideas in the treatment of linear and non-linear problems of elasticity theory and fluid flows.

6.3.2 Hyperbolic harmonic functions

In [168] is considered another very interesting generalization of CAUCHY–RIEMANN differential equations. Let G be a domain in $\mathbb{R}^3$.

$$t(\partial_x u - \partial_y v - \partial_t w) + w = 0 \tag{6.11}$$

$$\partial_y u = -\partial_x v, \quad \partial_t u = -\partial_x w, \quad \partial_t v = \partial_y w. \tag{6.12}$$

Let now $f = u + iv + jw$ $(ij + ji = 0)$ be a reduced quaternion. Any solution of the LAPLACE–BELTRAMI *equation*

$$t\,\Delta u - \partial_t u = 0 \quad \text{with} \quad \Delta = \partial_x^2 + \partial_y^2 + \partial_t^2 \tag{6.13}$$

is associated with the hyperbolic metric (cf. [134])

$$ds = \frac{(dx^2 + dy^2 + dt^2)^{\frac{1}{2}}}{t} \qquad (t > 0).$$

Note that the components u and v fulfil equation (6.13) and the third component w satisfies the equation

$$t^2 \Delta w - t\partial_t w + w = 0.$$

Let $q = u + iv + it$. H. LEUTWILER found that any power q^n $(n \in \mathbb{N})$ is a solution of (6.11). He was describing a system of fundamental polynomials and found out a large number of characteristic properties. Nevertheless, it is unclear up to now if a CAUCHY-type formula can be verified. The same problem is also open in the case of n dimensions (cf. [258]).

6.3.3 Intrinsic functions

In his well-known paper [243] R.F. RINEHART describes a method of extending complex functions to functions defined in a finite dimensional associative algebra. He introduced the notions of the so-called intrinsic sets and intrinsic functions.

Definition 6.11. *A subset $G' \subset G$ is called an intrinsic set of G if $h(G') = G'$ for any $h \in H$, where H denoted the group of all automorphisms and antiautomorphisms. Further, a function u which is defined on G' and maps* *into G is said to be an intrinsic function, if G' is an intrinsic set of G and for all $x \in G'$ the identity*

$$u(h(x)) = h(u(x)) \quad (\forall\, h \in G).$$

holds.

In the complex plane intrinsic sets are just those which are symmetric to the real axis, and intrinsic functions u have to fulfil the relation $\overline{u}_{\mathbb{C}}(z) = u_{\mathbb{C}}(\overline{z})$. Corresponding intrinsic sets in $\mathbb{H}$ are those which are symmetric to the real axis in any complex plane which is generated by 1 and $\xi \in S_4$. The induced function of an intrinsic function is called *stem function*. The intrinsic function is often called *primary function*. Let now

$$e(x) = \frac{\overline{x}}{|x|^4}$$

be the so-called CAUCHY–FUETER kernel. We now wish to extend functions defined on intrinsic domains in $\mathbb{C}$ to such in $\mathbb{H}$. First let

$$i_\xi : u_{\mathbb{C}} = u_1 + iu_2 \to u_1 + \xi u_2 = u_{\mathbf{H}} \quad (\xi \in S_4).$$

Furthermore, let τ be the extension operator, which is given by

$$\tau : u_{\mathbb{C}} \to -\frac{1}{4}\Delta i_\xi u_{\mathbb{C}} \quad \text{(see also Section 2.3)},$$

where Δ is the Laplacian. In this way one gets for instance

$$\tau(\log(\cdot)) = \frac{1}{|x|^2} \quad \text{and} \quad \tau((\cdot)^{-1}) = \frac{\overline{x}}{|x|^4}.$$

Proposition 6.12. [238] *Let $\overline{D} = \partial_0 - \sum_{i=1}^{3} \partial_i e_i$ and $\partial_z = 1/2(\partial_x - i\partial_y) \in \mathbb{C}$ then we have the relation*

$$\frac{1}{2}\overline{D}_x \tau(u_{\mathbb{C}}) = \tau(\partial_z(u_{\mathbb{C}})).$$

It follows that

$$P^{(-k)} = \tau((\cdot)^{-k}) \quad (k = -1, -2, ...).$$

Introducing KELVIN's inversion mapping I defined by

$$I(u)(x) = e(x)u(x^{-1}),$$

which transforms (left) $\mathbb{H}$-regular functions inside of S_4 to (left) $\mathbb{H}$-regular functions outside of S_4 and vice versa. We find then also

$$P^{(k-1)} = \tau((\cdot)^{k+1}) \quad (k = 1, 2, ...),$$

where

$$P^{(k-1)}(x) = \frac{(-\frac{1}{2})^{(k-1)}}{(k-1)!} D^{k-1} e(x). \tag{6.14}$$

Moreover, one can see that the relations

$$\frac{1}{2}\overline{D}P^{(k)} = kP^{(k-1)} \quad (k = -1, -2, ...)$$

$$\frac{1}{2}\overline{D}P^{(k)} = (k+2)P^{(k-1)} \quad (k = 1, 2, ...)$$

are valid. More generally we obtain

$$\sum_{k=1}^{\infty} P^{(k)}(x) + \sum_{k=-\infty}^{-1} -P^{(k)}(x) = \tau\left(\sum_{k=1}^{\infty} z^k + \sum_{k=-\infty}^{-1} z^k\right)$$

$$= \tau\left(\frac{1+z}{1-z}\right) = \tau\left(\frac{2}{1-z}\right) = 2e(1-x)$$

Note that $\tau(c) = 0$ for $c =$ const. These ideas are used in [238] to construct classes of star-shaped LIPSCHITZ bounded domains starting from a parametrization of a LIPSCHITZ curve in $\mathbb{C}$.

6.3.4 A multiplier problem

Let Γ be a star-shaped LIPSCHITZ surface, which means that for any points $x, y \in \Gamma$ there is a constant L' with

$$\frac{|\ln|y^{-1}x||}{\arg(x,y)} \leq L'$$

and that Γ has to be star-shaped relative to the origin. Setting $\inf L' =: L$. Let $r_1 = \min\{|x|, x \in \Gamma\}$ and $r_2 = \max\{|x|, x \in \Gamma\}$. In [50] the following class of functions is studied

$$R_A = \{u : \; u \; (left) \; regular \; in \; r_1 - r < |x| < r_2 + r, \; for \; some \; r\}.$$

Proposition 6.13. *[238] The subspace R_A is dense in $L_2(\Gamma)$.*

Proposition 6.14. *[238] Any $u \in R_A$ permits the representation*

$$u(x) = \sum_{-\infty}^{\infty} \frac{1}{2\pi^2} \int_{\Gamma} P^{(k)}(y^{-1}x)e(y)\alpha(y)u(y)dy$$

for $r_1 - r < |x| < r_2 + r$.

Already in [313] is shown that for $u \in R_A$

$$u(x) = \sum_{k=0}^{\infty} \sum_{\nu} \{P_\nu(x)a_\nu + E_\nu(x)b_\nu\}.$$

where

$$P_\nu(x) = \frac{1}{n!} \sum_{i_1,i_2,\dots i_n,\, i_k \neq i_\ell} (x_0 e_{i_1} - x_{i_1})\dots(x_0 e_{i_n} - x_{i_n})$$

and

$$|\nu| = n = n_1 + n_2 + n_3,\ (1 \le i_k \le 3).$$

Additionally, we have

$$\partial_\nu = \frac{\partial^n}{\partial_1^{n_1}\partial_2^{n_2}\partial_3^{n_3}}, \qquad E_\nu = \partial_\nu e,$$

$$a_\nu = \frac{1}{2\pi^2}\int_\Gamma E_\nu(y)\alpha(y)u(y)dy,$$

and

$$b_\nu = \frac{1}{2\pi^2}\int_\Gamma P_\nu(y)\alpha(y)u(y)dy.$$

Let now

$$m \in S_\omega^{\mathbb{C}} = \{z \in \mathbb{C} : |\arg(z)| \quad \text{or} \quad |\arg(-z)| < \omega\} \text{ with } \omega \in (\arctan L, \frac{\pi}{2}).$$

Now we are able to introduce the multiplier operator $M_m : R_A \to R_A$ by

$$(M_m)(x) := \sum_{k=1}^{\infty} m(k) \sum_\nu P_\nu(x) a_\nu + \sum_{k=1}^{\infty} m(-k) \sum_\nu E_\nu(x) b_\nu$$
$$(r_1 - r < |x| < r_2 + r).$$

A straightforward calculation using the notations (cf. 6.14)

$$\phi^+ := \sum_{k=1}^{\infty} m(k) P^{(k)} \quad \text{and} \quad \phi^- := \sum_{k=-1}^{-\infty} m(k) P^{(k)}$$

yields finally (see [238]) the integral representation

$$(M_m u)(x) = \lim_{t \to 1^-} \frac{1}{2\pi^2} \int_\Gamma (\phi^+(y^{-1}tx) + \phi^-(y^{-1}t^{-1}x))e(y)\alpha(y)u(y)dy.$$

After extension of M_m onto $L_2(\Gamma, \mathbb{H})$ one can obtain

$$||M_m||_{L(L_2)} \le C_\nu ||m||_{L_\infty(S_\nu^{\mathbb{C}})} \quad (\nu \in (L, \omega)).$$

As for

$$\begin{aligned} m_1, m_2 &\in H^\infty(S_\omega^{\mathbb{C}}), \quad \lambda_1, \lambda_2 \in \mathbb{C}, \\ M_{m_1 m_2} &= M_{m_1} \circ M_{m_2}, \\ M_{\lambda_1 m_1 + \lambda_2 m_2} &= \lambda_1 M_{m_1} + \lambda_2 M_{m_2}, \end{aligned}$$

a bounded functional calculus of the spherical DIRAC operator can be constituted.

6.3.5 On a chain rule

Chain rules belong to the basic properties in any differential calculus. In [161] rules of this type are deduced in the case of real quaternions. We will describe one of them in the following. Let $x = x_0 + \sum_{k=1}^{3} e_k x_k$ and $D = \partial_0 + \sum_{k=1}^{3} \partial_k e_k$ as usual. Further, let w be an arbitrary element in $\mathbb{H}$. We introduce the following new notation:

$$w^{(k)} := e_k w(-e_k) \quad (k = 1, 2, 3).$$

Let $\sigma : x \to y, u : y \to u$ be C^1-mappings, where the mapping

$$\sigma = \sigma_0 + \sigma_1 e_1 + \sigma_2 e_2 + \sigma_3 e_3 : x \to y$$

is an $\mathbb{R}$-linear isomorphism in $\mathbb{H}$. For any quaternionic function $u = u_0 + u_1 e_1 + u_2 e_2 + u_3 e_3$ we can now define the derivative

$$\frac{\partial u}{\partial \sigma_\nu(x)} := \sum_\mu \frac{\partial x_\mu}{\partial \sigma_\nu(x)} \frac{\partial u}{\partial x_\mu}.$$

Then one gets [161]

$$\frac{\partial u}{\partial \sigma_\nu(y)} = \sum_\mu \frac{\partial \sigma_\mu(x)}{\partial \sigma_\nu(y)} \frac{\partial u}{\partial \sigma_\mu(x)}.$$

Theorem 6.15. *Let G_1, G_2 be domains in $\mathbb{H}$, $u : \mathbb{H} \ni G_2 \to \mathbb{H}$, $v : G_1 \to G_2$ C^1- mappings in $\mathbb{H}$, where $v = v_0 + e_1 v_1 + e_2 v_2 + e_3 v_3$. Then it holds that*

$$D(u \circ v) = (Dv)(Du) \circ v) + \sum_{k=1}^{3} (Dv^{(k)})(\frac{\partial u}{\partial x^{(k)}} \circ v).$$

More recently also M.V. SHAPIRO found an interesting chain rule in the quaternionic case. It would be very good to have a similar chain rule also in more general CLIFFORD algebras.

6.3.6 Clifford-valued wavelets

Meanwhile there exists a comprehensive literature on wavelets (cf. [57], [175]). We will shed a little light on the notion of CLIFFORD-valued wavelets.

We introduce the inner product

$$(u, v) := \int_{R^m} u(x) b(x) v(x) dx,$$

where $b : \mathbb{R}^m \to \mathbb{R}^{n+1}$ is an essentially bounded function the scalar part of which is Sc $b(x) \geq \delta > 0$.

Definition 6.16. *Let $\{\psi^{\ell}_{jk}\}_{j,k}$ and $\{\psi^{r}_{jk}\}_{j,k}$ be a pair of systems of functions which belong to $C^p(R^m, Cl_{0,n} \otimes \mathbb{C})$. Here ℓ stands for left and r for right. Assume the following properties:*

$$\text{(i)} \quad (\psi^{\ell}_{jk}, \psi^{r}_{jk}) = \delta_{jj'}\delta_{kk'}e_0 \tag{6.15}$$

$$\text{(ii)} \quad u = \sum_{j,k}(u, \psi^{r}_{jk})\psi^{\ell}_{jk} = \sum_{j,k}\psi^{r}_{jk}(\psi^{\ell}_{jk}, u)$$

$$\text{(iii)} \quad ||u|| \approx \sum |(u, \psi^{r}_{jk})|^2 \approx \sum |(\psi^{\ell}_{jk})|^2 \quad (u \in L_2).$$

For a fixed $p \in \mathbb{R}^+$ we suppose

$$\text{(iv)} \quad |\partial^{\alpha}\psi^{\ell}_{jk}(x)| + |\partial^{\alpha}\psi^{r}_{jk}(x)| \leq 2^{k(m/2+|\alpha|)}e^{-\nu|2^k x-j|)} \quad \forall\, |\alpha| \leq p.$$

We will call functions which fulfil the properties (i)–(iv) CLIFFORD *wavelets.*

Our aim consists in the explicit construction of a pair of systems of CLIFFORD wavelets. We will do this very roughly only in the case of HAAR–CLIFFORD wavelets. Following [204] we have to introduce

$$L_2^{(n)} := L_2 \otimes Cl_{2,n} = \{x = \sum_A x_A \otimes e_A;\ x_A \in L_2\},$$

where

$$A \in \{i_1 i_2 ... i_k, 1 \leq k \leq n, i_s < i_t \text{ for } s < t\}.$$

A norm is given by $||x||^2_{L_2^{(n)}} = \sum_A ||x_A||^2_{L_2}$. Let further $V_k := \{u \in L_2^{(n)}(\mathbb{R}^m) :$ u piecewise constant on dyadic cubes of the side length $2^{-k}\}$.
Let $\mathcal{Q}_k$ be the collection of all dyadic cubes with side length 2^{-k}. Furthermore, $\mathcal{Q} := \cup_k \mathcal{Q}_k$. Each dyadic cube Q permits a new refinement. It arises 2^m new cubes with half side length. Suppose for the function b, additionally, the inequality

$$\left| \frac{1}{|Q|} \int_Q b(x)dx \right| \geq \delta$$

for any $Q \in \mathbb{R}^m$, where with $|Q|$ is denoted the EUCLIDean volume of Q. Moreover, we set

$$\mu(Q) := \int_Q b(x)dx \in \mathbb{R}^{n+1}.$$

One can easily verify that now $|\mu(Q)| \approx |Q|$. As usual we denote by χ_{Q^i} $(i = 1, ..., 2^m - 1)$ the characteristic function for the cube $Q^i \in \mathcal{Q}_{k+1}$. Define a new function

$$\Theta_{Q,i} := \left(\sum_{j=1}^{i} \mu(Q^j) \right)^{-1} \left(\sum_{j=1}^{i} \chi_{Q^j} \right) - \mu(Q^{i+1})^{-1} \chi_{Q^{i+1}}.$$

Remark 6.17. It can be always realized that

$$\sum_{j=1}^{i} \mu(Q^j) \neq 0 \text{ for }, i = 1, 2, ..., 2^m - 1$$

(see [203] p.271).

One has now to check that

$$\int_{\mathbf{R}^m} \Theta_{Q,i}(x) b(x) \Theta_{Q,j}(x) dx = \delta_{ij} e_0.$$

A direct computation shows

$$\begin{aligned} M(\mathcal{Q}, i) &:= \int_{\mathbf{R}}^{m} \Theta_{Q,i} b(x) \Theta_{Q,i}(x) dx \\ &= \left(\sum_{j=1}^{i} \mu(Q^j) \right)^{-1} \left(\sum_{j=1}^{i+1} \mu(Q^j) \right) \mu(Q^{i+1})^{-1} \in \mathbb{R}^{n+1}. \end{aligned}$$

Then we found the wanted pair of systems of CLIFFORD wavelets. For $i = 1, 2, ..., 2^m - 1$ we get

$$\psi_{Qi}^{\ell} := \sqrt{|M(\mathcal{Q}, i)|} M(\mathcal{Q}, i)^{-1} \Theta_{Q,i}$$

$$\psi_{Qi}^{r} := \frac{1}{\sqrt{|M(\mathcal{Q}, i)|}} \Theta_{Q,i},$$

where $Q \in \mathcal{Q}_k$ and $k \in Z$.

Definition 6.18. *The system $\{\psi_{Qi}^{\ell}\}_{Q,i}$ is called left* RIESZ *basis for $L^{(n)}(\mathbb{R}^m)$. The system $\{\psi_{Qi}^{r}\}$ is called right* RIESZ *basis for $L_2^{(n)}(\mathbb{R}^m)$.*

The following properties are valid.

Corollary 6.19. (i) *All* RIESZ *basic elements belong to V_k ($\forall\, k \in Z, Q \in \mathcal{Q}_k$).*

(ii) *We have*

$$\text{supp}\, \psi_{Qi}^{\ell(r)} \subset Q.$$

(iii) *Here* $\psi^{\ell}_{Qi}(x)$ *and* $\psi^{r}_{Qi}(x)$ *are orthogonal to 1 relative to the inner product introduced above.*

(iv) *Relations (6.15) are fulfilled.*

6.3.7 $\mathbb{H}$-regular splines

In [2] is introduced the notion of complex analytic spline. We will give a proposal for a definition of quaternionic regular splines. Let $G \subset \mathbb{R}^n$ be a bounded domain. Its boundary is a piecewise smooth LIAPUNOV surface. Let $\mathcal{P} = \{\Gamma_1, ..., \Gamma_\ell\}$ be a partition of the surface Γ.

Definition 6.20. *A function* $s_{\mathcal{P}} \in C^{k-1}(\Gamma, \mathbb{H})$ *is called* $\mathbb{H}$*-spline of degree* k *with respect to the partition* $\mathcal{P}$*, if on each part* Γ_j $s_{\mathcal{P}}$ *is an* $\mathbb{H}$*-regular polynomial of degree at most* k *in the quaternionic variable* x.

Definition 6.21. *Each* $\mathbb{H}$*-spline* $s_{\mathcal{P}}$ *generates by* CAUCHY*'s formula an* $\mathbb{H}$*-regular spline*

$$S_{\mathcal{P}}(x) = \int_{\Gamma} e(y-x)\alpha(y)s_{\mathcal{P}}(y)d\Gamma_y.$$

We are convinced that $\mathbb{H}$-regular splines can play an essential role in interpolation theory.

6.3.8 On generalized Cauchy kernels

We will introduce here a generalized CAUCHY kernel which has a better behaviour near to infinity. We consider the following kernel:

$$k(\xi, x) = \frac{1}{\omega_n}\left(\frac{\xi - x}{|\xi - x|^n} - \frac{\xi + x}{|\xi + x|^n}\right).$$

This kernel works on unbounded domains G lying in a half space with sufficiently smooth boundary Γ and satisfying the condition

$$\text{dist}(G, -G) \geq d^* > 0$$

where $\text{dist}(G, -G)$ is the usual distance between the domains G and $-G := y : (-y) \in G$. Defining now the modified operators

$$\tilde{T}_G u := \int_G k(x,y)u(y)dy \quad \text{and} \quad \tilde{F}_\Gamma u := \int_\Gamma k(y,x)\alpha(y)u(y)dy$$

many basic results from [110] hold also for this kernel. For a wide class of unbounded domains analogues to the BOREL–POMPEIU formula, the PLEMELJ–SOKHOTZKI formulae and also a corresponding orthogonal decomposition of the CLIFFORD-valued $L_2(G)$ can be obtained. For more detailed information and further references we refer to [77] and [106].

Much progress has been made in the treatment of convolution singular integrals on LIPSCHITZ surfaces by the help of CLIFFORD analysis. For a large class of right $Cl_{0,n}$- regular kernels $k(x)$ which satisfy the condition

$$|k(x)| \leq C|x|^{-n}$$

on a sector

$$S_\mu^0 = \{x = x_0 + x \in \mathbb{R}^{n+1} : |x_0| < |x| \tan \mu\}.$$

The LIPSCHITZ surface Γ is defined by

$$\Gamma := \{g(x)e_0 + x \in \mathbb{R}^{n+1} : x \in \mathbb{R}^n\}$$

with a LIPSCHITZ function g satisfying the inequality

$$||g||_\infty \leq \tan \omega < \infty.$$

Between μ and ω we have the relation $\mu > \omega$. The main result (cf. [172]) can formulated as follows

Theorem 6.22. *In the case of existence of an essential bounded function k^* such that*

$$k^*(R) - k^*(r) = \int_{\substack{r<|x|<R \\ x \in \mathbb{R}^n}} k(x)dx$$

the corresponding convolution singular integral operator

$$(F_{(k,k^*)}(x) = \lim_{\varepsilon \to 0+} \left(\int_{\substack{y \in \Gamma \\ |x-y| \geq \varepsilon}} k(x-y)\alpha y u(y) d\Gamma_y + k^*(\varepsilon\alpha(x))u(x) \right)$$

is bounded in $L_p(\Gamma)$ for $1 < p < \infty$.

Moreover, formulae of the PLEMELJ–SOKHOTZKI type and a related HARDY decomposition can be verified.

6.3.9 Quaternionic multi-layer perceptron (QMLP)

Neural network models have been used in a wide variety of applications. Nowadays, the multi-layer perceptron structure is used in a quaternionic domain denoted by QMLP. In order to explain this application we have to introduce the following denotations (cf. [76]): Let M be the number of layers in the network, l denotes the layer. For $l = 0$ we say input layer and for $l = M$ we say output layer. N_l is the number of neurons of the l-th layer, n describes the neuron index. Furthermore, with

$$x_n^{(l)} = \sum_{i=0}^{3} x_{in}^{(l)} e_i$$

is designed the quaternion output of the n-th neuron of the l-th layer. The quaternion

$$w_{nm}^{(l)} = \sum_{i=0}^{3} w_{inm}^{(l)} e_i$$

describes the sinaptic weight of the n-th neuron of the l-th layer relative to the m-th output of the $(l-1)$th-layer, the quaternion

$$S_n^{(l)} = \sum_{i=0}^{3} S_{in}^{(l)}$$

denotes the net function relative to the n-th neuron of the l-th layer and the bias of the n-th neuron of the l-th layer is given by

$$\Theta_n^{(l)} = \sum_{i=0}^{3} \Theta_{in}^{(l)} e_i .$$

The forward phase of the QMLP is covered by the equations:

$$S_n^{(l)} = \sum_{m=0}^{N_{l-1}} w_{nm}^{(l)} x_m^{l-1} + \Theta_n^{(l)},$$

$$x_n^{(l)} = \sigma(S_n^{(l)}),$$

where the activation function σ is defined by

$$\sigma(y) := \sum_{i=0}^{3} \sigma(y_i) e_i$$

and σ is a real sigmoidal function, which is a function defined on $\mathbb{R}$ and tends to one if the argument tends to ∞ and tends to zero if the argument tends to $-\infty$. It can be seen that the quaternionic language allows a compact description and, consequently, a convenient treatment of these problems. For more information see [76] and more recent papers of these authors.

6.3.10 Concluding remarks

It would not be difficult to continue to present recently progress in the application of CLIFFORD analysis. In this way we see good chances of a complete treatment of a special class of mixed boundary value problems, the so-called *window problems* (cf. [305]). Note also that previously problems of optimal estimation of a rotating body attitude with both differential equations of motion and observations of stars has been very successfully studied in [197], using the language of quaternions. As is known, the classical CAUCHY–RIEMANN operator is a so-called spectral operator. It could be proved that the MOISIL–TEODORESCU operator is also of this type. We are sure that it is not difficult to generalize this result to the corresponding differential operators in $Cl_{0,n}$. Moreover, it is a very important task to check the validity of function theoretic theorems in CLIFFORD algebras of the general type $Cl_{p,q}$ ([176], [164]), especially in CLIFFORD algebras over KREIN spaces. In his talk at the Seiffen Symposium (1996), E. MEISTER (Darmstadt) emphasized the importance of factorization problems in different applications. Therefore, we will at least hint at an unsolved factorization problem (cf. [259]):

Is it possible to find (explicitly) a factorization of functions which are given on a smooth boundary of a domain in $\mathbb{R}^n$ of the type

$$u(x) = u_+(x)x^\kappa u_-(x),$$

where u_+ is in some sense CLIFFORD *regular into the interior domain, and u_- is in some sense* CLIFFORD *regular into the outer domain?*

Together with new researches concerning some singular integral operators such a factorization seems to be important for the development of a self-contained systematic theory of quaternionic singular integral operators.

Appendix A

Exact computations of the Teodorescu transform

Assume that G is the unit ball in $\mathbf{R}^3$ and we write briefly T instead of T_G. Furthermore $x = \underline{x} = x_1e_1 + x_2e_2 + x_3e_3$ and computation take place within the algebra of real quaternions.

1. Integrals of rational functions

1.1. Integrals of the form x^n, n integer

1. $T(1) = -\frac{1}{3}x$

2. $T(x) = \frac{1}{2}(-x^2 - 1)$

3. $T(x^2) = -\frac{1}{5}x^3$

4. $T(x^3) = \frac{1}{4}(-x^4 + 1)$

5. $T(x^4) = -\frac{1}{7}x^5$

6. $T(x^5) = \frac{1}{6}(-x^6 - 1)$

7. $T(x^{-1}) = -\ln|x|$

8. $T(x^{-2}) = -x^{-1}$

9. $T(x^{-3}) = \frac{1}{2}(x^{-2} + 1)$

10. $T(x^{-4}) = x^{-3}$

11. $T(x^{-5}) = \frac{1}{4}(x^{-4} - 1)$

In general:

$$T(x^{n-1}) = \begin{cases} -\ln|x| & : \; n = 0 \\ \frac{-1}{n+2}x^n & : \; n \text{ odd} \\ \frac{-1}{n}(x^n - 1) & : \; n \text{ even and integer multiple of 4; } n \neq 0 \\ \frac{-1}{n}(x^n + 1) & : \; n \text{ even; } n \neq 0 \end{cases}$$

1.2. Integrals of the form $|x|^n$

12. $T(|x|) = -\frac{1}{4}(x|x|)$

13. $T(|x|^2) = -\frac{1}{5}(x|x|^2)$

14. $T(|x|^3) = -\frac{1}{6}(x|x|^3)$

15. $T(|x|^4) = -\frac{1}{7}(x|x|^4)$

16. $T(|x|^5) = -\frac{1}{8}(x|x|^5)$

17. $T(|x|^{-1}) = -\frac{1}{2}x|x|^{-1}$

18. $T(|x|^{-2}) = -x|x|^{-2}$

19. $T(|x|^{-4}) = x|x|^{-4}$

20. $T(|x|^{-5}) = \frac{1}{2}x|x|^{-5}$

In general:

$$T(|x|^n) = \frac{-1}{n+3}(x|x|^n),\ n \neq -3$$

for rational exponents respectively:

$$T(|x|^{\frac{n}{m}}) = \frac{-m}{n+3m}x|x|^{\frac{n}{m}}$$

1.3. Integrals of the form $x|x|^{\frac{n}{m}-2}$, $m \neq 0$, n, m integer

21. $T(x|x|) = \frac{1}{3}(|x|^3 - 1)$

22. $T(x|x|^2) = \frac{1}{4}(|x|^4 - 1)$

23. $T(x|x|^3) = \frac{1}{5}(|x|^5 - 1)$

24. $T(x|x|^4) = \frac{1}{6}(|x|^5 - 1)$

25. $T(x|x|^{3/4}) = \frac{4}{11}(|x|^{11/4} - 1)$

26. $T(x|x|^{-1}) = |x| - 1$

27. $T(x|x|^{-2}) = \ln|x|$

28. $T(x|x|^{-3}) = -|x|^{-1} + 1$

29. $T(x|x|^{-4}) = \frac{1}{2}(-|x|^{-2} + 1)$

30. $T(x|x|^{-3/4}) = \frac{4}{5}(|x|^{5/4} - 1)$

In general:

$$T(x|x|^{\frac{n}{m}-2}) = \begin{cases} \frac{m}{n}(|x|^{\frac{n}{m}} - 1) & : \quad m, n \neq 0 \\ \ln|x| & : \quad m \neq 0, n = 0 \end{cases}$$

1.4. Integrals of the form $x^n|x|^m$; n, m integer

31.	$T(x|x|)$	$=$	$-\frac{1}{3}(x^2|x| + 1)$
32.	$T(x^2|x|)$	$=$	$-\frac{1}{6}x^3|x|$
33.	$T(x^3|x|)$	$=$	$-\frac{1}{4}(x^4|x| - 1)$
34.	$T(x^4|x|)$	$=$	$-\frac{1}{8}x^5|x|$
35.	$T(x|x|^{-1})$	$=$	$-(x^2|x|^{-1} + 1)$
36.	$T(x^2|x|^{-1})$	$=$	$-\frac{1}{4}x^3|x|^{-1}$
37.	$T(x^3|x|^{-1})$	$=$	$-\frac{1}{3}(x^4|x|^{-1} + 1)$
38.	$T(x^4|x|^{-1})$	$=$	$-\frac{1}{6}x^5|x|^{-1}$
39.	$T(x^{-1}|x|)$	$=$	$-|x| + 1$
40.	$T(x^{-2}|x|)$	$=$	$-\frac{1}{2}|x|^{-1}|x|$
41.	$T(x^{-3}|x|)$	$=$	$x^{-2}|x| + 1$
42.	$T(x^{-5}|x|)$	$=$	$\frac{1}{3}(|x|^{-4} - 1)$
43.	$T(x^{-1}|x|^{-1})$	$=$	$|x|^{-1} - 1$
44.	$T(x^{-3}|x|^{-1})$	$=$	$\frac{1}{3}x^{-2}|x|^{-1} + 1$
45.	$T(x^{-4}|x|^{-1})$	$=$	$\frac{1}{2}x^{-2}|x|^{-1}$
46.	$T(x^{-5}|x|^{-1})$	$=$	$\frac{1}{5}(x^{-4}|x|^{-1} - 1)$
47.	$T(x^5|x|^8)$	$=$	$-\frac{1}{14}(x^6|x|^8 + 1)$
48.	$T(x^{-5}|x|^8)$	$=$	$-\frac{1}{4}(x^{-4}|x|^8 - 1)$
49.	$T(x^5|x|^{-8})$	$=$	$\frac{1}{2}(x^6|x|^{-8} + 1)$
50.	$T(x^{-5}|x|^{-8})$	$=$	$\frac{1}{12}x^{-4}(|x|^{-8} - 1)$

In general:

$$T(x^{n-1}|x|^m) = \begin{cases} -\ln|x| & : \; n = m = 0 \\ \frac{-1}{n+m+2}|x|^n|x|^m & : \; n \text{ odd and } n+m \neq -2 \\ \frac{-1}{n+m}(x^n|x|^m - 1) & : \; n \text{ even and integer multiple of 4, } n+m \neq 0 \\ \frac{-1}{n+m}(x^n|x|^m + 1) & : \; n \text{ even, } n+m \neq 0, n \neq 0 \end{cases}$$

2. Integrals of trigonometrical functions

51. $T(\frac{x}{|x|}\tan^2|x|) = \tan|x| - \tan 1 - |x| + 1$

52. $T(\frac{x}{|x|}\cot^2|x|) = -\cot|x| + \cot 1 - |x| + 1$

53. $T(\frac{x}{|x|}\frac{1}{\sqrt{1-|x|^2}}) = \begin{cases} \arcsin|x| - \pi/2 \\ -\arccos|x| \end{cases}$

54. $T(\frac{x}{|x|}\frac{1}{1+|x|^2}) = \begin{cases} \arctan|x| - \pi/4 \\ \mathrm{arccot}|x| + \pi/4 \end{cases}$

55. $T(\frac{x}{|x|}\frac{1}{\sin^2|x|}) = -\cot|x| + \cot|x|$

56. $T(\frac{x}{|x|}\frac{1}{\cos^2|x|}) = \tan|x| - \tan 1$

2.1. Integrals of the form $\frac{x}{|x|}\sin^n|x|\cos|x|$

57. $T(\frac{x}{|x|}\sin|x|\cos|x|) = \frac{1}{2}(\sin^2|x| - \sin^2 1)$

58. $T(\frac{x}{|x|}\sin^2|x|\cos|x|) = \frac{1}{3}(\sin^3|x| - \sin^3 1)$

59. $T(\frac{x}{|x|}\sin^3|x|\cos|x|) = \frac{1}{4}(\sin^4|x| - \sin^4 1)$

60. $T(\frac{x}{|x|}\sin^4|x|\cos|x|) = \frac{1}{5}(\sin^5|x| - \sin^5 1)$

In general:

$$T\left(\frac{x}{|x|}\sin^n|x|\cos|x|\right) = \frac{1}{n+1}(\sin^{n+1}|x| - \sin^{n+1} 1)$$

2.2. Integrals of the form $\frac{x}{|x|}\cos^n|x|\sin|x|$

61. $T(\frac{x}{|x|}\cos|x|\sin|x|) = \frac{1}{2}(-\cos^2|x| + \cos^2 1)$

62. $T(\frac{x}{|x|}\cos^2|x|\sin|x|) = \frac{1}{3}(-\cos^3|x| + \cos^3 1)$

63. $T(\frac{x}{|x|}\cos^3|x|\sin|x|) = \frac{1}{4}(-\cos^4|x| + \cos^4 1)$

64. $T(\frac{x}{|x|}\cos^4|x|\sin|x|) = \frac{1}{5}(-\cos^5|x| + \cos^5 1)$

In general:

$$T\left(\frac{x}{|x|}\cos^n|x|\sin|x|\right) = \frac{1}{n+1}(-\cos^{n+1}|x| + \cos^{n+1} 1)$$

2.3. Integrals of the form $\frac{x}{|x|}\sin^n|x|$

65. $T(\frac{x}{|x|}\sin|x|) = -\cos|x| + \cos 1$

66. $T(\frac{x}{|x|}\sin^2|x|) = \frac{1}{2}(-\sin|x|\cos|x| + \sin 1\cos 1 + |x| + 1)$

67. $T(\frac{x}{|x|}\sin^3|x|) = \frac{1}{3}(-\sin^2|x|\cos|x| + \sin^2 1\cos 1 - 2\cos|x| + 2\cos 1)$

68. $T(\frac{x}{|x|}\sin^4|x|) = \frac{1}{4}(-\sin^3|x|\cos|x| + \sin^3 1\cos 1) + \frac{3}{8}(-\sin|x|\cos|x| + \sin 1\cos 1 + |x| - 1)$

69. $T(\frac{x}{|x|}\sin^5|x|) = \frac{1}{5}(-\sin^4|x|\cos|x| + \sin^4 1\cos 1)$
$+ \frac{4}{15}(-\sin^2|x|\cos|x| + \sin^2 1\cos 1 - 2\cos|x| + 2\cos 1)$

In general:

$$T\left(\frac{x}{|x|}\sin^{n+1}|x|\right) = \frac{1}{n+1}(-\sin^n|x|\cos|x| + \sin^n 1\cos 1) + \frac{n}{n+1}T\left(\frac{x}{|x|}\sin^{n+1}|x|\right)$$

2.4. Integrals of the form $\frac{x}{|x|}\cos^n|x|$

70 $T(\frac{x}{|x|}\cos|x|) = \sin|x| - \sin 1$

71. $T(\frac{x}{|x|}\cos^2|x|) = \frac{1}{2}(\cos|x|\sin|x| - \cos 1\sin 1 + |x| - 1)$

72. $T(\frac{x}{|x|}\cos^3|x|) = \frac{1}{3}(\cos^2|x|\sin|x| - \cos^2 1\sin 1 + 2\sin|x| - 2\sin 1)$

73. $T(\frac{x}{|x|}\cos^4|x|) = \frac{1}{4}(\cos^3|x|\sin|x| - \cos^3 1\sin 1) + \frac{3}{8}(\cos|x|\sin|x| - \cos 1\sin 1 + |x| - 1)$

74. $T(\frac{x}{|x|}\cos^5|x|) = \frac{1}{5}(\cos^4|x|\sin|x| - \cos^4 1\sin 1)$
$+ \frac{4}{15}(\cos^2|x|\sin|x| - \cos^2 1\sin 1 + 2\sin|x| - 2\sin 1)$

In general:

$$T\left(\frac{x}{|x|}\cos^{n+1}|x|\right) = \frac{1}{n+1}(\cos^n|x|\sin|x| - \cos^n 1\sin 1) + \frac{n}{n+1}T\left(\frac{x}{|x|}\cos^{n+1}|x|\right)$$

2.5. Integrals of the form $\frac{x}{|x|}\sin^n|x|\cos^2|x|$

75. $T(\frac{x}{|x|}\sin|x|\cos|x|) = \frac{1}{3}(\sin^2|x|\cos|x| - \sin^2 1\cos 1 - \cos|x| + \cos 1)$

76. $T(\frac{x}{|x|}\sin^2|x|\cos^2|x|) = \frac{1}{4}(\sin^3|x|\cos|x| - \sin^3 1\cos 1) + \frac{1}{8}(-\sin|x|\cos|x| + \sin 1\cos 1 + |x| - 1)$

77. $T(\frac{x}{|x|}\sin^3|x|\cos^2|x|) = \frac{1}{5}(\sin^4|x|\cos|x| - \sin^4 1\cos 1)$
$+ \frac{1}{15}(-\sin^2|x|\cos|x| + \sin^2 1\cos 1 - 2\cos|x| + 2\cos 1)$

78. $T(\frac{x}{|x|}\sin^4|x|\cos^2|x|) = \frac{1}{6}(\sin^5|x|\cos|x| - \sin^5 1\cos^1) + \frac{1}{24}(-\sin^3|x|\cos|x| + \sin^3 1\cos 1)$
$+ \frac{1}{16}(-\sin|x|\cos|x| + \sin 1\cos 1 + |x| - 1)$

79. $T(\frac{x}{|x|}\sin^5|x|\cos^2|x|) = \frac{1}{7}(\sin^6|x|\cos|x| - \sin^6 1\cos 1) + \frac{1}{35}(-\sin^4|x|\cos|x| + \sin^4 1\cos 1)$
$+ \frac{4}{105}(-\sin^2|x|\cos|x| + \sin^2 1\cos 1 - 2\cos|x| + 2\cos 1)$

In general:

$$T\left(\frac{x}{|x|}\sin^n|x|\cos^2|x|\right) = T\left(\frac{x}{|x|}\sin^n|x|\right) - T\left(\frac{x}{|x|}\sin^{n+2}|x|\right)$$

2.6. Integrals of the form $\frac{x}{|x|}\cos^n|x|\sin^2|x|$

80. $T(\frac{x}{|x|}\cos|x|\sin|x|) = \frac{1}{3}(-\cos^2|x|\sin|x| + \cos^2 1\sin 1 + \sin|x| - \sin 1)$

81. $T(\frac{x}{|x|}\cos^2|x|\sin^2|x|) = \frac{1}{4}(-\cos^3|x|\sin|x| + \cos^3 1\sin 1) + \frac{1}{8}(\cos|x|\sin|x| - \cos 1\sin 1 + |x| - 1)$

82. $T(\frac{x}{|x|}\cos^3|x|\sin^2|x|) = \frac{1}{5}(-\cos^4|x|\sin|x| + \cos^4 1\sin 1) + \frac{1}{15}(\cos^2|x|\sin|x|\cos^2 1\sin 1 + 2\sin|x| - 2\sin 1)$

83. $T(\frac{x}{|x|}\cos^4|x|\sin^2|x|) = \frac{1}{6}(-\cos^5|x|\sin|x| + \cos^5 1\sin^1) + \frac{1}{24}(\cos^3|x|\sin|x| - \cos^3 1\sin 1) + \frac{1}{16}(\cos|x|\sin|x| - \cos 1\sin 1 + |x| - 1)$

84. $T(\frac{x}{|x|}\cos^5|x|\sin^2|x|) = \frac{1}{7}(-\cos^6|x|\sin|x| + \cos^6 1\sin 1) + \frac{1}{35}(\cos^4|x|\sin|x|\cos^4 1\sin 1) + \frac{4}{105}(\cos^2|x|\sin|x| - \cos^2 1\sin 1 + 2\sin|x| - 2\sin 1)$

In general:

$$T\left(\frac{x}{|x|}\cos^n|x|\sin^2|x|\right) = T\left(\frac{x}{|x|}\cos^n|x|\right) - T\left(\frac{x}{|x|}\cos^{n+2}|x|\right)$$

2.7. Integrals of the form $\frac{x}{|x|}\sin^n|x|\cos^3|x|$

85. $T(\frac{x}{|x|}\sin|x|\cos^3|x|) = \frac{1}{4}(-\sin^4|x| + \sin^4 1) + \frac{1}{2}(\sin^2|x| - \sin^2 1)$

86. $T(\frac{x}{|x|}\sin^2|x|\cos^3|x|) = \frac{1}{5}(-\sin^5|x| + \sin^5 1) + \frac{1}{3}(\sin^3|x| - \sin^3 1)$

87. $T(\frac{x}{|x|}\sin^3|x|\cos^3|x|) = \frac{1}{6}(-\sin^6|x| + \sin^6 1) + \frac{1}{4}(\sin^4|x| - \sin^4 1)$

In general:

$$T\left(\frac{x}{|x|}\sin^n|x|\cos^3|x|\right) = \frac{1}{n+3}\left(-\sin^{n+3}|x| + \sin^{n+3} 1\right) + \frac{1}{n+1}\left(\sin^{n+1}|x| - \sin^{n+1} 1\right)$$

2.8. Integrals of the form $\frac{x}{|x|}\cos^n|x|\sin^3|x|$

88. $T(\frac{x}{|x|}\cos|x|\sin^3|x|) = \frac{1}{4}(\cos^4|x| - \cos^4 1) + \frac{1}{2}(-\cos^2|x| + \cos^2 1)$

89. $T(\frac{x}{|x|}\cos^2|x|\sin^3|x|) = \frac{1}{5}(\cos^5|x| - \cos^5 1) + \frac{1}{3}(-\cos^3|x| + \cos^3 1)$

90. $T(\frac{x}{|x|}\cos^3|x|\sin^3|x|) = \frac{1}{6}(\cos^6|x| - \cos^6 1) + \frac{1}{4}(-\cos^4|x| + \cos^4 1)$

In general:

$$T\left(\frac{x}{|x|}\cos^n|x|\sin^3|x|\right) = \frac{1}{n+3}\left(\cos^{n+3}|x| - \cos^{n+3}1\right) + \frac{1}{n+1}\left(-\cos^{n+1}|x| + \cos^{n+1}1\right)$$

2.9. Integrals of the form $\frac{x}{|x|}\sin^n|x|\cos^4|x|$

91. $T(\frac{x}{|x|}\sin|x|\cos^4|x|) = \frac{1}{5}(-\sin^4|x|\cos|x| + \sin^4 1\cos 1) + \frac{2}{5}(\sin^2|x|\cos|x| - \sin^2 1\cos 1) + \frac{1}{5}(-\cos|x| + \cos 1)$

92. $T(\frac{x}{|x|}\sin^2|x|\cos^4|x|) = \frac{1}{6}(-\sin^5|x|\cos|x| + \sin^5 1\cos 1) + \frac{7}{24}(\sin^3|x|\cos|x| - \sin^3 1\cos 1) + \frac{1}{16}(-\sin|x|\cos|x| + \sin 1\cos 1 + |x| - 1)$

93. $T(\frac{x}{|x|}\sin^3|x|\cos^4|x|) = \frac{1}{7}(-\sin^6|x|\cos|x| + \sin^6 1\cos 1) + \frac{8}{35}(\sin^4|x|\cos|x| - \sin^4 1\cos 1) + \frac{1}{35}(-\sin^2|x|\cos|x| + \sin^2 1\cos 1 - 2\cos|x| + 2\cos 1)$

94. $T(\frac{x}{|x|}\sin^4|x|\cos^4|x|) = \frac{1}{8}(-\sin^7|x|\cos|x| + \sin^7 1\cos 1) + \frac{3}{16}(\sin^5|x|\cos|x| - \sin^5 1\cos 1) + \frac{1}{64}(-\sin^3|x|\cos|x| + \sin^3 1\cos 1) + \frac{3}{128}(-\sin|x|\cos|x| + \sin 1\cos 1 + |x| - 1)$

In general:

$$T\left(\frac{x}{|x|}\sin^n|x|\cos^4|x|\right) = T\left(\frac{x}{|x|}\sin^{n+4}|x|\right) - 2T\left(\frac{x}{|x|}\sin^{n+2}|x|\right) + T\left(\frac{x}{|x|}\sin^n|x|\right)$$

2.10. Integrals of the form $\frac{x}{|x|}\cos^n|x|\sin^4|x|$

95. $T(\frac{x}{|x|}\cos|x|\sin^4|x|) = \frac{1}{5}(\cos^4|x|\sin|x| - \cos^4 1\sin 1) + \frac{2}{5}(-\cos^2|x|\sin|x| + \cos^2 1\sin 1) + \frac{1}{5}(\sin|x| - \sin 1)$

96. $T(\frac{x}{|x|}\cos^2|x|\sin^4|x|) = \frac{1}{6}(\cos^5|x|\sin|x| - \cos^5 1\sin 1) + \frac{7}{24}(-\cos^3|x|\sin|x| + \cos^3 1\sin 1) + \frac{1}{16}(\cos|x|\sin|x| - \cos 1\sin 1 + |x| - 1)$

97. $T(\frac{x}{|x|}\cos^3|x|\sin^4|x|) = \frac{1}{7}(\cos^6|x|\sin|x| - \cos^6 1\sin 1) + \frac{8}{35}(-\cos^4|x|\sin|x| + \cos^4 1\sin 1) + \frac{1}{35}(\cos^2|x|\sin|x| - \cos^2 1\sin 1 - 2\sin|x| + 2\sin 1)$

98. $T(\frac{x}{|x|}\cos^4|x|\sin^4|x|) = \frac{1}{8}(\cos^7|x|\sin|x| - \cos^7 1\sin 1) + \frac{3}{16}(-\cos^5|x|\sin|x| + \cos^5 1\sin 1) + \frac{1}{64}(\cos^3|x|\sin|x| - \cos^3 1\sin 1) + \frac{3}{128}(\cos|x|\sin|x| - \cos 1\sin 1 + |x| - 1)$

In general:

$$T\left(\frac{x}{|x|}\cos^n|x|\sin^4|x|\right) = T\left(\frac{x}{|x|}\cos^{n+4}|x|\right) - 2T\left(\frac{x}{|x|}\cos^{n+2}|x|\right) + T\left(\frac{x}{|x|}\cos^n|x|\right)$$

2.11. Integrals of the form $x\sin^n|x|$

99. $T(x\sin|x|) = \sin|x| - \sin 1 - |x|\cos|x| + \cos 1$

100. $T(x\sin^2|x|) = \frac{1}{4}(\sin^2|x| - \sin^2 1 + |x|^2 - 1) + \frac{1}{2}(-|x|\sin|x|\cos|x| + \sin 1\cos 1)$

101. $T(x\sin^3|x|) = \frac{1}{9}(\sin^3|x| - \sin^3 1) + \frac{1}{3}(-|x|\sin^2|x|\cos|x| + \sin^2 1\cos 1 + 2\sin|x| - 2\sin 1 - 2|x|\cos|x| + 2\cos 1)$

102. $T(x\sin^4|x|) = \frac{1}{16}(\sin^4|x| - \sin^4 1) + \frac{1}{4}(-|x|\sin|x|\cos|x| + \sin^3 1\cos 1) + \frac{3}{16}(\sin^2|x| - \sin^2 1 + |x|^2 - 1) + \frac{3}{8}(-|x|\sin|x|\cos|x| + \sin 1\cos 1)$

In general:

$$T(x\sin^n|x|) = \frac{1}{n+1}(-\sin^n|x|\cos|x| + \sin^n 1\cos 1) + \frac{n}{n+1}T(x\sin^{n-1}|x|)$$
$$+ \frac{1}{n+1}T\left(\frac{x}{|x|}\sin^n|x|\cos|x|\right)$$
$$= \frac{1}{(n+1)^2}(\sin^{n+1}|x| - \sin^{n+1}1) + \frac{1}{n+1}(-|x|\sin^n|x|\cos|x| + \sin^n 1\cos 1)$$
$$+ \frac{n}{n+1}T(x\sin^{n-1}|x|)$$

2.12. Integrals of the form $x\cos^n|x|$

103. $T(x\cos|x|) = \cos|x| - \cos 1 + |x|\sin|x| - \sin 1$

104. $T(x\cos^2|x|) = \frac{1}{4}(\cos^2|x| - \cos^2 1 + |x|^2 - 1) + \frac{1}{2}(|x|\sin|x|\sin|x| - \cos 1\sin 1)$

105. $T(x\cos^3|x|) = \frac{1}{9}(\cos^3|x| - \cos^3 1)$
$+ \frac{1}{3}(|x|\cos^2|x|\sin|x| - \cos^2 1\sin 1 + 2\cos|x| - 2\cos 1 + 2|x|\sin|x| - 2\sin 1)$

106. $T(x\cos^4|x|) = \frac{1}{16}(\cos^4|x| - \cos^4 1) + \frac{1}{4}(|x|\cos^2|x|\sin|x| - \cos^3 1\sin 1)$
$+ \frac{3}{16}(\cos^2|x| - \cos^2 1 + |x|^2 - 1) + \frac{3}{8}(+|x|\cos|x|\sin|x| - \cos 1\sin 1)$

In general:

$$T(x\cos^n|x|) = \frac{1}{n+1}(\cos^n|x|\sin|x| - \cos^n 1\sin 1) + \frac{n}{n+1}T(x\cos^{n-1}|x|)$$
$$-\frac{1}{n+1}T\left(\frac{x}{|x|}\cos^n|x|\sin|x|\right)$$
$$= \frac{1}{(n+1)^2}(\cos^{n+1}|x| - \cos^{n+1}) + \frac{1}{n+1}(|x|\cos^n|x|\sin|x| - \cos^n 1\sin 1)$$
$$+\frac{n}{n+1}T(x\cos^{n-1}|x|)$$

2.13. Integrals of the form $x\sin^n|x|\cos|x|$

107. $T(x\sin|x|\cos|x|) = \frac{1}{2}(|x|\sin^2|x| - \sin^2 1) + \frac{1}{4}(\sin|x|\cos|x| - \sin 1\cos 1 - |x| + 1)$

108. $T(x\sin^2|x|\cos|x|) = \frac{1}{3}(|x|\sin^3|x| - \sin^3 1) + \frac{1}{9}(\sin^2|x|\cos|x| - \sin^2 1\cos 1 + 2\cos|x| - 2\cos$

109. $T(x\sin^3|x|\cos|x|) = \frac{1}{4}(|x|\sin^4|x| - \sin^4 1) + \frac{1}{16}(\sin^3|x|\cos|x| - \sin^3 1\cos 1)$
$+ \frac{3}{32}(\sin|x|\cos|x| - \sin 1\cos 1 - |x| + 1)$

110. $T(x\sin^4|x|\cos|x|) = \frac{1}{5}(|x|\sin^5|x| - \sin^5 1) + \frac{1}{25}(\sin^4|x|\cos|x| - \sin^4 1\cos 1)$
$+ \frac{4}{75}(\sin^2|x|\cos|x| - \sin^2 1\cos 1 + 2\cos|x| - 2\cos 1)$

In general:

$$T(x\sin^n|x|\cos|x|) = \frac{1}{n+1}(|x|\sin^{n+1}|x| - \sin^{n+1}1) - \frac{1}{n+1}T\left(\frac{x}{|x|}\sin^{n+1}|x|\right)$$

2.14. Integrals of the form $x \cos^n |x| \sin |x|$

111. $T(x \cos |x| \sin |x|) = \frac{1}{2}(-|x| \cos^2 |x| + \cos^2 1) + \frac{1}{4}(\cos |x| \sin |x| - \cos 1 \sin 1 + |x| - 1)$

112. $T(x \cos^2 |x| \sin |x|) = \frac{1}{3}(-|x| \cos^3 |x| + \cos^3 1) + \frac{1}{9}(\cos^2 |x| \sin |x| - \cos^2 1 \sin 1 + 2 \sin |x| - 2 \sin 1)$

113. $T(x \cos^3 |x| \sin |x|) = \frac{1}{4}(-|x| \cos^4 |x| + \cos^4 1) + \frac{1}{16}(\cos^3 |x| \sin |x| - \cos^3 1 \sin 1)$
$+ \frac{3}{32}(\cos |x| \sin |x| - \cos 1 \sin 1 + |x| - 1)$

114. $T(x \cos^4 |x| \sin |x|) = \frac{1}{5}(-|x| \cos^5 |x| + \cos^5 1) + \frac{1}{25}(\cos^4 |x| \sin |x| - \cos^4 1 \sin 1$
$+ \frac{4}{75}(\cos^2 |x| \sin |x| - \cos^2 1 \sin 1 + 2 \sin |x| - 2 \sin 1)$

In general:

$$T(x \cos^n |x| \sin |x|) = \frac{1}{n+1}(-|x| \cos^{n+1} |x| + \cos^{n+1} 1) + \frac{1}{n+1} T\left(\frac{x}{|x|} \cos^{n+1} |x|\right)$$

2.15. Integrals of the form $x \sin^n |x| \cos^2 |x|$

115. $T(x \sin |x| \cos^2 |x|) = \frac{1}{9}(-\sin^3 |x| + \sin^3 1) + \frac{1}{3}(|x| \sin^2 |x| \cos |x| - \sin^2 1 \cos 1$
$+ \sin |x| - \sin 1 - |x| \cos |x| + \cos 1)$

116. $T(x \sin^2 |x| \cos^2 |x|) = \frac{1}{16}(-\sin^4 |x| + \sin^4 1) + \frac{1}{4}(|x| \sin^3 |x| \cos |x| - \sin^3 1 \cos 1)$
$+ \frac{1}{16}(\sin^2 |x| - \sin^2 1 + |x|^2 - 1 - 2|x| \sin |x| \cos |x| + 2 \sin 1 \cos 1)$

117. $T(x \sin^3 |x| \cos^2 |x|) = \frac{1}{25}(-\sin^5 |x| + \sin^5 1)$
$+ \frac{1}{5}(|x| \sin^4 |x| \cos |x| - \sin^4 1 \cos 1) + \frac{1}{45}(\sin^3 |x| - \sin^3 1)$
$+ \frac{1}{15}(-|x| \sin^2 |x| \cos |x| + \sin^2 1 \cos 1$
$+ 2 \sin |x| - 2 \sin 1 - 2|x| \cos |x| + 2 \cos 1)$

118. $T(x \sin^4 |x| \cos^2 |x|) = \frac{1}{36}(-\sin^6 |x| + \sin^6 1) + \frac{1}{6}(|x| \sin^5 |x| \cos |x| - \sin^5 1 \cos 1)$
$+ \frac{1}{96}(\sin^4 |x| - \sin^4 1) + \frac{1}{24}(-|x| \sin^3 |x| \cos |x| + \sin^3 1 \cos 1)$
$+ \frac{1}{32}(\sin^2 |x| - \sin^2 + |x|^2 - 1 - 2|x| \sin |x| \cos |x| + 2 \sin 1 \cos 1)$

In general:

$$T(x \sin^n |x| \cos^2 |x|) = T(x \sin^n |x|) - T(x \sin^{n+2} |x|)$$

2.16. Integrals of the form $x \cos^n |x| \sin^2 |x|$

$$119.\quad T(x\cos|x|\sin^2|x|) = \tfrac{1}{9}(-\cos^3|x| + \cos^3 1) + \tfrac{1}{3}(-|x|\cos^2|x|\sin|x| + \cos^2 1\sin 1 + \cos|x| - \cos 1 + |x|\sin|x| - \sin 1)$$

$$120.\quad T(x\cos^2|x|\sin^2|x|) = \tfrac{1}{16}(-\cos^4|x| + \cos^4 1) + \tfrac{1}{4}(1 - |x|\cos^3|x|\sin|x| + \cos^3 1\sin 1) + \tfrac{1}{16}(\cos^2|x| - \cos^2 1 + |x|^2 - 1 + 2|x|\cos|x|\sin|x| - 2\cos 1\sin 1)$$

$$121.\quad T(x\cos^3|x|\sin^2|x|) = \tfrac{1}{25}(-\cos^5|x| + \cos^5 1) + \tfrac{1}{5}(-|x|\cos^4|x|\sin|x| - \cos^4 1\sin 1) + \tfrac{1}{45}(\cos^3|x| - \cos^3 1) + \tfrac{1}{15}(|x|\cos^2|x|\sin|x| - \cos^2 1\sin 1 + 2\cos|x| - 2\cos 1 + 2|x|\cos|x| - 2\cos 1)$$

$$122.\quad T(x\cos^4|x|\sin^2|x|) = \tfrac{1}{36}(-\cos^6|x| + \cos^6 1) + \tfrac{1}{6}(-|x|\cos^5|x|\sin|x| + \cos^5 1\sin 1) + \tfrac{1}{96}(\cos^4|x| - \cos^4 1) + \tfrac{1}{24}(|x|\cos^3|x|\sin|x| - \cos^3 1\sin 1) + \tfrac{1}{32}(\cos^2|x| - \cos^2 1 + |x|^2 - 1 + 2|x|\cos|x|\sin|x| - 2\cos 1\sin 1)$$

In general:

$$T(x\cos^n|x|\sin^2|x|) = T(x\cos^n|x|) - T(x\cos^{n+2}|x|)$$

2.17. Integrals of the form $x \sin^n |x| \cos^3 |x|$

$$123.\quad T(x\sin|x|\cos^3|x|) = \tfrac{1}{4}(-|x|\sin^4|x| + \sin^4 1) + \tfrac{1}{16}(-\sin^3|x|\cos|x| + \sin^3 1\cos 1) + \tfrac{1}{2}(\sin^2|x| - \sin^2 1) + \tfrac{5}{32}(2|x|\sin|x|\cos|x| - 2\sin 1\cos 1 + |x| - 1)$$

$$124.\quad T(x\sin^2|x|\cos^3|x|) = \tfrac{1}{5}(-|x|\sin^5|x| + \sin^5 1) + \tfrac{1}{25}(-\sin^4|x|\cos|x| + \sin^4 1\cos 1) + \tfrac{1}{3}(|x|\sin^3|x| - \sin^3 1) + \tfrac{9}{225}(|x|\sin^2|x|\cos|x| - \sin^2 1\cos 1 + 2|x|\cos|x| - 2\cos 1)$$

$$125.\quad T(x\sin^3|x|\cos^3|x|) = \tfrac{1}{6}(-|x|\sin^6|x| + \sin^6 1) + \tfrac{1}{36}(-\sin^5|x|\cos|x| + \sin^5 1\cos 1) + \tfrac{1}{4}(|x|\sin^4|x| - \sin^4 1) + \tfrac{1}{36}(\sin^3|x|\cos|x| - \sin^3 1\cos 1) + \tfrac{1}{24}(\sin|x|\cos|x| - \sin 1\cos 1 - |x| - 1)$$

$$126.\quad T(x\sin^4|x|\cos^3|x|) = \tfrac{1}{7}(-|x|\sin^7|x| + \sin^7 1) + \tfrac{1}{49}(-\sin^6|x|\cos|x| + \sin^6 1\cos 1) + \tfrac{1}{5}(|x|\sin^5|x| - \sin^5 1) + \tfrac{19}{1225}(\sin^4|x|\cos|x| - \sin^4 1\cos 1) + \tfrac{76}{3675}(\sin^2|x|\cos|x| - \sin^2 1\cos 1 + 2\cos|x| - 2\cos 1)$$

In general:

$$T(x\sin^n|x|\cos^3|x|) = T(x\sin^n|x|\cos|x|) - T(x\sin^{n+2}|x|\cos|x|)$$

2.18. Integrals of the form $x \cos^n |x| \sin^3 |x|$

127. $T(x\cos^{l}|x|\sin^3|x|) = \frac{1}{4}(|x|\cos^4|x| - \cos^4 1) + \frac{1}{16}(-\cos^3|x|\sin|x| + \cos^3 1\sin 1)$
$+ \frac{1}{2}(-\cos^2|x| + \cos^2 1) + \frac{5}{32}(2|x|\cos|x|\sin|x| - 2\cos 1\sin 1 + |x| - 1)$

128. $T(x\cos^2|x|\sin^3|x|) = \frac{1}{5}(|x|\cos^5|x| - \cos^5 1)$
$+ \frac{1}{25}(-\cos^4|x|\sin|x| + \cos^4 1\sin 1) + \frac{1}{3}(-|x|\cos^3|x| + \cos^3 1)$
$+ \frac{9}{225}(|x|\cos^2|x|\sin|x| - \cos^2 1\sin 1 + 2|x|\sin|x| - 2\sin 1)$

129. $T(x\cos^3|x|\sin^3|x|) = \frac{1}{6}(+|x|\cos^6|x| - \cos^6 1) + \frac{1}{36}(-\cos^5|x|\sin|x| + \cos^5 1\sin 1)$
$+ \frac{1}{4}(-|x|\cos^4|x| + \cos^4 1) + \frac{1}{36}(\cos^3|x|\sin|x| - \cos^3 1\sin 1)$
$+ \frac{1}{24}(\cos|x|\sin|x| - \cos 1\sin 1 - |x| + 1)$

130. $T(x\cos^4|x|\sin^3|x|) = \frac{1}{7}(+|x|\cos^7|x| - \cos^7 1) + \frac{1}{49}(-\cos^6|x|\sin|x| + \cos^6 1\sin 1)$
$+ \frac{1}{5}(-|x|\cos^5|x| + \cos^5 1) + \frac{19}{1225}(\cos^4|x|\sin|x| - \cos^4 1\sin 1)$
$+ \frac{76}{3675}(\cos^2|x|\sin|x| - \cos^2 1\sin 1 + 2\sin|x| - 2\sin 1)$

In general:

$$T(x\cos^n|x|\sin^3|x|) = T(x\cos^n|x|\sin|x|) - T(x\cos^{n+2}|x|\sin|x|)$$

2.19. Integrals of the form $x\sin^n|x|\cos^4|x|$

131. $T(x\sin|x|\cos^4|x|) = \frac{1}{25}(\sin^5|x| - \sin^5 1)$
$+ \frac{1}{5}(-|x|\sin^4|x|\cos|x| + \sin^4 1\cos 1) + \frac{2}{15}(-\sin^3|x| + \sin^3 1)$
$+ \frac{2}{5}(|x|\sin^2|x|\cos|x| - \sin^2 1\cos 1) + \frac{1}{5}(+\sin|x| - \sin 1 - |x|\cos|x| + \cos 1)$

132. $T(x\sin^2|x|\cos^4|x|) = \frac{1}{36}(\sin^6|x| - \sin^6 1) + \frac{1}{6}(-|x|\sin^5|x|\cos|x| + \sin^5 1\cos 1)$
$+ \frac{7}{96}(-\sin^4|x| + \sin^4 1) + \frac{7}{24}(|x|\sin^3|x|\cos|x| - \sin^3 1\cos 1)$
$+ \frac{1}{32}(\sin^2|x| - \sin^2 1 + |x|^2 - 1 - 2|x|\sin|x|\cos|x| + 2\sin 1\cos 1)$

133. $T(x\sin^3|x|\cos^4|x|) = \frac{1}{49}(\sin^7|x| - \sin^7 1)$
$+ \frac{1}{7}(-|x|\sin^6|x|\cos|x| + \sin^6 1\cos 1) + \frac{8}{175}(-\sin^5|x| + \sin^5 1)$
$+ \frac{8}{35}(|x|\sin^4|x|\cos|x| - \sin^4 1\cos 1) + \frac{1}{105}(\sin^3|x| - \sin^3 1)$
$+ \frac{1}{35}(-|x|\sin^2|x|\cos|x| + \sin^2 1\cos 1 + 2\sin|x| - 2\sin 1 - 2|x|\cos|x| + 2\cos 1)$

134. $T(x\sin^4|x|\cos^4|x|) = \frac{1}{64}(\sin^8|x| - \sin^8 1) + \frac{1}{8}(-|x|\sin^7|x|\cos|x| + \sin^7 1\cos 1)$
$+ \frac{1}{32}(-\sin^6|x| + \sin^6 1) + \frac{3}{16}(|x|\sin^5|x|\cos|x| - \sin^5 1\cos 1)$
$+ \frac{1}{256}(\sin^4|x| - \sin^4 1) + \frac{1}{64}(-|x|\sin^3|x|\cos|x| + \sin^3 1\cos 1)$
$+ \frac{3}{256}(\sin^2|x| - \sin^2 + |x|^2 - 1 - 2|x|\sin|x|\cos|x| + 2\sin 1\cos 1)$

In general:

$$T(x\sin^n|x|\cos^4|x|) = T(x\sin^{n+4}|x|) - 2T(x\sin^{n+2}|x|) + T(x\sin^n|x|)$$

2.20. Integrals of the form $x\cos^n|x|\sin^4|x|$

135. $$\begin{aligned} T(x\cos|x|\sin^4|x|) &= \tfrac{1}{25}(\cos^5|x| - \cos^5 1) \\ &+ \tfrac{1}{5}(|x|\cos^4|x|\sin|x| - \cos^4 1\sin 1) + \tfrac{2}{15}(-\cos^3|x| + \cos^3 1) \\ &+ \tfrac{2}{5}(-|x|\cos^2|x|\sin|x| + \cos^2 1\sin 1) + \tfrac{1}{5}(\cos|x| - \cos 1 + |x|\sin|x| - \sin 1) \end{aligned}$$

136. $$\begin{aligned} T(x\cos^2|x|\sin^4|x|) &= \tfrac{1}{36}(\cos^6|x| - \cos^6 1) + \tfrac{1}{6}(|x|\cos^5|x|\sin|x| - \cos^5 1\sin 1) \\ &+ \tfrac{7}{96}(-\cos^4|x| + \cos^4 1) + \tfrac{7}{24}(-|x|\cos^3|x|\sin|x| + \cos^3 1\sin 1) \\ &+ \tfrac{1}{32}(\cos^2|x| - \cos^2 1 + |x|^2 - 1 + 2|x|\cos|x|\sin|x| - 2\cos 1\sin 1) \end{aligned}$$

137. $$\begin{aligned} T(x\cos^3|x|\sin^4|x|) &= \tfrac{1}{49}(\cos^7|x| - \cos^7 1) \\ &+ \tfrac{1}{7}(|x|\cos^6|x|\sin|x| - \cos^6 1\sin 1) + \tfrac{8}{175}(-\cos^5|x| + \cos^5 1) \\ &+ \tfrac{8}{35}(-|x|\cos^4|x|\sin|x| + \cos^4 1\sin 1) + \tfrac{1}{105}(\cos^3|x| - \cos^3 1) \\ &+ \tfrac{1}{35}(|x|\cos^2|x|\sin|x| - \cos^2 1\sin 1 + 2\cos|x| - 2\cos 1 + 2|x|\cos|x| - 2\cos \end{aligned}$$

138. $$\begin{aligned} T(x\cos^4|x|\sin^4|x|) &= \tfrac{1}{64}(\cos^8|x| + \cos^8 1) + \tfrac{1}{8}(|x|\cos^7|x|\sin|x| - \cos^7 1\sin 1) \\ &+ \tfrac{1}{32}(-\cos^6|x| + \cos^6 1) + \tfrac{3}{16}(-|x|\cos^5|x|\sin|x| + \cos^5 1\sin 1) \\ &+ \tfrac{1}{256}(\cos^4|x| - \cos^4 1) + \tfrac{1}{64}(|x|\cos^3|x|\sin|x| - \cos^3 1\sin 1) \\ &+ \tfrac{3}{256}(\cos^2|x| - \cos^2 + |x|^2 - 1 + 2|x|\cos|x|\sin|x| - 2\cos 1\sin 1) \end{aligned}$$

In general:

$$T(x\cos^n|x|\sin^4|x|) = T(x\cos^{n+4}|x|) - 2T(x\cos^{n+2}|x|) + T(x\cos^n|x|)$$

2.21. Integrals of the form $x\sin^n|x|\cos^5|x|$

139. $$\begin{aligned} T(x\sin|x|\cos^5|x|) &= \tfrac{1}{6}(|x|\sin^6|x| - \sin^6 1) + \tfrac{1}{36}(\sin^5|x|\cos|x| - \sin^5 1\cos 1) \\ &+ \tfrac{1}{2}(-|x|\sin^4|x| + \sin^4 1) + \tfrac{13}{144}(-\sin^3|x|\cos|x| + \sin^3 1\cos 1) \\ &+ \tfrac{1}{2}(|x|\sin^2|x| - \sin^2 1) + \tfrac{11}{96}(\sin|x|\cos|x| - \sin 1\cos 1 - |x| + 1) \end{aligned}$$

140. $$\begin{aligned} T(x\sin^2|x|\cos^5|x|) &= \tfrac{1}{7}(|x|\sin^7|x| - \sin^7 1) \\ &+ \tfrac{1}{49}(\sin^6|x|\cos|x| - \sin^6 1\cos 1) + \tfrac{2}{5}(-|x|\sin^5|x| + \sin^5 1) \\ &+ \tfrac{68}{1225}(-\sin^4|x|\cos|x| + \sin^4 1\cos 1) + \tfrac{1}{3}(|x|\sin^3|x| - \sin^3 1) \\ &+ \tfrac{409}{11025}(\sin^2|x|\cos|x| - \sin^2 1\cos 1 + 2|x|\cos|x| - 2\cos 1) \end{aligned}$$

141. $$\begin{aligned} T(x\sin^3|x|\cos^5|x|) &= \tfrac{1}{8}(|x|\sin^8|x| - \sin^8 1) + \tfrac{1}{64}(\sin^7|x|\cos|x| - \sin^7 1\cos 1) \\ &+ \tfrac{1}{3}(-|x|\sin^6|x| + \sin^6 1) + \tfrac{43}{1152}(-\sin^5|x|\cos|x| + \sin^5 1\cos 1) \\ &+ \tfrac{1}{4}(|x|\sin^4|x| - \sin^4 1) + \tfrac{73}{4608}(\sin^3|x|\cos|x| - \sin^3 1\cos 1) \\ &+ \tfrac{73}{3072}(\sin|x|\cos|x| - \sin 1\cos 1 - |x| + 1) \end{aligned}$$

142. $$\begin{aligned} T(x\sin^4|x|\cos^5|x|) &= \tfrac{1}{9}(|x|\sin^9|x| - \sin^9 1) + \tfrac{1}{81}(\sin^8|x|\cos|x| - \sin^8 1\cos 1) \\ &+ \tfrac{2}{7}(-|x|\sin^7|x| + \sin^7 1) + \tfrac{106}{3969}(-\sin^6|x|\cos|x| + \sin^6 1\cos 1) \\ &+ \tfrac{1}{5}(|x|\sin^5|x| - \sin^5 1) + \tfrac{263}{33075}(\sin^4|x|\cos|x| - \sin^4 1\cos 1) \\ &+ \tfrac{1052}{99225}(\sin^2|x|\cos|x| - \sin^2 1\cos 1 + 2\cos|x| - 2\cos 1) \end{aligned}$$

In general:

$$T(x\sin^n|x|\cos^5|x|) = T(x\sin^{n+4}|x|\cos|x|) - 2T(x\sin^{n+2}|x|\cos|x|) + T(x\sin^n|x|\cos|x|)$$

2.22. Integrals of the form $x\cos^n|x|\sin^5|x|$

143. $T(x\cos|x|\sin^5|x|) = \frac{1}{6}(-|x|\cos^6|x| + \cos^6 1) + \frac{1}{36}(\cos^5|x|\sin|x| - \cos^5 1\sin 1)$
$+ \frac{1}{2}(|x|\cos^4|x| - \cos^4 1) + \frac{13}{144}(-\cos^3|x|\sin|x| + \cos^3 1\sin 1)$
$+ \frac{1}{2}(-\cos^2|x| + \cos^2 1) + \frac{11}{96}(\cos|x|\sin|x| - \cos 1\sin 1 + |x| - 1)$

144. $T(x\cos^2|x|\sin^5|x|) = \frac{1}{7}(-|x|\cos^7|x| + \cos^7 1)$
$+ \frac{1}{49}(\cos^6|x|\sin|x| - \cos^6 1\sin 1) + \frac{2}{5}(|x|\cos^5|x| - \cos^5 1)$
$+ \frac{68}{1225}(-\cos^4|x|\sin|x| + \cos^4 1\sin 1) + \frac{1}{3}(-|x|\cos^3|x| + \cos^3 1)$
$+ \frac{409}{11025}(\cos^2|x|\sin|x| - \cos^2 1\sin 1 + 2\sin|x| - 2\sin 1)$

145. $T(x\cos^3|x|\sin^5|x|) = \frac{1}{8}(-|x|\cos^8|x| + \cos^8 1) + \frac{1}{64}(\cos^7|x|\sin|x| - \cos^7 1\sin 1)$
$+ \frac{1}{3}(|x|\cos^6|x| - \cos^6 1) + \frac{43}{1152}(-\cos^4|x|\sin|x| + \cos^4 1\sin 1)$
$+ \frac{1}{4}(-|x|\cos^4|x| + \cos^4 1) + \frac{73}{4608}(\cos^3|x|\sin|x| - \cos^3 1\sin 1)$
$+ \frac{73}{3072}(\cos|x|\sin|x| - \cos 1\sin 1 + |x| - 1)$

146. $T(x\cos^4|x|\sin^5|x|) = \frac{1}{9}(-|x|\cos^9|x| + \cos^9 1) + \frac{1}{81}(\cos^8|x|\sin|x| - \cos^8 1\sin 1)$
$+ \frac{2}{7}(|x|\cos^7|x| - \cos^7 1) + \frac{106}{3969}(-\cos^6|x|\sin|x| + \cos^6 1\sin 1)$
$+ \frac{1}{5}(-|x|\cos^5|x| + \cos^5 1) + \frac{263}{33075}(\cos^4|x|\sin|x| - \cos^4 1\sin 1)$
$+ \frac{1052}{99225}(\cos^2|x|\sin|x| - \cos^2 1\sin 1 + 2\sin|x| - 2\sin 1)$

In general:

$$T(x\cos^n|x|\sin^5|x|) = T(x\cos^{n+4}|x|\sin|x|) - 2T(x\cos^{n+2}|x|\sin|x|) + T(x\cos^n|x|\sin|x|)$$

3. Integrals in base of exponential function

147. $T(xe^{|x|}) = e^{|x|}(2|x| - 4) - 6e$

148. $T(\frac{x}{|x|}e^{|x|}) = 2(e^{|x|} - e)$

4. Integrals with the natural logarithm

4.1. Integrals of the form $x(\ln|x|)^n$

149. $T(x\ln|x|) = \frac{1}{2}|x|^2\ln|x| + \frac{1}{4}(-|x|^2 + 1)$

150. $T(x(\ln|x|)^2) = \frac{1}{2}\left(|x|^2(\ln|x|)^2 - |x|^2\ln|x|\right) + \frac{1}{4}(|x|^2 - 1)$

151. $T(x(\ln|x|)^3) = \frac{1}{2}|x|^2(\ln|x|)^3 - \frac{3}{4}\left(|x|^2(\ln|x|)^2 - |x|^2\ln|x|) + \frac{3}{8}(|x|^2 - 1\right)$

In general:

$$T(x\ln|x|^n) = \frac{1}{2}|x|^2(\ln|x|)^n - \frac{n}{2}T(x\ln|x|^{n-1})$$

4.2. Integrals of the form $\frac{x}{|x|}(\ln|x|)^n$

152. $T(\frac{x}{|x|}\ln|x|) = |x|\ln|x| - |x| + 1$

153. $T(\frac{x}{|x|}(\ln|x|)^2) = |x|(\ln|x|)^2 - 2|x|\ln|x| - 2|x| + 2$

154. $T(\frac{x}{|x|}(\ln|x|)^3) = |x|(\ln|x|)^3 - 3|x|(\ln|x|)^2 + 6|x|\ln|x| + 6|x| - 6$

In general:

$$T\left[\frac{x}{|x|}(\ln|x|)^n\right] = |x|(\ln|x|)^n - nT\left[\frac{x}{|x|}(\ln|x|)^{n-1}\right]$$

4.3. Integrals of the form $x|x|^n(\ln|x|)^m$

155. $T(x|x|\ln|x|) = \frac{1}{3}|x|^3\ln|x| - \frac{1}{9}|x|^3 + \frac{1}{9}$

156. $T(x|x|^2\ln|x|) = \frac{1}{4}|x|^4(\ln|x|) - \frac{1}{16}|x|^4 + \frac{1}{16}$

157. $T(x|x|(\ln|x|)^2) = \frac{1}{3}|x|^3(\ln|x|)^2 - \frac{2}{9}|x|^3\ln|x| + \frac{2}{27}(|x|^3 - 1)$

158. $T(x|x|^2(\ln|x|)^2) = \frac{1}{4}|x|^4(\ln|x|)^2 - \frac{1}{8}|x|^4\ln|x| + \frac{1}{32}(|x|^4 - 1)$

In general:

$$T[x|x|^{n-2}(\ln|x|)^m] = \frac{1}{n}|x|^n(\ln|x|)^m - \frac{m}{n}T[x|x|^{n-2}(\ln|x|)^{m-1}],\ n \neq 0$$

4.4. Further examples

159. $T(\ln|x|) = \frac{-1}{3}x\ln|x| + \frac{1}{9}x$

160. $T(|x|\ln|x|) = \frac{-1}{4}x|x|\ln|x| + \frac{1}{12}x|x|$

161. $T(x^{-2}\ln|x|) = \frac{1}{3x}(\ln|x| - 1)$

162. $T(|x|^{-2}\ln|x|) = \frac{-x}{3|x|^2}(\ln|x| - 1)$

Appendix B

Discrete fundamental solutions

Discrete fundamental solutions in $\mathbf{R}^2$ and $\mathbf{R}^3$

I) **in $\mathbf{R}^3$:**

Problem: $-\Delta_h E_h(x,y) = \delta_h(x,y) = \begin{cases} \frac{1}{h^3} & x = y \\ 0 & \text{else} \end{cases}$

$$\begin{aligned}
F_h\left(-\Delta_h E_h(x,y)\right) &= d^2 F_h E_h(x,y) \\
F_h\delta_h = \frac{h^3}{\sqrt{2\pi}^3}\frac{1}{h^3}\mathrm{e}^{\mathrm{i}(y,\xi)} &= \frac{1}{\sqrt{2\pi}^3}\mathrm{e}^{\mathrm{i}(y,\xi)} \quad \xi \in Q_h \\
\Rightarrow\ d^2 F_h E_h(x,y) &= \frac{1}{\sqrt{2\pi}^3}\mathrm{e}^{\mathrm{i}(y,\xi)} \\
F F_h E_h(x,y) &= \frac{1}{\sqrt{2\pi}^3} F\left(\frac{\mathrm{e}^{\mathrm{i}(y,\xi)}}{d^2}\right) \\
E_h(x,y) &= \frac{1}{(2\pi)^3} R_h \int\limits_{[-\pi/h,\pi/h]^3} \frac{\mathrm{e}^{\mathrm{i}(y,\xi)}}{d^2}\mathrm{e}^{-\mathrm{i}(x,\xi)}\mathrm{d}\xi
\end{aligned}$$

II) **in $\mathbf{R}^2$:** $\quad (h = 1, y = \vec{0})$

$$\begin{aligned}
E_h(x,y) &= \frac{1}{(2\pi)^2} R_h \int\limits_{[-\pi,\pi]^2} \frac{\mathrm{e}^{-\mathrm{i}(x,\xi)} - 1}{d^2}\mathrm{d}\xi \\
&= \frac{1}{(\pi)^2} R_h \int\limits_{[0,\pi]^2} \frac{\cos x_1\xi_1 \cos x_2\xi_2 - 1}{d^2}\mathrm{d}\xi
\end{aligned}$$

on the diagonals we obtain

$$E_h(n,n,0,0) = -\frac{1}{\pi}\left[1 + \frac{1}{3} + \frac{1}{5} + \frac{1}{7} + \ldots + \frac{1}{2|n| - 1}\right]$$

$$E_h(0,0,0,0) = 0$$

Special discrete fundamental solutions

1. Half space and boundary condition of Dirichlet type

$$\begin{aligned} -\Delta_h G_h^+(x,y) &= \delta_h(x,y) \\ G_h^+(x_1,x_2,0,y_1,y_2,y_3) &= 0 \\ \textbf{solution: } G_h^+(x,y) &= E_h(x-y) - E_h(x-\bar{y}) \end{aligned}$$

2. One quadr. and boundary conditions of Dirichlet type

$$\begin{aligned} -\Delta_h G_h^{++}(x,y) &= \delta_h(x,y) \\ G_h^{++}(0,x_2,y_1,y_2) &= 0, \\ G_h^{++}(x_1,0,y_1,y_2) &= 0 \\ \textbf{solution: } G_h^{++}(x,y) &= E_h(x-y) - E_h(x-\bar{y}) - E_h(x-y') + E_h(x-\bar{y}') \end{aligned}$$

3. One quadr. and central diff. in the direction of the outer normal

$$\begin{aligned} -\Delta_h G_h^{++}(x,y) &= \delta_h(x,y) \\ G^{++}_{\overset{o}{h,x_1}}(0,x_2,y_1,y_2) &= 0, \\ G^{++}_{\overset{o}{h,x_2}}(x_1,0,y_1,y_2) &= 0 \\ \textbf{solution: } G_h^{++}(x,y) &= E_h(x-y) + E_h(x-\bar{y}) + E_h(x-y') + E_h(x-\bar{y}') \end{aligned}$$

4. One quadr. and central diff. in the direction of the tangent

$$\begin{aligned} -\Delta_h G_h^{++}(x,y) &= \delta_h(x,y) \\ G^{++}_{\overset{o}{h,x_1}}(x_1,0,y_1,y_2) &= 0, \\ G^{++}_{\overset{o}{h,x_2}}(0,x_2,y_1,y_2) &= 0 \\ \textbf{solution: } G_h^{++}(x,y) &= E_h(x-y) - E_h(x-\bar{y}) - E_h(x-y') + E_h(x-\bar{y}') \end{aligned}$$

5. One quadr. + forward diff. in the direction of the outer normal and the tangent

$$\begin{aligned} -\Delta_h G_h^{++}(x,y) &= \delta_h(x,y) \\ G_{h,\bar{x}_1}^{++}(0,x_2,y_1,y_2) &= 0 \\ G_{h,\bar{x}_1}^{++}(x_1,0,y_1,y_2) &= 0 \\ \textbf{solution: } G_h^{++}(x,y) &= E_h(x-y) - E_h(x-\bar{y}) + E_h(x-y') - E_h(x-\bar{y}') + \\ & \quad hE_{h,x_1}(x-\bar{y}') - hE_{h,x_1}(x-y') \end{aligned}$$

Discrete fundamental solution of the Lamé–system

$$\Delta_h u + \alpha \operatorname{grad}_h^+ \operatorname{div}_h^- u = \begin{cases} \frac{1}{h^3} & \text{for } x = 0 \\ 0 & \text{else} \end{cases}$$

Solution:

$$u =$$

$$\begin{Bmatrix} F\left[\frac{1}{d}\right] - \frac{\alpha}{\alpha+1} F\left[\frac{\xi_1 \xi_{-1}}{d^2}\right] & -\frac{\alpha}{\alpha+1} F\left[\frac{\xi_1 \xi_{-2}}{d^2}\right] & -\frac{\alpha}{\alpha+1} F\left[\frac{\xi_1 \xi_{-3}}{d^2}\right] \\ -\frac{\alpha}{\alpha+1} F\left[\frac{\xi_2 \xi_{-1}}{d^2}\right] & F\left[\frac{1}{d}\right] - \frac{\alpha}{\alpha+1} F\left[\frac{\xi_2 \xi_{-2}}{d^2}\right] & -\frac{\alpha}{\alpha+1} F\left[\frac{\xi_2 \xi_{-3}}{d^2}\right] \\ -\frac{\alpha}{\alpha+1} F\left[\frac{\xi_3 \xi_{-1}}{d^2}\right] & -\frac{\alpha}{\alpha+1} F\left[\frac{\xi_3 \xi_{-2}}{d^2}\right] & F\left[\frac{1}{d}\right] - \frac{\alpha}{\alpha+1} F\left[\frac{\xi_3 \xi_{-3}}{d^2}\right] \end{Bmatrix}$$

with

$$d = \sum_{j=1}^{3}(2 - 2\cos hy_j)$$

$$\xi_{\pm j} = \frac{1}{h}\left[e^{\mp ihy_j} - 1\right]$$

$$F\left[\frac{1}{d}\chi_{\pi/h}\right] \quad - \quad \text{fundamental solution of } -\Delta_h$$

Fundamental solution of discrete Cauchy–Riemann operators

define discrete Cauchy–Riemann operators by D_h^+ and D_h^-

$$D_h^+ u_h^{k,j} = \frac{1-i}{2h}\left(u_h^{k+1,j+1} - u_h^{k,j}\right) + \frac{1+i}{2h}\left(u_h^{k,j+1} - u_h^{k+1,j}\right)$$

$$D_h^- u_h^{k,j} = \frac{1+i}{2h}\left(u_h^{k,j} - u_h^{k-1,j-1}\right) + \frac{1-i}{2h}\left(u_h^{k-1,j} - u_h^{k,j-1}\right).$$

i imaginary unit and $u_h^{k,j} = u_h(kh, jh)$

$$D_h^+ \to (-i)\left(\frac{\partial}{\partial x_1} + i\frac{\partial}{\partial x_2}\right) \quad \text{and} \quad D_h^- \to i\left(\frac{\partial}{\partial x_1} - i\frac{\partial}{\partial x_2}\right) \quad \text{for } h \to 0$$

We investigate the difference equation

$$D_h^+ E_h^+(x) = \delta_h(x) \qquad \text{for } x = (kh, jh)$$

Computation of $E_h^+(kh, jh)$ by the help of

$$E_h^+(kh, jh) = -D_h^- \tilde{E}_h(kh, jh),$$

where $\tilde{E}_h(kh, jh)$ solves the equation

$$-\tilde{\Delta}_h \tilde{E}_h^{\,k,j} = \frac{1}{2h^2}\left(4\tilde{E}_h^{\,k,j} - \tilde{E}_h^{\,k+1,j+1} - \tilde{E}_h^{\,k+1,j-1} - \tilde{E}_h^{\,k-1,j+1} - \tilde{E}_h^{\,k-1,j-1}\right)$$

$$= -D_h^+ D_h^- \tilde{E}_h^{\,k,j} = \begin{cases} h^{-2} & k = j = 0 \\ 0 & \text{otherwise} \end{cases}$$

$$h = 1 \Longrightarrow E_1^+(k,j) = E_{11}^+(k,j) + E_{12}^+(k,j)$$

with

$$E_{11}^+(k,j) = -\frac{1+i}{16\pi^2}\int_{\xi_1=-\pi}^{\pi}\int_{\xi_2=-\pi}^{\pi}\frac{\cos k\xi_1 \cos j\xi_2 - \cos(k-1)\xi_1 \cos(j-1)\xi_2}{1 - \cos\xi_1 \cos\xi_2}\,d\xi_1\,d\xi_2$$

and

$$E_{12}^+(k,j) = -\frac{1-i}{16\pi^2}\int_{\xi_1=-\pi}^{\pi}\int_{\xi_2=-\pi}^{\pi}\frac{\cos(k-1)\xi_1 \cos j\xi_2 - \cos k\xi_1 \cos(j-1)\xi_2}{1 - \cos\xi_1 \cos\xi_2}\,d\xi_1\,d\xi_2\,.$$

$$h > 0 \Longrightarrow t^{-1}E_{h/t}^+(x_1/t, x_2/t) = E_h^+(x_1, x_2)$$

Efficient computation of $\tilde{E}_1(k,j)$.

Substitute

$k = \alpha - \beta$ and $j = \alpha + \beta$ for $k + j = 2m,\ m \in \mathbf{Z}$

Then,

$$\tilde{E_1}(k,j) = 2\,E_1((k+j)/2\,,\,(j-k)/2) = 2\,E_1(\alpha,\beta).$$

Define

$\tilde{E_1}(k,j) = 0$ for $k + j = 2m + 1\,,\, m \in \mathbf{Z}$

For $k + j = 2m$ we have

$$E_1^+(k,j) = -\frac{1+i}{2}\Big(\tilde{E_1}(k,j) - \tilde{E_1}(k-1,j-1)\Big) = E_{11}^+(k,j)$$

and for $k + j = 2m + 1$ it follows that

$$E_1^+(k,j) = E_{12}^+(k,j)\,.$$

Remember:

$$\begin{aligned} E_1(n,n) &= -\frac{1}{\pi}\left[1 + \frac{1}{3} + \frac{1}{5} + \frac{1}{7} + \ldots + \frac{1}{2|n|-1}\right] \\ E_1(0,0) &= 0 \end{aligned}$$

Fundamental solution of the explicit heat conduction operator
Let $\mathbf{R}_h^2 = \{x = (x_1,x_2) : x_1 = kh$ and $x_2 = jh\ (k,j \in \mathbf{Z})\}$ and $\mathbf{R}_{h_t} = \{t = lh_t : l \in \mathbf{N})\}$. Further, we use the notations

$$\delta_h(x) = \begin{cases} \frac{1}{h^2} & \text{if } x = (0,0) \\ 0 & \text{if } x \in \mathbf{R}_h^2 \setminus \{(0,0)\} \end{cases} \quad \text{and} \quad \delta_{h_t}(t) = \begin{cases} \frac{1}{h_t} & \text{if } t = 0 \\ 0 & \text{if } t \in \mathbf{R}_{h_t} \setminus \{0\} \end{cases}$$

for the discrete delta function. The HEAVISIDE function will be denoted by $\Theta = \Theta(t)$. We study the explicit difference equation

$$\Big((-a^2\Delta_h + D_{h_t}^+)E_h\Big)(x,t) \;= \delta_{h,h_t}(x,t) = \delta_h(x)\delta_{h_t}(t),$$

where

$$\begin{aligned} (\Delta_h E_h)(x,t) \;&=\; \frac{1}{h^2}\Big[-4E_h(x_1,x_2,t) + E_h(x_1+h,x_2,t) \\ &\quad + E_h(x_1-h,x_2,t) + E_h(x_1,x_2+h,t) + E_h(x_1,x_2-h,t)\Big] \end{aligned}$$

and

$$(D_{h_t}^+ E_h)(x,t) \;=\; \frac{1}{h_t}\Big[E_h(x,t+h_t) - E_h(x,t)\Big].$$

The fundamental solution is represented by [103]

$$E_h(x,t) = \Theta(t)\,\Big((1 + a^2 h_t\Delta_h)^{t/h_t - 1}\delta_h\Big)(x).$$

Fundamental solution of the biharmonic operator $(n = 2, (\Delta_h)^2)$.

$$E_1(k,0) = \sum_{j=0}^{[k+2]} (-1)^{j+1}\binom{k}{2j} 2^{3j-3} J_{k,j} + \frac{k^2\pi}{4} - \frac{2k^2-1}{8}\ln 2 - \frac{k^2}{2} - \frac{3}{16}$$

with

$$J_{k,j} = \sum_{l=0}^{k-2_j} \sum_{m=0}^{j} (-1)^{l+m} \binom{k-2_j}{l} \binom{j}{m} 3^{-2j-l} \left(3K_{k,j,l,m} - L_{k,j,l,m}\right)$$

$$K_{k,j,l,m} = V_{k-1,k-j-l+m} \; L_{k,j,l,m} = V_{k-1,k-j-l+m+1}$$

$$V_{i,n} = \frac{i+1}{2^{i+2}} \left(-\sum_{p=1}^{2n} \frac{1}{p} \right) + \frac{1}{2^{i+2}} \left(\sum_{p=1}^{2n-1} \frac{p}{2n-p} - 1 \right) + \frac{i+1}{2^{i+2}} \left(\sum_{p=1}^{2n} \frac{(-1)^p}{p} + \ln 2 \right)$$

$$+ \frac{1}{2^{i+2}} \left(-\sum_{p=1}^{2n-1} \frac{p(-1)^{2n-p}}{2n-p} + \frac{1}{2} - 2n \ln 2 \right) + \sum_{s=1}^{i} \frac{i+1-5}{2^i+2-s} W_{s,n}$$

$$W_{i,k} = -\frac{1}{(2n+1)(2^{i-1})} \sum_{k=1}^{n} \prod_{j=1}^{n} \frac{2n+3-2j}{2i-2n-3+2j} + \prod_{j=1}^{n} \frac{2n-2j+1}{2i-2n-2+2j-1} \cdot Z_i$$

$$Z_i = \frac{1}{2i-1} \sum_{k=1}^{i-1} \frac{(2i-1) \cdot \ldots \cdot (2i-2k+1)}{2(i-1)(i-2) \cdot \ldots \cdot (i-k)} + \frac{2i-3)!!}{2^{i-1}(i-1)!} \cdot \frac{\pi}{4}$$

Remark:

$$Z_i = \int_0^1 \frac{1}{(1+t^2)^i} dt; \; W_{i,n} = \int_0^1 \frac{t^{2n}}{(1+t^2)^i} dt$$

These formulas can be used for approximation calculations.

Fundamental solution of discrete generalized Cauchy-Fueter operators

1. Fundamental solution of $D_h^+ D_h^- = \widetilde{\Delta} \to .\widetilde{E_h}$
2. Fundamental solution of $D_h^+ \to D_h^- \widetilde{E_h} = e_h^+$.

$$\widetilde{\Delta_h} f = \begin{pmatrix} -\Delta_h f_0 - (\partial_2\partial_{-3} - \partial_3\partial_{-2}) f_1 + (\partial_3\partial_{-1} - \partial_1\partial_{-3}) f_2 - (\partial_1\partial_{-2} - \partial_2\partial_{-1}) f_3 \\ -\Delta_h f_1 + (\partial_2\partial_{-3} - \partial_3\partial_{-2}) f_0 + (\partial_3\partial_{-1} - \partial_1\partial_{-3}) f_3 - (\partial_1\partial_{-2} - \partial_2\partial_{-1}) f_2 \\ -\Delta_h f_2 + (\partial_3\partial_{-1} - \partial_1\partial_{-3}) f_0 - (\partial_2\partial_{-3} - \partial_3\partial_{-2}) f_3 + (\partial_1\partial_{-2} - \partial_2\partial_{-1}) f_1 \\ -\Delta_h f_3 + (\partial_1\partial_{-2} - \partial_2\partial_{-1}) f_0 + (\partial_2\partial_{-3} - \partial_3\partial_{-2}) f_2 - (\partial_3\partial_{-1} - \partial_1\partial_{-3}) f_1 \end{pmatrix}$$

where $\partial_1 = D_{1,h}^+$, $\partial_{-i} = D_{1,h}^-$ and with

$$\widehat{\xi}_j = \frac{1}{h}[e^{-ihx_j} - 1], \widehat{\xi}_{-j} = \frac{1}{h}[1 - e^{-ihx_j}]$$

$$a = \widehat{\xi}_1\widehat{\xi}_{-3} - \widehat{\xi}_3\widehat{\xi}_{-2}, \quad b = \widehat{\xi}_3\widehat{\xi}_{-1} - \widehat{\xi}_1\widehat{\xi}_{-3}, \quad c = \widehat{\xi}_1\widehat{\xi}_{-2} - \widehat{\xi}_2\widehat{\xi}_{-1}, \quad d = \sum_{j=1}^{3} \widehat{\xi}_j\widehat{\xi}_{-j}$$

$$\Longrightarrow \widetilde{E_h} = R_h \widehat{F}\{[de_0 + ae_1 + be_2 + ce_3]^{-1}\}$$

$$\Longrightarrow e_h^+ = D_h^- R_h \widehat{F}\{[de_0 + ae_1 + be_2 + ce_3]^{-1}]\}$$

Appendix C

Discrete Teodorescu transform

As announced in Section 5.2 we will now describe the index sets for the explicit definition of the discrete operators. First we define some subsets of the discrete domain G_h and of its boundary, respectively.

$$
\begin{aligned}
G_h &= \{x \in G_h : dist(x, coG_h) \leq \sqrt{3}h\} \\
\partial G_{h,l} &= \{x \in G_h : \exists i \in \{1,2,3\}, \text{ with } V_{i,h}^{-} x \notin G_h\} \\
\partial G_{h,r} &= \{x \in \partial G_h : \exists\, i \in \{1,2,3\}, \text{ with } V_{i,h}^{+} x \notin G_h\} \\
\partial G_{h,l,i} &= \{x \in \partial G_h : V_{i,h}^{-} x \notin G_h\}, i = 1,2,3 \\
\partial G_{h,r,i} &= \{x \in \partial G_h : V_{i,h}^{+} x \notin G_h\}, i = 1,2,3 \\
\partial G_{h,l,i,j} &= \partial G_{h,l,i} \cap \partial G_{h,l,j},\ i,j \in \{1,2,3\} \\
\partial G_{h,l,i,j,k} &= \partial G_{h,l,i} \cap \partial G_{h,l,j} \cap \partial G_{h,l,k},\ i,j,k \in \{1,2,3\}
\end{aligned}
$$

We introduce the discrete analogues to the operator T_G by

$$
\begin{aligned}
(T_h^{+} f)(x) &:= \sum_{\overset{\circ}{G}_h \cup \partial G_{h,l}} e_h^{+}(x-y) f(y) h^3 - \sum_{\substack{k,j=1 \\ k<j}}^{3} \sum_{y \in \partial G_{h,l,j,k}} e_h^{+}(x-y) f(y) h^3 + \\
&\quad + \sum_{\substack{y \in \partial G_{h,l,i,j,k} \\ i \neq j \neq k}} e_h^{+}(x-y) f(y) h^3
\end{aligned}
$$

$$
\begin{aligned}
(T_h^{-} f)(x) &:= \sum_{\overset{\circ}{G}_h \cup \partial G_{h,r}} e_h^{-}(x-y) f(y) h^3 - \sum_{\substack{k,j=1 \\ k<j}}^{3} \sum_{y \in \partial G_{h,r,j,k}} e_h^{-}(x-y) f(y) h^3 + \\
&\quad + \sum_{\substack{y \in \partial G_{h,r,i,j,k} \\ i \neq j \neq k}} e_h^{-}(x-y) f(y) h^3 .
\end{aligned}
$$

Remark: In the theory of finite difference methods, sometimes boundary points are introduced in a different way by $x \in \partial G_h$ iff there exists an index i with $V_{h,i}^{+} x \notin G_h$ ($V_{h,i}^{-} x \notin G_h$). Because we also have to consider internal corner-points or rather internal edges, this approach is not possible for our purposes, otherwise we are not able to construct a discrete BOREL–POMPEIU formula.

We carry out further investigations only for the operators which are marked by "+" . Operators signified by "-" can be dealt in the same way.

Theorem: *We have*

$$(D_h^+ T_h^+ f)(x) = f(x) \quad x \in \text{ int } G_h.$$

Proof. Denoting the action of D_h^+ with respect to x by $D_{h,x}^+$, it follows that

$$(D_h^+ T_h^+ f)(x) = D_{h,x}^+ e_h^+(x-y) f(y) h^3 = f(x)$$

holds. #

Using the structure of the BOREL–POMPEIU formula, we introduce the following discrete analogue F_h^+ of the operator F_Γ.

$$(F_h^+ f)(x) = f(x) - (T_h^+ D_h^+ f)(x)$$

We have to verify that the operator F_h^+ depends only on the boundary values of the function f. Because the determination of the domain of the operator F_h^+ is necessary for theoretical considerations, we shall carry out the proof in a detailed manner.

Theorem: *Let $\alpha_i(y)$ be the components of the outward-pointing normal unit vector to Γ at the point y,*

$$\alpha(y) = \sum_{i=1}^{3} \alpha_i(y) e_i,$$

$$R_{ij} = V_{i,h}^+ \partial G_{h,l,j} \setminus \partial G_{h,l,j}, \; S_{ij} = \partial G_{h,l,j} \setminus V_{i,h}^+ \partial G_{h,l,j},$$

$$R_{ijk} = V_{i,h}^+ \partial G_{h,l,j,k} \setminus \partial G_{h,l,j,k}$$

and

$$S_{ijk} = \partial G_{h,l,j,k} \setminus V_{i,h}^+ \partial G_{h,l,j,k}.$$

The operator F_h^+ allows the following representation:

$$\begin{aligned}(F_h^+ f)(x) = & -\sum_{i=1}^{3} \sum_{\partial G_{h,r,i} \cup \partial G_{h,l,i}} e_h^+(x - V_{i,h}^- y) n f(y) h^2 \\ & + \sum_{i=1}^{3} \sum_{\substack{j=1 \\ j \neq i}} \sum_{y \in S_{ij}} e_h^+(x - V_{i,h}^- y) e_i f(y) h^2 \\ & - \sum_{i=1}^{3} \sum_{\substack{j,k=1 \\ j>k}}^{3} \sum_{y \in S_{ijk}} e_h^+(x - V - h, iy) e_i f(y) h^2 \\ & + \sum_{i=1}^{3} \sum_{\substack{y \in \partial G_{h,l,m,j,k} \\ m \neq j \neq k}} e_h^+(x-y) e_i f(y) h^2. \text{ for } x \in \text{ int } G_h\end{aligned}$$

Proof.

$$(F_h^+ f)(x) = f(x) - (T_h^+ D_h^+ f)(x)$$

$$= f(x) - \left(\sum_{\overset{\circ}{G}_h \cup \partial G_{h,l}} - \sum_{j>k} \sum_{\partial G_{h,l,j,k}} + \sum_{y \in \partial G_{h,l,m,j,k}} \right) \sum_{i=1}^{3} e_h^+(x-y) e_i [f(V_{i,h}^+ y) - f(y)] h^2$$

$$= f(x) - \sum_{i=1}^{3} \left(\sum_{V_{i,h}^+ (\overset{\circ}{G}_h \cup \partial G_{h,l})} - \sum_{j>k} \sum_{V_{i,h}^+ \partial G_{h,l,j,k}} + \sum_{V_{i,h}^+ \partial G_{h,l,m,j,k}} \right) e_h^+(x - V_{i,h}^- y) e_i f(y) h^2$$

$$+ \sum_{i=1}^{3} \left(\sum_{(\overset{\circ}{G}_h \cup \partial G_{h,l})} - \sum_{j>k} \sum_{\partial G_{h,l,j,k}} + \sum_{\partial G_{h,l,m,j,k}} \right) e_h^+(x-y) e_i f(y) h^2$$

$$= f(x) -$$

$$- \sum_{i=1}^{3} \left(\sum_{V_{i,h}^+ (\overset{\circ}{G}_h \cup \partial G_{h,l,i}} + \sum_{j \neq i} \sum_{V_{h,i}^+ \partial G_{h,l,j}} - \sum_{j>k} \sum_{V_{h,i}^+ \partial G_{h,l,j,k}} \right.$$

$$\left. + \sum_{V_{i,h}^+ \partial G_{h,l,m,j,k}} \right) e_h^+(x - V_{i,h}^- y) e_i f(y) h^2$$

$$+ \sum_{i=1}^{3} \left(\sum_{\overset{\circ}{G}_h \cup \partial G_{h,l}} - \sum_{j>k} \sum_{\partial G_{h,l,j,k}} + \sum_{\partial G_{h,l,m,j,k}} \right) e_h^+(x-y) e_i f(y) h^2$$

$$= - \sum_{i=1}^{3} (\sum_{\partial G_{h,l,j}} + \sum_{j \neq i} \sum_{V_{h,i}^+ \partial G_{h,l,j}} - \sum_{j>k} \sum_{V_{h,i}^+ \partial G_{h,l,m,j,k}} \bigg) e_h^+(x - V_{i,h}^- y) e_i f(y) h^2$$

$$+ \sum_{i=1}^{3} \left(\sum_{\overset{\circ}{G}_h \cup \partial G_{h,l}} - \sum_{j>k} \sum_{\partial G_{h,l,j,k}} + \sum_{\partial G_{h,l,m,j,k}} \right) e_h^+(x-y) e_i f(y) h^2 \qquad (0.1)$$

To obtain this formula we have used the following formula:

$$\sum_{i=1}^{3} \sum_{y \in \overset{\circ}{G}_h} [e_h^+(x - V_{i,h}^- y) - e_h^+(x-y)] e_i f(y) h^2$$

$$= \sum_{y \in \overset{\circ}{G}_h} [(e_h^+(x-y)) D_{h,x}^+] f(y) h^3 = f(x) x \in \text{ int } G_h,$$

because $e_h^+ = D_h^- E_h = (E_h D_h^-)$ holds. With the relation $V_{i,h}^+ \partial G_{h,l,j} = (\partial G_{h,l,j} \setminus S_{ij}) \cup R_{ij}$ formula (0.1) can be simplified. It follows that

$$\sum_{\substack{j=1 \\ j \neq i}}^{3} \sum_{V_{h,i}^+ \partial G_{h,l,j}} e_h^+(x - V_{i,h}^- y) e_i f(y) h^2 - \sum_{i=1}^{3} \sum_{\partial G_{h,l}} e_h^+(x-y) e_i f(y) h^2$$

$$= \sum_{\substack{j=1\\ j\neq i}}^{3} \sum_{\partial G_{h,l,j}} e_h^+(x - V_{i,h}^- y) e_i f(y) h^2 - \sum_{i,j=1}^{3} \sum_{\partial G_{h,l,j}} e_h^+(x - y) e_i f(y) h^2 -$$

$$- \sum_{\substack{j=1\\ j\neq i}}^{3} \sum_{S_{ij}} e_h^+(x - V_{i,h}^- y) e_i f(y) h^2 + \sum_{\substack{j=1\\ j\neq i}}^{3} \sum_{R_{ij}} e_h^+(x - V_{i,h}^- y) e_i f(y) h^2$$

$$= - \sum_{i=1}^{3} \sum_{\partial G_{h,l,i}} e_h^+(x - V_{i,h}^- y) e_i f(y) h^2 + \sum_{j=1}^{3} \sum_{\partial G_{h,l,j}} [e_h^+(x - y) D_{h,x}^+] f(y) h^3$$

$$- \sum_{\substack{j=1\\ j\neq i}}^{3} \sum_{S_{ij}} e_h^+(x - V_{i,h}^- y) e_i f(y) h^2 + \sum_{\substack{j=1\\ j\neq i}}^{3} \sum_{R_{ij}} e_h^+(x - V_{i,h}^- y) e_i f(y) h^2$$

$$= \sum_{\partial G_{h,l}} e_h^+(x - V_{i,h}^- y) n(y) f(y) h^2 - \sum_{\substack{j=1\\ j\neq i}}^{3} \Big(\sum_{S_{ij}} - \sum_{R_{ij}}\Big) e_h^+(x - V_{i,h}^- y) e_i f(y) h^2.$$

Using (0.1) we obtain

$$(F_h^+ f)(x) = \sum_{i=1}^{3} \left(- \sum_{\partial G_{h,l}} - \sum_{\partial G_{h,r}} \right) e_h^+(x - V_{i,h}^- y) \alpha(y) f(y) h^2 +$$

$$+ \sum_{i=1}^{3} \left(\sum_{j>k} \sum_{V_{i,h}^+ \partial G_{h,l,j,k}} - \sum_{V_{i,h}^+ \partial G_{h,l,m,j,k}} + \sum_{j\neq i} \sum_{S_{ij}} - \sum_{j\neq i} \sum_{R_{ij}} \right) e_h^+(x - V_{i,h}^- y) e_i f(y) h^2$$

$$+ \sum_{i=1}^{3} \left(- \sum_{j>k} \sum_{\partial G_{h,l,j,k}} + \sum_{\substack{\partial G_{h,l,m,j,k}\\ j\neq m\neq k}} \right) e_h^+(x - y) e_i f(y) h^2.$$

Using $V_{i,h}^+ G_{h,l,j,k} = (G_{h,l,j,k} \setminus S_{ijk}) \cup R_{ijk}$ and taking into consideration that

$$\sum_{i=1}^{3} \left[\sum_{V_{i,h}^+ G_{h,l,j,k}} + \sum_{\substack{j=1\\ j\neq i}}^{3} \sum_{R_{ij}} - \sum_{\substack{i,k=1\\ j>k}}^{3} \sum_{R_{ijk}} \right] e_h^+(x - V - h, iy) e_i f(y) h^2 = 0$$

we get the formula

$$(F_h^+ f)(x) = \sum_{i=1}^{3} \left(- \sum_{\partial G_{h,l,i}} - \sum_{\partial G_{h,r,i}} \right) e_h^+(x - V_{i,h}^- y) \alpha(y) f(y) h^2$$

$$+ \sum_{i=1}^{3} \sum_{j>k} \sum_{\partial G_{h,l,j,k}} - \sum_{j>k} \sum_{S_{ijk}} + \sum_{\substack{j=1\\ j\neq i}}^{3} \sum_{S_{ij}} e_h^+(x - V_{i,h}^- y) e_i f(y) h^2$$

$$
\begin{aligned}
&+\sum_{i=1}^{3}\left(-\sum_{j>k}\sum_{\partial G_{h,l,j,k}}+\sum_{\partial G_{h,l,m,j,k}}\right)e_h^+(x-y)e_i f(y)h^2\\
=\ &-\sum_{i=1}^{3}\left(\sum_{\partial G_{h,l,i}}+\sum_{\partial G_{h,r,i}}\right)e_h^+(x-V_{h,i}^- y)\alpha(y)f(y)h^2\\
&+\sum_{i=1}^{3}\left(-\sum_{j>k}+\sum_{j=1}^{3}\sum_{S_{ij}}\right)e_h^+(x-V_{h,i}^- y)e_i f(y)h^2\\
&+\sum_{i=1}^{3}\sum_{\substack{\partial G_{h,l,m,j,k}\\ m\neq j\neq k}} e_h^+(x-y)e_i f(y)h^2. \qquad \#
\end{aligned}
$$

Appendix D

Bibliography

Bibliography

[1] Agmon S. (1965) *Lectures on elliptic boundary value problems. Van Nostrand, Princeton.*

[2] Ahlberg J.H., Nilson E.N. and Walsh J.L. (1967) Complex cubic splines. *Trans. Amer. Math. Soc.* 129: 391–413.

[3] Ahlfors L.V. (1966) *Lectures on quasiconformal mappings. Van Nostrand, Princeton.*

[4] Ahlfors L.V. (1984) Old and new in Möbius groups. *Ann. Acad. Sci. Fenn. Ser. A 1, Math.* 9: 93–105.

[5] Ahlfors L.V. (1986) Möbius transformations in $\mathbf{R}^n$ expressed through 2×2-matrices of Clifford numbers. *Complex Variables, Theory and Appl.* 5: 215–224.

[6] Alekzidse M.A. (1978) *On the solution of boundary value problems by non-orthogonal decomposition methods. Nauka, Moscow* (Russian)

[7] Altmann S.L. (1986) *Rotations, quaternions and double groups. Clarendon Press, Oxford.*

[8] Anglés P. (1985) Construction de revêtements du groupe conforme d'un espace vectoriel muni d'une "métrique" du type (p, q). *Ann. Inst. H. Poincaré Sect.A* 33: 33–51.

[9] Aramanowitsch (Meister) L.I. (1995) Quaternion Non-linear filter for estimation of rotating body attitude. *Mathematical Meth. in the Appl. Sciences* 18: 1239–1255.

[10] Arnold W.I. (1973) *Ordinary differential equations. MIT Press, Cambridge, Mass.*

[11] Aronszajn N. (1950) Theory of reproducing kernels. *Trans. Amer. Math. Soc.* 68: 337–404

[12] Aschmjanskij V.V. (1984) Boundary value problems for an 8-dimensional analogue of the Cauchy–Riemann system. *Diff. Uravnenija* 20, 1: 172–174 (Russian).

[13] Atiyah M.F. Hitchin N.J. and Singer I.M. (1978) Selfduality in four-dimensional geometry. *Proc. Roy. Soc. London* A 362: 425–461.

[14] Bandle C. (1980) Isoperimetric inequalities and applications. *Pitman Research Notes in Math. Ser., Boston, London, Melbourne.*

[15] Bartik V., Ferreira A.V., Markl M. and Souček V. (1985) Index and Cauchy integral formula in complex-quaternionic analysis. *Simon Stevin, Wisk.en Natuurkunding Tijdschrift* 59, No. 3: 321–329.

[16] Begehr H. and Xu Z. (1992) Nonlinear half-Dirichlet problems for first-order elliptic equations in the unit ball of $\mathbf{R}^m$ $(m \geq 3)$. *Appl. Anal.* 45: 3–18.

[17] Bergman S. (1970) The kernel function and conformal mapping. *Mathematical Surveys, 5. Providence, R.I..: Amer. Math. Soc.* X: 257p.

[18] Bernstein S. (1991) Operator calculus for elliptic boundary value problems in unbounded domains. *Zeitschrift f. Analysis u. Anwend.* 10, 4: 447–460.

[19] Bernstein S. (1993) Elliptic boundary value problems in unbounded domains. *In: F. Brackx, et al (eds), Clifford algebras and their applications in mathematical physics, Kluwer, Dordrecht* : 45–53.

[20] Bernstein S. (1993) Analytische Untersuchungen in unbeschränkten Gebieten mit Anwendungen auf quaternionische Operatorentheorie und elliptische Randoperatoren. *Dissertation, Freiberg.*

[21] Bernstein S. (1996) Fundamental solutions of Dirac type operators. *Banach Center Publications, Banach Center Symposium: Generalizations of Complex Analysis, May 30 – July 1 1994, Warsaw* 37: 159-172.

[22] Bernstein S. On the index of Clifford valued singular integral operators and the left-linear Riemann problem. *Complex Variables, Theory and Appl.*: to appear.

[23] Bers L. (1953) *Theory of pseudo-analytic functions. Institute of Mathematics and Mechanics, New York University.*

[24] Bers L. and A. Gelbart (1943) On a class of differential equations in mechanics of continua. *Quarterly Appl. Math.* 1: 168–188.

[25] Beyer U. (1996) Eine erste Integraltafel zur Teodorescu-Transformation über der Einheitskugel und deren Anwendung. *Diplomarbeit, Freiberg.*

[26] Bitsadze A.W. (1955) On two-dimensional integrals of Cauchy type. *Akademii Nauk Grus. SSR* 16: 177–184 (Russian).

[27] Bitsadze A.W. (1973) *Grundlagen der Theorie der Analytischen Funktionen. Akademie-Verlag, Berlin.*

[28] Bjørken J.D. and Drell S.D. (1965) *Relativistic quantum fields. McGraw-Hill, New York.*

[29] Bloom F. (1981) Ill-posed problems for integro-differential equations in mechanics and electromagnetic theory. *SIAM Studies in Applied Mathematics, Philadelphia.*

[30] Bojarski B. (1966) Theory of generalized analytic vectors. *Ann. Polonici Math.* 17: 281–320 (Russian).

[31] Bojarski B. (1989) Conformally covariant differential operators. *Proc. 20th Iranian Math. Congress, Teheran.*

[32] Bolinder E.F. (1957) The classical electromagnetic equations expressed as complex four-dimensional quantities. *J. Franklin Inst.* 263: 213–223.

[33] Booss B. (1977) *Eine Einführung in die Atiyah-Singer-Indexformel. Springer.*

[34] Boor C., Höllig K. and Riemenschneider S. (1989) Fundamental solutions for multivariate difference equations. *Amer. J. Math.* 111: 403–415.

[35] Brackx F. and Van Acker N. (1993) A conjugate Poisson kernel in Euclidean space. *Simon Stevin* 67, No. 1–2: 3–13.

[36] Brackx F. Delanghe R. and Sommen F. (1982) Clifford analysis. *Pitman Research Notes in Math., Boston, London, Melbourne.*

[37] Browder F.E. (1960) Approximation by solutions of partial differential equations. *Amer. J. of Math.* 84: 137–160.

[38] Bunse-Gerstner A., Byers R. and Mehrmann R. (1989) A quaternion QR-algorithm. *Numerische Mathematik* 55: 83–95.

[39] Bures J. and Souček V. (1985) Generalized hypercomplex analysis and its integral formulas. *Complex Variables, Theory and Appl.* 1: 53–70.

[40] Calderón A.P. (1977) Cauchy integrals on Lipschitz curves and related operators. *Proc. Nat. Acad. Sci. USA* 74: 1324–1327.

[41] Cartan É. (1908) Nombres complexes. *In: Molk J. (ed.), Encyclopédie des sciences mathématiques, Tome 11*, Fasc. 4, art I5: 329–468.

[42] Casanova G. (1976) *L'algèbre vectorielle. Presses Universitaires de France, Que sais-je?, Paris* .

[43] Chevalley Cl. (1954) *The algebraic theory of spinors. Columbia U.P., New York.*

[44] Chisholm J.S.R. and Farwell R.S. (1990) Unified spin gauge theories of the four fundamental forces. *In: Quillen D.G. et al. (eds), The interface of mathematics and particle physics, Clarendon Press, Oxford* : 193–202.

[45] Clifford W.K. (1978) Applications of Grassmann's extensive algebra. *Americ. J. of Math. Pure and Appl.* 1: 350–358.

[46] Clifford W.K. (1882, 1969) Collected mathematical papers. *Chelsea reprint.*

[47] Coifman R.R., McIntosh A. and Meyer Y. (1982) L'intégrale de Cauchy définit un opérateur borné sur L2 pour les courbes lipschitziennes. *Annals of Mathematics* 116: 361–387.

[48] Cnops J. (1989) Orthogonal functions associated with the Dirac operator. *Thesis, State University of Ghent, Academic year 1988–89*, 129 pages.

[49] Cnops J. and Malonek H.R. (1997) Introduction to Clifford analysis *Lecture Notes, 1. European Intensive Course "Complex analysis and its generalizations", Dep. de Matematica da Universidade de Coimbra, Textos de Matematica, Série B.*

[50] Coifman R. and Meyer Y. (1980) Fourier analysis of multilinear convolution, Calderons theorem and analysis on Lipschitz curves. *Lecture Notes in Mathematics, Springer, New York, Berlin, Heidelberg* 779: 104–122.

[51] Common A.K. and Sommen F. (1993) Axial monogenic functions from holomorphic functions. *J. of Mathematical Anal. and Appl.* 179, No. 2: 610–628.

[52] Common A.K. and Sommen F. (1994) Special bi-axial monogenic functions. *J. of Math. Anal. and Appl.* 185, No. 1: 189–206.

[53] Costabel M. (1984) Starke Elliptizität von Randintegraloperatoren erster Art. *Thesis, TH Darmstadt.*

[54] Courant R., Friedrichs K.O. and Lewy H. (1928) Über die partiellen Differenzengleichungen der mathematischen Physik. *Math. Ann.* 100: 32–74.

[55] Crawford J. (1985) On the algebra of Dirac bispinor densities: factorization and inversion theorems. *J. Math Phys.* 26: 1439–1441.

[56] Crumeyrolle A. (1990) *Orthogonal and symplectic Clifford algebras, Spinor Structures. Kluwer, Dordrecht.*

[57] David G. (1991) Wavelets and singular integrals on curves and surfaces. *Lectures Notes in Mathematics, Springer, New York, Berlin, Heidelberg* 1465.

[58] Deavours C.A. (1973) The quaternionic calculus. *Am. Math. Monthly* 80: 995–1008.

[59] Deeter C.R. and Lord M.E. (1969) Further theory of operational calculus on discrete analytic functions. *J. Math. Anal. Appl.* 26: 92–113.

[60] Deeter C.R. and Springer G. (1965) Discrete harmonic kernels. *J. Math. Mech.* 14: 41 3–438.

[61] Delanghe R. (1970) On regular-analytic functions with values in a Clifford algebra. *Math. Ann.* 185: 91–111.

[62] Delanghe R. and Souček V. (1992) On the structure of spinor-valued differential forms. *Complex Variables, Theory and Appl.* 18: 223–236.

[63] Delanghe R. Sommen F. and Souček V. (1992) Clifford algebra and spinor-valued functions *Kluwer, Dordrecht.*

[64] Delanghe R. Sommen F. and Souček V. (1992) Residues in Clifford analysis. *In: Begehr H. and Jeffrey A. (eds), Partial differential equations with complex analysis, Pitman Res. Notes in Math. Ser.* 262: 61–92.

[65] Desin A.A (1960) Invariant elliptic systems. *Sib. Mat. Shurnal* 1, Nr. 4: 578–608 (Russian).

[66] Dix D. (1995) Applications of Clifford analysis to inverse scattering for the linear hierarchy in several space dimensions. *In: Ryan J.(ed.), Clifford algebras in analysis and related topics. CRC Press, Boca Raton, New York, London, Toronto*: 260–282.

[67] Dixon A.C. (1904) On the newtonian potential. *Quarterly Journal of Mathematics* 35: 283–296.

[68] Doppel K. and Sprößig W. *Einführung in die mehrdimensionale Analysis*: to prepare.

[69] Duffin R.J. (1953) Discrete potential theory. *Duke Math. J.* 20: 233–251.

[70] Duffin R.J. (1956) Basic properties of discrete analytic functions. *Duke Math. J.* 31: 335–363.

[71] Duffin R.J. (1971) Yukawa potential theory. *Journal of Math. Anal. and Appl.* 35: 105–130.

[72] Duffin R.J. and Duris C.S. (1964) A convolution product for discrete function theory. *Duke Math. J.* 31: 199–220.

[73] Dzuraev A. (1992) On the Moisil-Teodorescu system. *In: Begehr H. and Jeffrey A. (eds) Partial differential equations with complex analysis. Pitman Res. Notes in Math. Ser.* 262: 186–202.

[74] Elsrodt J., Grunewald F. and Mennicke J. (1987) Vahlen's group of Clifford matrices and spin-groups *Math. Zeitschrift* 196: 369–390.

[75] Ferrand J. (1944) Fonctions préharmonique et fonctions préholomorphes. *Bulletin des Sciences Mathématiques, sec. series* 68: 152–180.

[76] Fortuna L., Muscato G. and Xibilia M. (1997) Neural Attitude Control in Quaternion Algebra. *In Sprößig W. and Gürlebeck K. (eds) Proceedings of the Symposium "Analytical and Numerical Methods in Quaternionic and Clifford Analysis" held at Seiffen, Germany 1996, University of Mining and Technology Freiberg*, 207–216.

[77] Franks E. and Ryan J. Bounded monogenic functions on unbounded domains: to appear.

[78] Fueter R. (1932) Analytische Theorie einer Quaternionenvariablen. *Comment. Math. Helv.* 4: 9–20.

[79] Fueter R. (1935) Die Funktionentheorie der Differentialgleichungen $\Delta u = 0$ und $\Delta\Delta u = 0$ mit vier reellen Variablen. *Comm. Math. Helv.* 7: 307–330.

[80] Fueter R. (1936-37) Die Singularitäten der eindeutigen regulären Funktionen einer Quaternionenvariablen I. *Comment. Math. Helv.* 9: 320–334.

[81] Fueter R. (1949) Reguläre Funktionen einer Quaternionenvariablen. *Math. Inst. d. Universität Zürich.*

[82] Fuzhen Z. (1997) Quaternions and Matrices of Quaternions. *Linear Algebra and its Applications* 251: 21–57

[83] Gaier D. (1980) Vorlesungen über Approximation im Komplexen. *Birkhäuser, Basel, Boston, Stuttgart.*

[84] Gegelia T.G. (1961) On properties of multidimensional singular integrals in $\mathbf{R}^3$. *Doklady Akademii Nauk* 139, Nr. 2: 279–282 (Russian).

[85] Gentili G., Mariconda C. and Tarallo M. (1993) Quaternionic regular maps and $\overline{\partial}$-type operators. *Simon Stevin* 67, No. 3-4: 333–559.

[86] Gilbert J.P. and Buchanan J. (1983) First-order ellipitic systems: A function theoretic approach. *Academic Press, San Diego.*

[87] Gilbert J.E., Hogan J. and Lakey J. (1995) Frame decomposition of form-valued Hardy spaces. *In: Ryan J. (ed.), Clifford algebras in analysis and related topics. CRC Press, Boca Raton, New York, London, Toronto*: 237–259.

[88] Gilbert J.E. and Murray M.A.M. (1990) Clifford algebras and Dirac operators in harmonic analysis. *Cambr. Studies in advanced mathematics, Cambridge 26.*

[89] Goerisch F. and Zimmermann S. (1985) Complementary variational principles and bounds to eigenvalues. *Forschungsbericht Mathematik-Informatik der TU Clausthal 85/1.*

[90] Goldschmidt B. (1981) Properties of generalized analytic vectors in $\mathbf{R}^n$. *Math. Nachrichten* 103: 245–254.

[91] Goldschmidt B. (1982) Existence and representation of solutions of a class of elliptic systems of partial differential equations of first order in the space. *Math. Nachr.* 108: 159–166.

[92] Grassmann H. (1854) Sur les différents genres de multiplication. *J. für die Reine und Angew. Mathematik* 49: 123–141.

[93] Grassmann H. (1877) Der Ort der Hamiltonschen Quaternionen in der Ausdehnungslehre. *Math. Ann.* 12: 375–386.

[94] Grötsch H. (1928) Über die Verzerrung bei schlichten nichtkonformen Abbildungen und eine damit zusammenhängende Erweiterung des Picardschen Satzes. *Ber. Verh. Sächs. Akad. Wiss. Leipzig* 80: 503–507.

[95] Gürlebeck K. (1984) Über die optimale Interpolation verallgemeinert analytischer quaternionenwertiger Funktionen und ihre Anwendung zur näherungsweisen Lösung wichtiger räumlicher Randwertaufgaben der mathematischen Physik. *Diss., TH Karl-Marx-Stadt.*

[96] Gürlebeck K. (1988) Grundlagen einer diskreten räumlich verallgemeinerten Funktionentheorie und ihrer Anwendungen. *Habilitationsschrift, TU Karl-Marx-Stadt.*

[97] Gürlebeck K. (1988) Interpolation and best approximation in spaces of monogenic functions. *Wissenschaftliche Zeitschrift der TU Karl-Marx-Stadt* 30, H 1: 38–41.

[98] Gürlebeck K. (1990) Approximative solution of the stationary Navier–Stokes equations. *Math. Nachr.* 145: 297–308.

[99] Gürlebeck K. (1991) Lower and upper bounds for the first eigenvalue of the Lamé system *In: Kühnau R. and Tutschke W. (eds) Boundary value and initial value problems in complex analysis. Pitman Research Notes in Mathematics, Series* 256: 184–192.

[100] Gürlebeck K. (1994) Zur Berechnung von Fundamentallösungen diskreter Cauchy–Riemann Operatoren. *ZAMM* 74, 6: T 625–T 627.

[101] Gürlebeck K. On some operators in Clifford analysis. *Proceedings of the International Conference on Complex and Hypercomplex Analysis and Operator theory, held in Mexico City, December 1994*: to appear.

[102] Gürlebeck K. (1995) Zur Theorie der Differenzenpotentiale. *ZAMM* 75, S 2: 461–462.

[103] Gürlebeck K. and Hommel A. (1994) On fundamental solutions of the heat conduction difference operator. *Zeitschrift für Analysis und Anwendungen.* 13, No. 3: 425–441.

[104] Gürlebeck K. and Kähler U. (1996) On a spatial generalization of the complex Π-operator. *J. Anal. Appl.* 15: 283–297.

[105] Gürlebeck K. and Kähler U. On a boundary value problem of the biharmonic equation. *Math. Meth. in the Applied Sciences*: to appear.

[106] Gürlebeck K., Kähler U., Ryan J. Sprößig W. (1996) Clifford analysis over unbounded domains. *Applic. Analysis*: to appear.

[107] Gürlebeck K. and Sprößig W. (1987) A unified approach to estimation of lower bounds for the first eigenvalue of several elliptic boundary value problems. *Math. Nachr.* 131: 183–199.

[108] Gürlebeck K. and Sprößig W. (1987) A generalized Leibniz rule and a discrete quaternionic analysis. *Suppl. Rend. Circ. Mat. Palermo, Serie 2* 18: 44–64.

[109] Gürlebeck K. and Sprößig W. (1987) An application of quaternionic analysis to the solution of time-independent Maxwell equations and of Stokes equations. *Suppl. Rend. Circolo Mat. Palermo, Serie 2* 16: 61–76.

[110] Gürlebeck K. and Sprößig W. (1990) *Quaternionic analysis and elliptic boundary value Problems. Birkhäuser, Basel.*

[111] Gürlebeck K. and Sprößig W. (1995) Clifford analysis and elliptic boundary value problems. *In: Ablamowicz R. and Lounesto P. (eds), Clifford algebras and spinor structures. Kluwer, Dordrecht* : 325–334.

[112] Gürlebeck K. and Sprößig W. (1997) *Introduction in analytical and numerical methods in Clifford algebras. Dep. de Matematica da Universidade de Coimbra, Textos de Matematica, Série B* to appear.

[113] Gürlebeck K., Sprößig W. and Tasche M. (1985) Numerical realization of boundary collocation methods. *Int. Ser. of Num. Math. Birkhäuser, Basel* 75: 206–217.

[114] Gürlebeck K., Sprößig W. and Wimmer U. (1993) Hypercomplex function theory for consideration of non-linear Stokes problems with variable viscosity. *Complex Variables, Theory and Appl.* 22: 195–202.

[115] Gürsey F. and Tze H.C. (1980) Complex and quaternionic analyticity in chiral and gauge theories I. *Annals of Physics* 128: 29–130.

[116] Habetha K. (1976) Eine Bemerkung zur Funktionentheorie in Algebren. *In: Meister E., Weck N. and Wendland W. (eds), Function theoretic methods of partial differential equations. Proc. of the Int. Symp. Darmstadt 1976, Lecture Notes in Mathematics, Vol. 561, Springer, Stuttgart*: 502–509.

[117] Habetha K. (1986) Eine Definition des Kroneckerindex in $\mathbf{R}^{n+1}$ mit Hilfe der Cliffordanalysis. *Zeitschrift f. Anal. und Anw.* 5: 133–137.

[118] Haefeli H.G. (1947) Hyperkomplexe differentiale. *Comment. Math. Helv.* 20: 382–420.

[119] Hamilton W.R. (1866, 1969) Elements of quaternions. *Chelsea* (reprint).

[120] Han W. (1994) A posteriori error analysis for linearization of nonlinear elliptic problems and their discretizations. *Math. Methods in the Applied Sciences* 17: 487–508.

[121] Harvey F.R. (1990) *Spinors and calibrations perspectives in mathematics. Academic Press, Inc., London* 9.

[122] Hayabara S. (1966) Operational calculus on the discrete analytic functions. *Math. Japon.* 11: 35–65.

[123] Hedberg L. (1972) On certain convolution inequalities. *Proc. Amer. Math. Soc.* 36: 505–510.

[124] Hermann P. and Wennemar L. (1994) Die Methode der Singularitätenfunktionen zur Lösung der Prae-Maxwellschen Gleichungen. *ZAMM* 74, 6: T 635–T 637.

[125] Hestenes D. (1966) *Space-time algebra. Gordon and Breach, New York.*

[126] Hestenes D. (1968) Multivector calculus. *J. Math. Anal. Appl.* 24: 313–325.

[127] Hestenes D. (1985) *New foundations for classical mechanics. Reidel, Dordrecht, Boston.*

[128] Hestenes D. (1991) A unified language for mathematics and physics. *Adv. in Applied Clifford Algebras* 1, No 1: 5–29.

[129] Hestenes D. and Sobzyk G. (1985) *Clifford algebras for mathematics and physics. Reidel, Dordrecht.*

[130] Helgason S. (1980) The radon transform. *Progress in Mathematics 5, Birkhäuser, Boston, Basel, Stuttgart.*

[131] Hile G.N. (1977) Representations of solutions of a special class of first-order systems. *J. of Differential Equations.* 25, No. 3: 410–424.

[132] Hillion P. (1995) Constitutive relations and Clifford algebra in electromagnetism. *Adv. in Appl. Clifford Algebras* 5, Nr. 2: 141–158.

[133] Hillion P. (1994) Beltrami–Moses spinors in electromagnetism. *Adv. in Appl. Clifford Algebras* 4, Nr. 1: 13–26.

[134] Hodge W.V.D. and Pedoe D. (1952) Methods of algebraic geometry. *Cambridge University Press, Cambridge.*

[135] Hommel A. Fundamentallösungen partieller Differenzenoperatoren und die Lösung diskreter Randwertprobleme mit Hilfe von Differenzenpotentialen. *Dissertation, Bauhaus-Universität Weimar* (in preparation).

[136] Imaeda I. (1976) A new formulation of classical electrodynamics. *Nuovo Cimento* 32 A: 138–162.

[137] Iftimie V. (1965) Fonctions hypercomplexes. *Bull. Roy. Math. Soc. Roumanie* 9: 279–332.

[138] Iwaniec T. and Martin G. (1993) Quasiregular mappings in even dimension. *Acta Math.* 170: 29–81.

[139] Jancewicz B. (1988) Multivectors and Clifford algebra in electrodynamics. *World Scientific Publ., Singapore.*

[140] Jank G. and Sommen F. (1990) Clifford analysis, biaxial symmetry, and pseudoanalytic functions. *Complex Variables, Theory and Appl.* 13: 195–212.

[141] Joos, G. (1964) *Lehrbuch der theoretischen Physik. Akademische Verlagsgesellschaft Geest & Portig K.G., Leipzig.*

[142] Juvet G. (1932) Les nombres de Clifford et leurs applications à la physique mathématique. *Congrès int. des mathématiciàns de Zurich*: 306–307.

[143] Juvet G. and Schidlof A. (1932) Sur les nombres hypercomplexes de Clifford et leurs applications à l'analyse vectorielle ordinaire, à l électromagnétisme de Minkowski et à la théorie de Dirac. *Bull. Soc. Neuchateloise Sc. Nat.* 57: 127–147.

[144] Kähler U. (1997) Applications of hypercomplex Π-operators to the solution of Beltrami systems. *In Sprößig W. and Gürlebeck K. (eds) Proceedings of the Symposium "Analytical and Numerical Methods in Quaternionic and Clifford Analysis" held at Seiffen, Germany 1996, University of Mining and Technology Freiberg,* 69–76.

[145] Kantorovich L.W. and Akilov G.P. (1964) *Funktionalanalysis in normierten Räumen. Akademie-Verlag, Berlin.*

[146] Kawohl B. and Sweers G. (1987) Remarks on eigenvalues and eigenfunctions of a special elliptic system. *J. Appl. Math. Phys.(ZAMP)* 38: 730–740.

[147] Keller J. (1991) Spinors and multivectors as a unified tool for spacetime geometry and for elementary particle physics. *Int. Journal of Theoretical Physics* 30, 2: 137-145.

[148] Keller J. and Rodrigues-Romo S. (1990) A multivectorial Dirac equation. *J. of Math. Phys.* 31: 2502–2511.

[149] Kippig F. (1997) Untersuchungen zu Randwert- und Anfangswertaufgaben für partielle Differentialgleichungen mit Methoden der Clifford Analysis. *Dissertation, Freiberg.*

[150] Kisil V.V. (1995) Connection between different function theories in Clifford analysis. *Advanc. in Applied Clifford algebras* 5, No. 1: 63–74.

[151] Khimshiashwili G.N. Elliptic boundary problems for Cauchy–Riemann systems. (1997) *In Sprößig W. and Gürlebeck K. (eds) Proceedings of the Symposium "Analytical and Numerical Methods in Quaternionic and Clifford Analysis" held at Seiffen, Germany 1996, University of Mining and Technology Freiberg*, 77–84.

[152] Korowkin P.P. (1973) Ungleichungen. *Deutscher Verlag der Wissenschaften, Berlin, 7. Auflage.*

[153] Kravshenko V.V. (1992) On the relation between holomorphic biquaternionic functions and time-harmonic electromagnetic fields. *Deposited in UKREINTEI 29.12.92 under Nr. 2073-Uk-92* (Russian).

[154] Kravshenko V.V. (1993) On a hypercomplex factorization of some equations of mathematical physics. *Preprint #1, Dep. of Math., Instituto Superior Tecnico, Lisboa, Portugal.*

[155] Kravshenko V.V., Malonek H.R. and Santana G. Biquaternionic integral representations for massive Dirac spinors in a magnetic field and generalized biquaternionic differentiability. *Math. Meth. in the Applied Sciences* 19, 17: 1415–1431.

[156] Kravshenko V.V. and Shapiro M.V. (1993) Helmholtz operator with a quaternionic wave number and associated function theory II. Integral representations. *Acta Applicandae Mathematicae* 32, No. 3: 243–265.

[157] Kravshenko V.V. and Shapiro M.V. (1993) On a generalized system of equations of Cauchy–Riemann type with quaternionic parameters. *Doklady Akademii Nauk* 329, Nr. 5: 547–549 (Russian).

[158] Kravshenko V.V. and Shapiro M.V. (1996) Integral representations for spatial models of mathematical physics. *Pitman Research Notes in Math. Series* 351.

[159] Kravshenko V.V., Ramirez de Arellano and Shapiro M.V. (1996) On integral representations and boundary properties of spinor fields. *Math. Meth. in the Appl. Sciences* 19: 977–989.

[160] Królokowski W. and Ramirez de Arellano E. (1992) Fueter–Hurwitz regular mappings and an integral representation. *In: Micali A., Boudet R. and Helmstetter J. (eds), Clifford algebras and their applications in mathematical physics, Kluwer, Dordrecht, Fund. Theories of Physics* 47: 221–237.

[161] Krolikowski W. and Porter R.M. (1993) Biregular quaternionic functions. *In: Brackx F., et.al. (eds), Clifford algebras and their applications in mathematical physics, Kluwer, Dordrecht* : 129–135.

[162] Kron G. (1959) *Tensors for circuits. Dover, New York.*

[163] Krüger H. (1991) New solutions of the Dirac equation for central fields. *In: The Electron, Kluwer, Dordrecht* : 49–81.

[164] Kusnetsov S.P. and Motshalov W.W. (1992) Automorphisms of the Clifford algebra, and strongly regular functions. *Izv. Vyssh. Uchebn. Zared. Mat.* 10:(Russian).

[165] Lawrentjew M.A. and Schabat B.W. (1967) *Methoden der komplexen Funktionentheorie. DVW, Berlin.*

[166] Lense J. (1950) *Kugelfunktionen. Geest & Portig, Leipzig.*

[167] Leray J. (1959) Le calcule différentiel et intégral sur une variété analytique complexe (Problème de Cauchy III). *Bull. Soc. Math. France* 87: 81–180.

[168] Leutwiler H. (1992) Modified Clifford analysis. *Complex Variables, Theory and Appl.* 17: 153–171.

[169] Leutwiler H. (1992) Modified quaternionic analysis in $\mathbf{R}^3$. *Complex Variables* 20: 19–51.

[170] Leutwiler H. (1995) More on modified quaternionic analysis in $\mathbf{R}^3$. *Forum Mathematicum* 7: 279–305.

[171] Li C., McIntosh A. and Qian T. (1994) Clifford algebras, Fourier transforms, and singular convolution operators on Lipschitz surfaces. *Revista Matemática Iberoamericana* 10: 665–721.

[172] Li C., McIntosh A. and Semmes S. (1992) Convolution singular integrals on Lipschitz surfaces *Journal of the American Math. Society* 5: 455–481.

[173] Lindelöf E. (1905) *Le Calcul des résidus et ses applications á la théorie des fonctions. Gauthier-Villars, Paris.*

[174] Lipschitz R. (1886) *Untersuchungen über die Summe von Quadraten. Max Cohen und Sohn, Bonn* 147 pages.

[175] Louis A.K., Maaß P. and Rieder A. (1994) *Wavelets. Teubner, Stuttgart.*

[176] Lounesto P. (1979) Spinor-valued regular functions in hypercomplex analysis. *Helsinki Univ. of Technology, Inst. of Math. Report HTKK-MAT-A154* 79 pages.

[177] Lounesto P. (1981) Scalar products of spinors and an extension of Brauer–Wall groups. *Foundations of Physics* 11, Nos 9/10: 721–740.

[178] Lounesto P. (1989) Möbius transformations and Clifford algebras of Euclidean and anti-Euclidean spaces. *In: Lawrynowicz J.(ed.), Deformations of mathematical structures. Kluwer, Dordrecht*: 79–90.

[179] Lounesto P. (1993) Clifford algebras and Hestenes spinors. *Foundation of Physics, Special issue dedicated to D. Hestenes* 23, No. 9: 1203–1237.

[180] Lounesto P. (1995) Möbius transformations, Vahlen matrices and their factorization. *In: Ryan J. (ed.), Clifford algebras in analysis and related topics. CRC Press, Boca Raton, New York, London, Toronto*: 355–359.

[181] Lounesto P. and Bergh P. (1983) Axially symmetric vector fields and their complex potentials. *Complex Variables, Theory and Appl.* 2: 139–150.

[182] Lounesto P., Mikkola R. Vierros V. (1987) CLICAL user manual. *Helsinki Univ. of Technology, Math.- Report A2480.*

[183] Maaß H. (1949) Automorphe Funktionen von mehereren Veränderlichen und Dirichletsche Reihen. *Abh. Math. Sem. Univ. Hamburg* 16, 3-4:72–100.

[184] Maeda F.-Y., Murakami A. and Yamasaki M. (1991) Discrete initial value problems and discrete parabolic potential theory. *Hiroshima Math. J.* 21: 285–299.

[185] Maks J. (1992) Clifford algebras and Möbius transformations. *In: Micali A. et al. (eds), Clifford Algebras and their Applications in Mathematical Physics. Kluwer, Dordrecht*: 57–63.

[186] Malonek H.R. (1987) *Zum Holomorphiebegriff in höheren Dimensionen. Habilitationsschrift, Pädagogische Hochschule Halle.*

[187] Malonek H.R. (1990) A new hypercomplex structure of the Euclidean space $\mathbf{R}^{m+1}$ and a concept of hypercomplex differentiability. *Complex Variables, Theory and Appl.* 14: 25–33.

[188] Malonek H.R. (1993) Hypercomplex differentiability and its applications. *Brackx F. et al. (eds), Clifford algebras and applications in mathematical physics. Kluwer, Dordrecht*: 141–150.

[189] Malonek H.R. and Müller B. (1992) Definition and properties of a hypercomplex singular integral operator. *Res. Math.* 22: 713–724.

[190] Mariconda C. (1989) Functions of quaternionic variables. *Magister Philosophiae Thesis, S.I.S.S.A. — I.S.A.S., Trieste.*

[191] Marinov M.S. (1988) On a class of hyperholomorphic functions *Comptes rendus de l'Acad. bulgare des Sciences* 41, No. 9:.

[192] Marinov M.S. (1994) Regeneration of regular-quaternionic functions. *Proc. 20-th Summer School "Appl. of Mathematics in Engineering" in Varna 1994, Sofia*: 85-101.

[193] Marinov M.S. (1995) On the S-regular functions. *Journal of Natural Geometry* 7, No. 1: 21–44.

[194] Martensen E. (1968) Potentialtheorie. *Leitfäden der Angewandten Mathematik und Mechanik, Bd. 12, Teubner, Stuttgart.*

[195] McIntosh A. (1989) *Clifford algebras and the higher dimensional Cauchy integral approximation theory and function spaces. Banach Center Publications* 22: 253–267.

[196] McIntosh A. (1995) Clifford algebras, Fourier theory, singular integrals, and harmonic functions on Lipschitz domains. *In: Ryan J. (ed.) Clifford algebras in analysis and related topics. CRC Press, Boca Raton, New York, London, Toronto*: 33–86.

[197] Meister L. (1997) Estimation of rotating body attitude by quaternion filter. *Report in AAS/AIAA Astrodynamics Conference Aug.14–17 1995, Halifax, I&II Advances in the Astronautic. Sciences* 90: to appear.

[198] Mehlikzon A.S. (1948) On the notion of monogeous quaternions. *Doklady Akademii Nauk SSSR* 59, 3: 431–434 (Russian).

[199] Mercier M. (1935) Expression des équations de électromagnétisme au moyen des nombres de Clifford. *Thesis Nr. 953, University of Geneve.*

[200] Michlin S.G. (1969) *Numerische Realisierung von Variationsaufgaben. Akademie-Verlag, Berlin.*

[201] Michlin S.G. (1981) Konstanten in einigen Ungleichungen der Analysis. *Teubner-Texte zur Mathematik 39, Teubner, Leipzig.*

[202] Michlin S.G. and Prößdorf S. (1986) *Singular integral operators. Akademie-Verlag Berlin, Berlin.*

[203] Mitrea M. (1994) Singular integrals, Hardy spaces and Clifford wavelets. *Lecture Notes in Mathematics, No. 1575, Springer-Verlag, Heidelberg.*

[204] Mitrea M. (1992) Construction of Clifford wavelets. *Advanc. in Appl. Clifford Algebras* 2, No.2: 249–276.

[205] Mitrea D., Mitrea M. and Pipher J. (1996) Vector potential theory on nonsmooth domains in $\mathbf{R}^3$ and applications to electromagnetic scattering. *Preprint of the University of Minneapolis.*

[206] Mitelman I.M. and Shapiro M.V. (1995) Differentiation of the Martinelli–Bochner integrals and the notion of hyperderivability. *Math. Nachr.* 172: 211–238.

[207] Mogomedov G.A. and Palamodov V.P. (1978) Generalized analytic functions of several variables. *Math. Sbornik* 106: 515–543.

[208] Moisil G. and Teodorescu N. (1931) Fonctions holomorphes dans l'espace. *Matematica, Cluj.* 5: 142–150.

[209] Murray A. (1985) The Cauchy integral, Calderon commutation and conjugation of singular integrals in $\mathbf{R}^m$. *Trans. of the A.M.S.* 298: 497–518.

[210] Mushelishwili N.I. (1966) *Einige Grundaufgaben der mathematischen Elastizitätstheorie. Nauka, Moskva* (Russian).

[211] Mushelischwili N.I. (1965) *Singuläre Integralgleichungen. Akademie-Verlag, Berlin.*

[212] Natterer F. (1986) *The mathematics of computersized tomography. Wiley, Stuttgart, Chichester.*

[213] Nečas J. (1967) *Les méthodes directes en théorie des équations elliptiques. Masson, Paris.*

[214] Nef W. (1942-43) über die singulären Gebilde der regulären Funktionen einer Quaternionenvariablen. *Comment. Math. Helv.* 15: 144–174.

[215] Nef W. (1944) Die Funktionentheorie der partiellen Differentialgleichungen zweiter Ordnung (Hyperkomplexe Funktionentheorie). *Bull. Soc. Fribourgoise Sc. Nat.* 37: 348–37.

[216] Nono K. (1993) Regularity of functions with values in Clifford algebras based on a generalized axially symmetric potential theory operator. *In: Brackx F. et al.(eds), Clifford algebras and their applications in mathematical physics. Kluwer, Fundamental Theories of Physics, Dordrecht* 55: 159–166.

[217] Norguet F. (1959) Sur la théorie des résidues. *C.R. Acad. Sci. sér. A* 248, 14: 2057–2059.

[218] Obolaschwili E.I. (1975) Multidimensional generalized holomorphic vectors. *Diff. Uravnenija.* tom 11, No. 1: 108–115.

[219] Obolaschwili E.I. (1988) Effective solutions of some boundary value problems in two and three dimensional cases. *Funct. Anal. Methods in Compl. Anal. and Applications to PO, Trieste*: 149–172.

[220] Obolaschwili E.I. (1997) Boundary and initial value problems in Clifford analysis. *In Sprößig W. and Gürlebeck K. (eds) Proceedings of the Symposium "Analytical and Numerical Methods in Quaternionic and Clifford Analysis" held at Seiffen, Germany 1996, University of Mining and Technology Freiberg,* 145-152.

[221] Parra M. (1992) On Dirac, Dirac–Darwin and Dirac–Hestenes equations. An example of the subtile relationship between mathematics and physics. *In: Micali A., Boudet R. and Helmstetter J.(eds), Clifford algebras and their applications in mathematical physics. Kluwer, Dordrecht, Fund. Theories of Physics* 47: 463–479.

[222] Parra M. (1992) Clifford algebras. Towards a common language for mathematicians and physicists. Some suggested notations. *Adv. in Appl. Clifford Algebras* 2, Nr. 1: 145–168.

[223] Pascali D. (1963) A homogenous Hilbert problem for holomorphic quaternions. *Rev. Math. Pures et Appl., Acad. R.P. de Roumanie 8* 4: 653–659.

[224] Peetre J. and Qian T. (1994) Möbius covariance of iterated Dirac operators *J. Australian Math. Soc., Ser. A* 56: 403–414.

[225] Pertici D. (1993) Quaternion regular functions and domains of regularity . *Bollettino U.M.I* 7, 7-B: 973–988.

[226] Pertici D. (1993) Regular extension theorems for quaternionic functions. *Riv. Mat. Univ. Parma* 5, 2: 291–300.

[227] Pezzaglia W.M. Jr. (1992) Clifford algebra, geometric-multispinor particles, and multivector-current gauge fields. *Found. Phys. Lett.* 5: 57–62.

[228] Pezzaglia W.M. Jr. (1995) Multivector solutions to the hyperholomorphic massive Dirac equation. *In: Ryan J. (ed.) Clifford algebras in analysis and related topics. CRC Press, Boca Raton, New York, London, Toronto*: 339–354.

[229] Pfaff U. (1985) Entwicklung einer Funktionentheorie in Algebren, die durch homogene partielle Differentialgleichungen induziert sind. *Dissertation, RWTH Aachen.*

[230] Pfeifer E. and Rauhöft A. (1984) Über Grundlösungen von Differenzenoperatoren. *Z. f. Anal. u. Anw.* 3: 227–236.

[231] Piskorek A. (1978) Über die Anfangsrandwertaufgabe der quasistationären Thermoelastizität. *ZAMM* 58: 230–231.

[232] Plemelj J. (1908) Ein Ergänzungssatz zur Cauhyschen Integraldarstellung analytischer Funktionen, Randwerte betreffend. *Monatsheft für Mathematik und Physik* 76: 205–210.

[233] Poloshi G.N. (1965) Generalized theory of analytic functions of one complex variable. *Naukova Dumka, Kiev* (Russian).

[234] Post E.J. (1962) Formal structure of electromagnetics. *North-Holland, Amsterdam.*

[235] Porteous I. (1969) Topological geometry. *Van Nostrand-Reinhold, London.*

[236] Porter R.M., Shapiro M.V. and Vasilevski N.L. (1994) Quaternionic differential and integral operators and the $\bar{\partial}$-problem. *Journal of Natural Geometry* 6: 101–124.

[237] Qian T. (1995) Singular integrals with monogenic kernels on the m-torus and their Lipschitz perturbations. *In: Ryan J. (ed.) Clifford algebras in analysis and related topics. CRC Press, Boca Raton, New York, London, Toronto* : 158–172.

[238] Qian T. (1996) Singular integrals on star-shaped Lipschitz surfaces in the quaternionic space. *Research and Technical Reports, Univ. of New England, Dep. of Math., Statist. and Computing Sciences 96-120*, 30 pages.

[239] Remmert R. (1995) *Funktionentheorie Springer, Berlin, Heidelberg.*

[240] Rathsfeld A. (1992) The invertibility of the double layer potential operator in the space of continuous functions defined on a polyhedron: The panel method.*Appl. Anal.* 45: 135-177.

[241] Ricci M.M.G and Levi-Cività T. (1902) Méthodes de calcul différential absolu et leurs applications. *Math. Ann.* 54: 125–201.

[242] Riesz M. (1958) *Clifford numbers and spinors. Lecture Series, Institute of Physical Science and Technology, Nr. 38, Maryland.*

[243] Rinehart R.F. (1965) Elements of a theory of intrinsic functions on algebras *Duke Math. Journal* 32: 1–19.

[244] Rodrigues O. (1840) Des lois geométriques qui régissent les déplacements d'un systéme solide dans l'espace, et de la variation des coordonnées provenant de produire. *J. de Mathématiques Pures et Appliquées* 5: 380–440.

[245] Rodriguez-Romo S. (1993) An analysis of Fierz identities, factorization and inversion theorems. *Foundations of Physics* 23, 11: 1535–1542.

[246] Rosch N. (1983) Time reversal-symmetry, Kramer's degeneracy and the algebraic eigenvalue problem. *Chemical Physics* 80: 1–5.

[247] Rose A. On the use of a complex (quaternion) velocity potential in three dimensions. *Comment. Math. Helv.* 24: 135–148.

[248] Rühs F. *Funktionentheorie, DVW, Berlin* (1962).

[249] Rwatschov W.L. (1982) *Theory of R-functions and some of its applications. Naukova Dumka, Kiew* (Russian).

[250] Ryabenskij V.S. (1987) *The method of difference potentials for some problems of continuum mechanics. Moscow, Nauka* (Russian).

[251] Ryan J. (1982) Complexified Clifford analysis. *Complex Variables, Theory and Appl.* 1: 119–149.

[252] Ryan J. (1982) Clifford analysis with generalised elliptic and quasifunctions. *Applicable Analysis* 13: 151–171.

[253] Ryan J. (1984) Properties of isolated singularities of some functions taking values in real Clifford algebras. *Math. Proc. Cambr. Philos. Soc.* 95: 277–298.

[254] Ryan J. (1990) Dirac operators, Schrödinger-type operators in C^n and Huygens' principle. *J. Funct. Anal.* 87: 295–318.

[255] Ryan J. (1993) Intertwining operators for iterated Dirac operators over Minkowski-type spaces. *Journal of Mathematical Analysis and Applications* 177: 1–23.

[256] Ryan J. (1993) Cauchy–Green type formulae in Clifford analysis. *University of Arkansas, Research Report UofA-R-57.*

[257] Ryan J. (1995) Cauchy–Green type formulae in Clifford analysis. *University of Arkansas, Transactions of the A.M.S.* 347:1331-1341.

[258] Ryan J. (1995) Some applications of conformal convariance in Clifford analysis. *In: Ryan J. (ed.), Clifford algebras in analysis and related topics. CRC Press, Boca Raton, New York, London, Toronto* 134–155.

[259] Ryan J. (ed.) (1995) *Clifford algebras in analysis and related topics. CRC Press, Boca Raton, New York, London, Toronto.*

[260] Ryan J. (1997) *Conformally covariant operators in Clifford analysis*: to appear.

[261] Ryan J. *Intrinsic Dirac operators in* $\mathbb{C}^n$. *Advances in Mathematics*: to appear.

[262] Ryan J. The Fourier Transform on the Sphere: to appear.

[263] Saak E.M. (1975) On the theory of multidimensional first order systems. *Doklady Academmii Nauk, SSSR* 222: 43–46 (Russian).

[264] Sce, M. (1957) Sulle serie di potenzi nei moduli quadratici. *Lincei Rend. Sci. Fis. Mat. et Nat.* 23: 220–225.

[265] Samarskij A.A. (1977) *The theory of difference methods. Nauka, Moscow* (Russian).

[266] Schischmarov I.A. (1979) *Introduction in the theory of elliptic equations. Moscow University Publ., Moscow* (Russian).

[267] Shapiro M.V. and Vasilevski N.L. (1993) On the Bergman kernel function in the Clifford analysis. *In: Brackx F., Delanghe R. and Serras H. (eds), Clifford algebras and their applications in mathematical physics. Kluwer, Dordrecht*: 183–192.

[268] Shapiro M.V. and Vasilevski N.V. (1994) Quaternionic ψ -hyperholomorphic functions, singular integral operators and boundary value problems, I. *Complex Variables, Theory and Applications* 27: 17–46.

[269] Shapiro M.V. and Vasilevski N.V. (1994) Quaternionic ψ -hyperholomorphic functions, singular integral operators and boundary value problems, II. *Complex Variables, Theory and Applications* 27: 67–96.

[270] Silberstein L. (1914) *The theory of relativity. Macmillan, London.*

[271] Sili A. (1992) Existence d'états stationnaires pour l'équation de Dirac non lineaire. *Portugaliae Mathematica* 49, Fasc. 1: 52–67.

[272] Smirnov W.I. (1961) *Lehrgang der Höheren Mathematik*, Teil 4. *DVW, Berlin.*

[273] Smirnov W.I. (1967) *Lehrgang der Höheren Mathematik*, Teil 5. *DVW, Berlin.*

[274] Smolitzki Ch.L. (1950) Estimation of the derivative of fundamental functions. *Doklady Akademii Nauk SSSR n. Ser.* 74, 205-208 (Russian).

[275] Sobczyk G.E. (1992) Simplicial calculus with geometric algebra. *In: Micali A., Boudet R. and Helmstetter J.(eds), Clifford algebras and their applications in mathematical physics. Kluwer, Dordrecht, Fundamental Theories of Physics* 47: 279–292.

[276] Sobolev S.L. (1952) On the uniqueness of solutions of elliptical difference equations. *Doklady Akad. Nauk SSSR* 87, No.2: 179–182 (Russian).

[277] Sobolev S.L. (1952) On a difference equation. *Doklady Akad. Nauk SSSR* 87, No. 3: 341–343 (Russian).

[278] Sokhotzki J.W. (1873) Über bestimmte Integrale und Funktionen, die für Reihenentwicklung benötigt werden , S. Petersburg (1873) (Russian).

[279] Solomjak M.S. (1963) Linear elliptic systems of first order. *Doklady Akademii Nauk SSSR* 150, Nr. 1: 48–51 (Russian).

[280] Sommen F. (1981) A product and an exponential function in hypercomplex function theory. *Appl. Anal.* 12: 13–26.

[281] Sommen F. (1984) Formale randwaarden van monogene functies en elliptischen systemen in het vlak. *Thesis, Ghent.*

[282] Sommen F. (1984) Monogenic differential forms and homology theory. *Proc. Royal Irish Academy* 84 A: 87–109.

[283] Sommen F. (1985) Plane waves, biregular functions and hypercomplex Fourier analysis. *Proceedings of the 13th Winter School on abstract analysis, Srni, 5–12 January 1985, Suppl. ai Rendiconti del Circolo Matematico di Palermo, Serie 2* 9: 205–219.

[284] Sommen F. (1987) An extension of the radon transform to Clifford analysis. *Complex Variables, Theory and Appl.* 8: 243–266.

[285] Sommen F. (1987) A Clifford–radon transform in Euclidean space. *Proc. XV.-DGM Conference Clausthal, World Scientific, Singapore* : 377–389.

[286] Sommen F. (1988) Special functions in Clifford analysis and axial symmetry. *J. Math. Anal. Appl.* 130: 100–133.

[287] Sommen F. (1988) Clifford–radon transform applied to massless fields. *Simon Stevin* 62: 293–319.

[288] Sommen F. (1988) Plane wave decomposition of monogenic functions *Ann. Polon. Math.* 49: 101–114.

[289] Sommen F. (1989) Radon and X-ray transforms in Clifford analysis. *Complex Variables* 11: 49–70.

[290] Sommen F. (1997) Curved Radon transforms in Clifford analysis. *In: Proceedings of the 4th International Conference Clifford algebras and their applications in mathematical physics. Kluwer, Dordrecht*: to appear.

[291] Sommen F. and Souček V. (1985) Hypercomplex differential forms applied to the de Rham and the Dolbeault complex. *Sem. Geom. 1984, Univ. Bologna* : 177–192.

[292] Sommen F. and Souček V. (1992) Monogenic differential forms. *Complex Variables* 19: 81–90.

[293] Sommen F. and Xu Z. (1992) Fundamental solutions for operators which are polynomials in the Dirac operator. *In: Micali A., Boudet R. and Helmstetter J. (eds), Clifford algebras and their applications in mathematical physics. Kluwer, Dordrecht*: 313–326.

[294] Souček V. (1980) Regularni funkce quaternionve promenne. *Thesis, Charles University Prague.*

[295] Souček V. Quaternion valued differential forms in $\mathbf{R}^4$. *Suppl. Rend. Circ. Mat. Palermo* 33, (2): 293–300.

[296] Sprößig W. (1974) Über ein System zweidimensionaler singulärer Integralgleichungen nicht normalen Typs. *WZ der TH Karl-Marx-Stadt* 16, Heft 3: 433–446.

[297] Sprößig W. (1978) Räumliches Analogon zum komplexen T-Operator. *Beiträge zur Analysis* 12: (1978), 127–137.

[298] Sprößig W. (1979) Über eine mehrdimensionale Operatorenrechnung über beschränkten Gebieten des E_n. *Thesis, TH Karl-Marx-Stadt.*

[299] Sprößig W. (1981) Methode der harmonischen Approximation. *Beiträge zur Numerischen Mathematik* 9: 185–193.

[300] Sprößig W. (1995) Representation formulae of solutions of boundary value problems with Clifford analysis methods. *Preprint TU-Bergakademie Freiberg: Fakultät für Mathematik und Informatik, Nr. 95-06.*

[301] Sprößig W. (1995) On decompositions of the Clifford valued Hilbert space and their applications to boundary value problems. *Advances in Applied Clifford Algebras* 5, No. 2: 167–186.

[302] Sprößig W. On a class of non-linear boundary value problems *In: Proceedings of the 4th International Conference Clifford algebras and their applications in mathematical physics. Kluwer, Dordrecht*: to appear.

[303] Sprößig W. and Gürlebeck K. (1992) Application of quaternionic analysis on generalized nonlinear Stokes eigenvalue problems. *In: Begehr H. and Jeffrey A. (eds), Partial differential equations with complex analysis. Pitman Research Notes in Mathematics Series* 262: 52–61.

[304] Sprößig W. and Gürlebeck K. (1996) On the treatment of fluid problems by methods of Clifford analysis. *In: Deville M., Gavrilakis S. and Ryhming I.L.(eds), Computation three-dimensional complex flows Notes on Numerical Fluid Mechanics, Vieweg, Braunschweig, Wiesbaden* 53: 304–310.

[305] Sprößig W. and Venturino E. (1993) The treatment of window problems by transform methods. *Zeitschrift für Analysis und Anwendungen* 12: 643–654.

[306] Stein E.M. and Weiss G. (1974) *Einführung in die harmonische Analyse im Euklidischen Raum. Isd. Nauka, Moscow.*

[307] Stein E.M. and Weiss G. (1968) Generalization of the Cauchy–Riemann equations and representations of the rotation group. *Amer. J. Math.* 90: 163–196.

[308] Stern I. (1989) Randwertaufgaben für verallgemeinerte Cauchy–Riemann-Systeme im Raum. *Dissertation A, Martin-Luther-Universität Halle-Wittenberg.*

[309] Stern I. (1991) Boundary value problems for generalized Cauchy–Riemann systems in the space. *In: Kühnau R. and Tutschke W. (eds), Boundary value and initial value problems in complex analysis. Pitman Res. Notes in Math.* : 159–183.

[310] Stern I. (1993) Direct methods for generalized Cauchy–Riemann systems in the space. *Complex Variables, Theory and Appl.* 23: 73–100.

[311] Stummel F. (1967) Elliptische Differenzenoperatoren unter Dirichletrandbedingungen. *Math. Zeitschrift* 97: 169–211.

[312] Suchomlinov G.A. (1938) On the extension of linear functionals in linear normed spaces and linear quaternionic spaces. *Mat. Sbornik* (2), 3: 353–358 (Russian).

[313] Sudbery A. (1979) Quaternionic analysis. *Math. Proc. Cambr. Phil. Soc.* 85: 199–225.

[314] Temam T. (1979) Navier–Stokes equations. *Studies in Math. and its Applic., Vol. 2, North-Holland, Amsterdam, New York, Oxford.*

[315] Thomée V. (1968) Discrete interior Schauder estimates for elliptic difference operators *SIAM J. Numer. Anal.* 5: 626–645.

[316] Triebel H. (1980) Höhere Analysis. *Deutscher Verlag der Wissenschaften Berlin.*

[317] Tutschke W. (1983) Classical and modern methods of complex analysis. *In: Lanckau E. and Tutschke W. (eds), Complex analysis. Akademie-Verlag, Berlin* 15–58.

[318] Vahlen K.Th. (1902) Über Bewegungen und komplexe Zahlen. *Math. Ann.* 55: 585–593.

[319] Vekua I.N. (1959) *Generalized analytic functions. Nauka, Moskow* (Russian).

[320] Vekua I.N. (1963) *Verallgemeinerte analytische Funktionen. Akademie-Verlag, Berlin.*

[321] Vladimirov V.S. (1972) *Gleichungen der mathematischen Physik. DVW, Berlin.*

[322] van der Waerden B.L. (1948) *Spinor analysis. Nachr. Akad. Wiss. Göttingen, Mathem. Physik, Kl.100.*

[323] Vinogradov V.S. (1971) An elliptic system, which des not permit Noetherian boundary data. *Doklady Akademii Nauk SSSR* 199, Nr. 5: 1008–1010 (Russian).

[324] von Mises R. (1944) Integral theorems in three-dimensional potential flow. *Bull. Americ. Math. Soc.* 50: 599–611.

[325] von Wolfersdorf L. (1997) Der Beitrag J. Radons für die Entwicklung der Integralgeometrie und Computer-Tomographie. *Wege und Perspektiven der Wissenschaft, Jubiläumsband der Sächsischen Akademie der Wissenschaften zu Leipzig, Akademie-Verlag*: to appear.

[326] Walsh J.L. (1969) Interpolation and approximation by rational functions in the complex domain. *Amer. Math. Soc., Providence, R.I.*.

[327] Wimmer U. (1992) Hyperkomplexe Methoden bei der Betrachtung nichtlinearer Strömungsprobleme mit veränderlicher Viskosität. *Diplomarbeit, Freiberg.*

[328] Wimmer U. (1994) Orthogonal Decomposition of the space $L_2(G, CH)$. *Proceedings of the 19-th Summer School "Applications of Mathematics in Engineering", August 24 – September 02, 1993, Techn. Univ. of Sofia, Sofia*, (1994), 203–210.

[329] Weyl H. (1921) *Raum, Zeit, Materie. Springer, Berlin.*

[330] Wloka J. (1982) *Partielle Differentialgleichungen. Teubner, Stuttgart.*

[331] Wu Zhijian (1996) Clifford Algebras, Hardy Spaces, and Compensated Compactness. *In: Ryan J. (ed.), Clifford algebras in analysis and related topics. CRC Press, Boca Raton, New York, London, Toronto* 215–236.

[332] Xu Z. (1986) On a Riemann boundary value problem for regular functions with values in a Clifford algebra. *Kexue Tongbao* 32: 1294–1295.

[333] Xu Z. (1989) On boundary value problems of Neumann type for hypercomplex functions with values in a Clifford algebra. *Rend. Circolo Math. Palermo, Serie 2* 22: 213–226.

[334] Xu Z. (1990) On linear and nonlinear Riemann problems for regular functions with values in a Clifford algebra. *Chin. Ann. Math.* 11B:3 :349–358.

[335] Xu Z. (1991) A function theory for the operator $D - \lambda$. *Complex Variables Theory and Appl.* 16: 27–42.

[336] Xu Z. and Zhou C. (1993) On boundary value problems of Riemann–Hilbert type for monogenic functions in a half space of $\mathbf{R}^m, (m \geq 2)$. *Complex Variables, Theory and Appl.* 22: 181–194.

[337] Yanushauskas A. (1995) On an elliptic system of three equations of partial derivations of second order. *Liet. Matem. Rink.* 35, No. 2: 190–197 (Russian).

[338] Zeilberger Z. (1980) The algebra of linear partial difference operators and its applications. *SIAM J. Math. Anal.* 11: 919–932.

[339] Zhou C. (1993) On boundary value problems of Neumann type for the Dirac operator in a half-space of $\mathbf{R}^m (m \geq 3)$. *Complex Variables, Theory and Appl.* 23: 1–16.

[340] Zöll G. (1987) Ein Residuenkalkül in der Clifford-Analysis und die Möbiustransformationen der Euklidischen Räume. *Dissertation, RWTH Aachen*, 141 pages.

Index

$$\log x := \ln|x| + \omega(x)\operatorname{arc\,cos}\frac{x_0}{|x|}, \quad |\underline{x}| \neq 0 \quad \text{or } |\underline{x}| = 0, x_0 > 0.$$

(ii) $1 - \frac{1}{|x|} - \operatorname{arc\,tan}\frac{|\underline{x}|}{|x_0|} \leq |\log x| \leq |x| - 1 + \operatorname{arc\,tan}\frac{|\underline{x}|}{|x_0|}.$

Proof. Let $u \in \operatorname{im} \mathbf{P}_{ab}$, then it is immediately clear that $u \in \operatorname{im} F_a$ and, therefore, $\operatorname{im} \mathbf{P}_{ab} \subset \ker D_a \cap W_2^k(G, C\ell_{0,n})$. Now let $u \in \ker D_a \cap W_2^k(G, C\ell_{0,n})$. It is clear, that then $u = F_a u$. Then there exists an element $w \in \operatorname{im} Q_b \cap W_2^{k-\frac{1}{2}}(\Gamma, C\ell_{0,n})$ with $u = (\operatorname{tr}_\Gamma T_b F_a)^{-1} w$. Hence, $w = Q_b g = \operatorname{tr}_\Gamma T_b D_b v$. We get now

$$u = F_a (\operatorname{tr}_\Gamma T_b F_a)^{-1} \operatorname{tr}_\Gamma T_b D_b v$$

and so $u \in \operatorname{im} \mathbf{P}_{ab}$. #

$$\mathbf{P}_{ab} D_b v = F_a (\operatorname{tr}_\Gamma T_b F_a)^{-1} \operatorname{tr}_\Gamma T_b D_b v = 0,$$

so $\operatorname{tr}_\Gamma T_b D_b v = \operatorname{tr}_\Gamma v = 0.$ #

$$\begin{aligned} D_b u &= F_a (\operatorname{tr}_\Gamma T_b F_a)^{-1} \mathbf{Q}_b g + \mathbf{Q}_{ab} T_a f \qquad \text{and} \\ D_a D_b u &= D_a \mathbf{Q}_{ab} T_a f = D_a T_a f - D_a \mathbf{P}_{ab} T_a f = f . \end{aligned}$$

$$\operatorname{tr}_\Gamma F_b g + \operatorname{tr}_\Gamma T_b F_a (\operatorname{tr}_\Gamma T_b F_a)^{-1} \mathbf{Q}_b g = P_b g + \mathbf{Q}_b g = g. \qquad \#$$

$$\begin{aligned} -\Delta u + \eta^{-1} \operatorname{grad} p &= \Lambda f(u) \quad \text{in } G & (4.70)\\ \operatorname{div} \eta^{-1} u &= 0 \qquad\quad \text{in } G & (4.71)\\ u &= 0 \qquad\quad \text{on } \Gamma. & (4.72) \end{aligned}$$

$$-\operatorname{div} \underline{u} + \psi(u, Du) = \operatorname{Sc} T_G f + \phi_0$$

$$-\Delta u + D(\operatorname{Sc} T_G f + \phi_0) + \operatorname{grad} \operatorname{div} \underline{u} = f$$

$$D(I - \operatorname{Sc})Du = f - D\operatorname{Sc} T_G f - D\phi_0.$$

Now denote $I - \operatorname{Sc} =: \operatorname{Vec}$ the projection onto the vector part and we get

$$D\operatorname{Vec} Du = D\operatorname{Vec} T_G f + \phi_1, \quad \phi_1 \in \ker D.$$

$$\operatorname{Vec} \Delta w = \Delta \operatorname{Vec} w = \operatorname{Vec} T_G f + T_G \phi_1 + \phi_2 \quad (\phi_k \in \ker D;\ k = 1, 2).$$

$$\Delta \operatorname{Sc} w = 0 \Leftrightarrow w_0 = 0$$

$$u = \phi^* + \mathbf{Q} T_G \operatorname{Vec} T_G f + \mathbf{Q} T_G H,$$

$$g = \operatorname{tr}_\Gamma T_G u = \operatorname{tr}_\Gamma T_G F \phi^* + \operatorname{tr}_\Gamma T_G \mathbf{Q} T_G (\operatorname{Vec} T_G f + H).$$

$$\operatorname{tr}_\Gamma \phi^* = (\operatorname{tr}_\Gamma T F)^{-1} \operatorname{tr}_\Gamma T_G u = (\operatorname{tr}_\Gamma T F)^{-1} g \quad (\text{cf. [110]})$$

$$\Rightarrow \quad u = F_\Gamma (\operatorname{tr}_\Gamma T F)^{-1} g + \mathbf{Q} T_G (\operatorname{Vec} T_G f + H).$$

$$\begin{aligned} \Lambda_1(m, G) &= \inf\{(DMDu, u)/(u, u), u \in \overset{\circ}{W}{}_2^1(G), \operatorname{Sc} u = 0, u \neq 0\} \\ &= \inf\{(MDu, Du)/(u, u), u \in \overset{\circ}{W}{}_2^1(G), \operatorname{Sc} u = 0, u \neq 0\} \\ &\leq \inf\{(MDu, Du)/(u, u), \hat{u} \in \overset{\circ}{W}{}_2^1(G) \cap \ker \operatorname{div}, \operatorname{Sc} u = 0, u \neq 0\} \\ &= \inf\{(Du, Du)/(u, u), \hat{u} \in \overset{\circ}{W}{}_2^1(G) \cap \ker \operatorname{div}, \operatorname{Sc} u = 0, u \neq 0\} \\ &= \Lambda_* . \end{aligned}$$

$$\begin{aligned} \psi^{\ell}_{Qi} &:= \sqrt{|M(\mathcal{Q}, i)|} M(\mathcal{Q}, i)^{-1} \Theta_{Q,i} \\ \psi^{r}_{Qi} &:= \frac{1}{\sqrt{|M(\mathcal{Q}, i)|}} \Theta_{Q,i}, \end{aligned}$$

$$\operatorname{supp} \psi^{\ell(r)}_{Qi} \subset Q.$$

31.	$T(x\|x\|)$	$=$	$-\frac{1}{3}(x^2\|x\| + 1)$	41.	$T(x^{-3}\|x\|)$	$=$	$x^{-2}\|x\| + 1$
32.	$T(x^2\|x\|)$	$=$	$-\frac{1}{6}x^3\|x\|$	42.	$T(x^{-5}\|x\|)$	$=$	$\frac{1}{3}(\|x\|^{-4} - 1)$
33.	$T(x^3\|x\|)$	$=$	$-\frac{1}{4}(x^4\|x\| - 1)$	43.	$T(x^{-1}\|x\|^{-1})$	$=$	$\|x\|^{-1} - 1$
34.	$T(x^4\|x\|)$	$=$	$-\frac{1}{8}x^5\|x\|$	44.	$T(x^{-3}\|x\|^{-1})$	$=$	$\frac{1}{3}x^{-2}\|x\|^{-1} + 1$
35.	$T(x\|x\|^{-1})$	$=$	$-(x^2\|x\|^{-1} + 1)$	45.	$T(x^{-4}\|x\|^{-1})$	$=$	$\frac{1}{2}x^{-2}\|x\|^{-1}$
36.	$T(x^2\|x\|^{-1})$	$=$	$-\frac{1}{4}x^3\|x\|^{-1}$	46.	$T(x^{-5}\|x\|^{-1})$	$=$	$\frac{1}{5}(x^{-4}\|x\|^{-1} - 1)$
37.	$T(x^3\|x\|^{-1})$	$=$	$-\frac{1}{3}(x^4\|x\|^{-1} + 1)$	47.	$T(x^5\|x\|^8)$	$=$	$-\frac{1}{14}(x^6\|x\|^8 + 1)$
38.	$T(x^4\|x\|^{-1})$	$=$	$-\frac{1}{6}x^5\|x\|^{-1}$	48.	$T(x^{-5}\|x\|^8)$	$=$	$-\frac{1}{4}(x^{-4}\|x\|^8 - 1)$
39.	$T(x^{-1}\|x\|)$	$=$	$-\|x\| + 1$	49.	$T(x^5\|x\|^{-8})$	$=$	$\frac{1}{2}(x^6\|x\|^{-8} + 1)$
40.	$T(x^{-2}\|x\|)$	$=$	$-\frac{1}{2}\|x\|^{-1}\|x\|$	50.	$T(x^{-5}\|x\|^{-8})$	$=$	$\frac{1}{12}x^{-4}(\|x\|^{-8} - 1)$

143. $$T(x \cos|x| \sin^5|x|) = \tfrac{1}{6}(-|x|\cos^6|x| + \cos^6 1) + \tfrac{1}{36}(\cos^5|x|\sin|x| - \cos^5 1 \sin 1) + \tfrac{1}{2}(|x|\cos^4|x| - \cos^4 1) + \tfrac{13}{144}(-\cos^3|x|\sin|x| + \cos^3 1 \sin 1) + \tfrac{1}{2}(-\cos^2|x| + \cos^2 1) + \tfrac{11}{96}(\cos|x|\sin|x| - \cos 1 \sin 1 + |x| - 1)$$

144. $$T(x \cos^2|x| \sin^5|x|) = \tfrac{1}{7}(-|x|\cos^7|x| + \cos^7 1) + \tfrac{1}{49}(\cos^6|x|\sin|x| - \cos^6 1 \sin 1) + \tfrac{2}{5}(|x|\cos^5|x| - \cos^5 1) + \tfrac{68}{1225}(-\cos^4|x|\sin|x| + \cos^4 1 \sin 1) + \tfrac{1}{3}(-|x|\cos^3|x| + \cos^3 1) + \tfrac{409}{11025}(\cos^2|x|\sin|x| - \cos^2 1 \sin 1 + 2\sin|x| - 2\sin 1)$$

145. $$T(x \cos^3|x| \sin^5|x|) = \tfrac{1}{8}(-|x|\cos^8|x| + \cos^8 1) + \tfrac{1}{64}(\cos^7|x|\sin|x| - \cos^7 1 \sin 1) + \tfrac{1}{3}(|x|\cos^6|x| - \cos^6 1) + \tfrac{43}{1152}(-\cos^4|x|\sin|x| + \cos^4 1 \sin 1) + \tfrac{1}{4}(-|x|\cos^4|x| + \cos^4 1) + \tfrac{73}{4608}(\cos^3|x|\sin|x| - \cos^3 1 \sin 1) + \tfrac{73}{3072}(\cos|x|\sin|x| - \cos 1 \sin 1 + |x| - 1)$$

146. $$T(x \cos^4|x| \sin^5|x|) = \tfrac{1}{9}(-|x|\cos^9|x| + \cos^9 1) + \tfrac{1}{81}(\cos^8|x|\sin|x| - \cos^8 1 \sin 1) + \tfrac{2}{7}(|x|\cos^7|x| - \cos^7 1) + \tfrac{106}{3969}(-\cos^6|x|\sin|x| + \cos^6 1 \sin 1) + \tfrac{1}{5}(-|x|\cos^5|x| + \cos^5 1) + \tfrac{263}{33075}(\cos^4|x|\sin|x| - \cos^4 1 \sin 1) + \tfrac{1052}{99225}(\cos^2|x|\sin|x| - \cos^2 1 \sin 1 + 2\sin|x| - 2\sin 1)$$

Index compiled by Geoffrey Jones